Pasture Management
in South Africa

Pasture Management
in South Africa

Edited by
N.M. Tainton

University of Natal Press
Pietermaritzburg
2000

ISBN 0 86980 960 1   Hardback
ISBN 0 86980 959 8   Softback

Cover design by Brett Armstrong
Original art by Olive Anderson

Typeset by Alpha Typesetters cc, New Germany
Printed and bound by Interpak Books, Pietermaritzburg

Contents

Preface

The year 1981 saw the publication of *Veld and pasture management in South Africa*, a book which aimed to 'synthesise the research of pasture scientists and workers in related disciplines in South Africa over the last 50 years'. Eight authors contributed to this book which was designed primarily as a text for use by university and college students studying subjects related to veld (range) and pasture production and management, and to also serve the needs of top farmers in this field.

Any document which draws heavily on ongoing research typically becomes outdated and ideas change as social, economic and technological conditions and attitudes change. In this way the need to update the information of the original became more and more apparent until a decision was taken to completely revise it. A large number of new authors was introduced for the revision, both to bring a wider perspective to the subject matter, as well as to expand some of the topics and introduce some new topics which members of the profession felt had been missing from the original.

As a result of its expansion, it was impractical to publish the new text as a whole, and it was decided to separate aspects of veld production and management from those relating more specifically to the management of the cultivated pasture. This book, then, covers the production and management of the second of these two major agricultural resources – that of the pasture. The veld material was published with the title *Veld management in South Africa* in 1999. New chapters have been introduced on such aspects as physiology, mineral and energy supplementation, fodder conservation and pasture production planning, while the material of many of the other chapters has been considerably expanded. Some 23 authors were invited to contribute to the revision of the pasture book which was then edited with the express intention of producing an integrated text covering the field of pasture production and management in South Africa.

Since the content of many of the revised sections, now mostly under new authorship, was substantially different from that of the previous text, the material of the new text could not logically be seen as merely a revision of the original. For this reason it is published independently of the original and under a completely new title. This text comprises some 14 chapters dealing with the plant/soil/animal complex which forms the basis of livestock production off the cultivated pasture in South Africa. Some of the topics dealing with fundamental material common to both veld and pasture production are repeated but have been modified where appropriate to cover aspects of pasture production. The material now deals with both summer and winter rainfall zones, and with humid, semi-arid and arid areas. It therefore provides a wider coverage of the ecological zones as well as a wide coverage of the aspects which relate to pasture production and utilisation in South Africa.

Acknowledgements

The following persons are gratefully acknowledged by the editor: members of the University of Natal Press, including Glenn Cowley, who played a prominent role in driving the process with considerable vigour; Mobbs Moberly, under whose guidance the programme was initiated; and Sally Hines, Trish Comrie, Ilze Gertenbach and Michelle Paterson for their patience and dedication to the onerous task of editing the material.

Mark Horan is thanked for producing the map and Olive Anderson undertook some of the illustrations to her usual high standards. Their assistance is much appreciated by the editorial team.

The editor and publishers also wish to thank most sincerely the contributors who are credited with their chapters and also Mark Hardy and Neil Murray, who unselfishly acted as consulting editors; and Pete Zacharias and his 3rd year students of the Department of Range and Forage Resources of the University of Natal who class tested the manuscript.

Thanks are also due to Dr Amie Aucamp who assisted in the initial planning of the publication and who, as Director of the Range and Forage Institute of the Agricultural Research Council, was in a position to apply gentle pressure on many members of his staff who contributed to this publication. Thanks are also due to the eight authors of the original publication, *Veld and pasture management in South Africa*, which formed the basis of this new publication and whose material was used where appropriate in some of the chapters.

Finally, I wish to thank, most sincerely, my wife Rina who transferred the original text of *Veld and pasture management in South Africa* to computer disk to serve as an aid to the various new authors, as well as for her considerable support during the many, many hours of editorial work on this book.

The place and role of cultivated pastures in South Africa

A.J. Aucamp

CONTENTS

1.1 HISTORICAL OVERVIEW

Interest in South Africa in the use of introduced plant species for the production of forage dates back to 1812. At this time a certain Mr Van Reenen is reported to have tested a number of European grass species as well as lucerne and to have reported that a number of these species had shown promise. In 1861, the Honourable F.W. Reitz suggested that research programmes be initiated at Worcester and George to test introduced grass species, with George being subsequently recommended as the site for these trials. In 1893, H.R. Rait from Dordrecht in the Eastern Cape reported to the Woodhouse Farmers Association that he had been impressed with the forage production potential of cocksfoot (*Dactylis glomerata*), clover (*Trifolium* spp.) and certain ryegrasses (*Lolium* spp.).

Official testing of pasture species was undertaken for the first time in South Africa in 1903 when 76 co-operative trials were established in the Cape. Over the next two years this number increased to 1 330 and in the following year to 4 492. No fewer than 1 491 individual farmers were involved in these trials in which the main emphasis fell on *Paspalum dilatatum*, although other species were also tested. During this time a prominent botanist, Burtt-Davy, initiated an impressive testing programme using pasture crops collected from various localities throughout the world. In this programme a total of 114 different species were tested in the 1904/05 season, with the number increasing to 212 by the 1907/08 season. In 1908 kikuyu (*Pennisetum clandestinum*) was first introduced to South Africa from central Africa and entered a testing programme at the University of Pretoria.

A variety of European temperate grass species were tested in the Underberg region in KwaZulu-Natal early in the 20th century, and some of these species (e.g. *D. glomerata* (cocksfoot)) persist in isolated high altitude humid areas to this day. Trials were also undertaken in the Harrismith district in the Free State at the turn of the 20th century.

Following this period of active testing little was done until the final report of the Drought Investigation Commission (1923) gave perspective to the potential role of cultivated pastures in a balanced farming system. It was probably this report that prompted the establishment of a number of studies on cultivated forage species at Potchefstroom during 1925. In 1928, *Digitaria pentzii* (now *Digitaria eriantha*) was planted out on a farm scale and showed great promise. In 1938, a test programme based mainly on legumes was initiated in the winter rainfall region. At around this time pasture species were being tested throughout the country and their value was being reported in a number of local publications.

Much has been written in recent years on the potential value of cultivated pastures in South African farming systems. In this chapter emphasis will be given to their importance to and role in the local livestock industry. In essence this value revolves around their potential to reduce the unacceptably high grazing pressures currently being applied to veld, the role they can play in providing forage for the periods when veld is unpalatable (as in the sourveld in winter), and the important contribution they can make to meet the increasing demand for animal products.

1.2 THE NEED FOR INCREASED ANIMAL PRODUCTION

With the increasing demand for food as human populations increase, there is a need to increase the productivity of each hectare of agricultural land without degrading the natural resources of the country. The most direct way of achieving this on land set aside for animal production, is to increase livestock numbers, but this provides only a short-term solution to the problem. The inevitable long-term consequence of this is veld degradation and, in many situations, an increase in the rate of soil erosion unless increased animal numbers can be accompanied by an increase in forage production from the land. It has been repeatedly shown that this is possible over large parts of South Africa by either reinforcing or completely replacing the veld with improved forage species. Such reinforcement or replacement can result in considerable increases in productivity. Not only can dry matter production be increased, but also the palatability, nutritive value and digestibility of the forage which is produced by the modified sward.

Such major modifications of the composition and general characteristics of veld through its replacement or reinforcement elicits often severe criticism from environmentalists in particular. Their argument revolves around the supposed loss of stability arising from such interventions, claiming that veld should not be replaced by pastures under any circumstances because of this. However, such an attitude would seem to be unrealistic provided appropriate steps are taken to stabilise the pasture ecosystem which is produced by such intervention to the extent that losses of soil, in particular, are strictly controlled.

As things stand at present (1999/2000), South Africa is confronted with the following situation:

(a) the rate of human population increase is extremely high;

(b) cattle numbers, which in 1995 stood at approximately 13 million, had not appreciably changed over the previous 40 years; and

(c) sheep numbers, which in 1995 stood at 29 million, were similarly not much different from the numbers in the mid-1950s and are likely to have declined in recent years.

In order to meet the increasing local demand for milk, beef and mutton, it is estimated that levels of production will need to increase by 34%, 22% and 56%, respectively, within the next 20 years. Veld cannot be expected to provide for such increases in production and so the most logical, and in the long term most profitable, means of doing this would seem to be to convert appropriate areas into pasture. The attraction of this is that it could be done within the constraints of acceptable patterns of land use and acceptable levels of resource conservation.

Currently, however, there is little economic incentive to increase production levels in South Africa, particularly since the introduction of large-scale importation of a number of agricultural products. To what extent the country will be prepared to continue importing large quantities of its food when it has the capacity to produce its own requirements is an open question. In any event, this situation may well change of its own accord as the population grows and the needed volumes of food increase. Any measures which need to be taken to increase food production locally will inevitably involve an increase in the areas under pasture.

1.3 THE INTEGRATION OF CULTIVATED PASTURES INTO THE FARMING SYSTEM

It is important to appreciate that veld and pastures can play complementary roles in providing fodder to our livestock. Before pastures are introduced into any system, however, an assessment should be made of the extent to which productivity is likely to be increased, the amount of capital needed, the livestock system which is envisaged, the availability of labour and management expertise, and perhaps most important of all, the attitude of each individual farmer to pasture development. There are perhaps few other farming ventures which can lead so rapidly to financial ruin than an unplanned or poorly planned headlong dash into cultivated pastures.

It is essential, before embarking on a pasture programme, to view the forage resources of a property holistically so that species can be selected which meet identifiable needs within the specific forage and livestock programme of the property concerned. So, for example, on a particular property it may not be profitable to maintain a cow herd all year round on cultivated pastures, but such pastures may be used to provide strategic grazing for cows or for weaners that have to be weaned prematurely. In other situations, legume-based pastures may be used to boost the protein concentration in the diet of livestock during the winter. A complete inventory of the forage situation on any property first needs to be undertaken to identify the weak points in the forage flow. The pasture programme then needs to be targeted specifically at these weak points.

Unfortunately, cultivated pastures will do little to reduce the risks associated with variable rainfall conditions unless they are irrigated. Also, any forage they produce will normally cost more than that produced by veld. A full economic evaluation is therefore essential before a decision is made to embark on a pasture programme, particularly in the semi-arid regions where pastures normally do not have a quality advantage over veld and where production may be extremely variable from year to year. This evaluation should provide an estimate of the amount that can reasonably be spent on establishing and maintaining the pasture. An example of such an analysis is presented by Van Niekerk *et al.* (1987) for the Free State region. They showed that only R43.93/ha could be spent on pastures used for a cow/calf production system compared with a necessary outlay of R60/ha/year for a lucerne pasture and between R80 and R102/ha/year for a grass pasture.

1.4   THE ROLE OF PASTURES IN LIVESTOCK SYSTEMS

1.4.1   Legumes

The energy crisis of the 1970's and the rapidly increasing costs of nitrogen fertilizers directed attention to the potential role of legumes in providing protein-rich forage and in their potential to reduce the nitrogen fertilizer needs of pastures. The re-awakening interest in legumes is based largely on the successes which have been achieved with their use in countries such as Australia and New Zealand and on their successful use in certain situations in this country. Clearly, for successful animal production, a balanced diet is necessary. It is precisely in this role that the legume can play an important part. A pure stand of legumes will normally more than satisfy the protein needs of highly productive animals but considerable management attention needs to be given to these pastures if they are to be used effectively in livestock systems.

Legume pastures can be integrated into feeding programmes in a number of ways, as for example, by using them:
(a)   to re-inforce veld;
(b)   to replace veld;
(c)   in a mixture with grasses; or
(d)   in a rotation with cash crops.
The first determining factor here is the environmental potential and the availability of adapted species for the area under consideration. There can nonetheless be no doubt that an increased use of legumes would lead to improved levels of animal production and a reduction in the need for high levels of nitrogen fertilizer. They have the potential, therefore, to have a major impact on net farm income.

There is a wide variation in the optimum soil conditions for the growth of different legumes and they generally have rather more specific requirements than do grasses. Perennial legumes in particular, tend to be extremely susceptible to disease in their post-establishment years. Great care therefore needs to be taken in choosing the legume species for use in any situation.

1.4.2   Grasses

South Africa is world renowned for the richness of its grassland flora. The value of this flora was recognised locally only when a number of species indigenous to South Africa were successfully adopted as pasture species elsewhere in the world. Selections from among these species can, under both dryland conditions and under irrigation, produce 2 to 15 times as much dry matter, respectively, as can veld. Because of their adaptability to the South African climate, grasses are likely to retain their pre-eminent role in the cultivated pastures of the country.

1.4.3 Drought-tolerant fodder crops

The climate of South Africa, particularly between longitudes 17 and 24 (representing arid and semi-arid regions), is such that for all practical purposes a year seldom passes in which farmers do not experience a drought period. It is here that drought-tolerant crops can play an extremely important role in feeding livestock. Three drought-tolerant forage crops have received a great deal of attention over the years. These are the American aloe (*Agave mexicana*), the spineless cactus (*Opuntia aurantiaca*) and Old Man Saltbush (*Atriplex nummularia*). The establishment, cultivation and use of these crops is discussed in Chapters 10 & 12.

1.5 THE POTENTIAL ROLE OF PASTURES IN DIFFERENT TYPES OF VELD

The major role of the cultivated pasture in the farming system is to satisfy the forage needs of animals during periods when the quantity or quality, or both, of forage produced by veld is inadequate. Their primary role should be to:
(a) provide forage during the periods of greatest food shortage in the forage flow programme; and
(b) increase the total amount of forage and digestible nutrients produced on any property.

The precise role of the cultivated pasture will depend, in any circumstance, on the exact nature of the livestock system and on the quality and quantity of forage which is available. Their role will, therefore, be very much veld type dependent.

In the sourveld, cultivated pastures can:
(a) increase the length of the growing season, particularly in the spring and autumn periods;
(b) increase the total amount and quality of forage produced on any property;
(c) provide high quality green forage during the winter months (particularly for the dairy industry, using irrigated temperate pastures);
(d) provide high quality forage for carry-over into the winter in the form of, for example, foggage, hay or silage; and
(e) increase the level of animal production per unit area of land.

In the sweetveld, on the other hand, the major role of the cultivated pasture is likely to be one of the following:
(a) to provide forage for summer use so that the veld can be rested out at this time;
(b) to provide hay for drought periods;
(c) to increase the total amount of forage available for animal feeding through the year; and
(d) to increase the level of animal production per unit area of land.

Finally, in mixedveld, the major role of the pasture is likely to be:

(a) the provision of forage in the early summer, when it is often in short supply in these areas;
(b) to provide relief to the veld during the summer by allowing it to rest;
(c) to provide high quality forage during the winter months; and
(d) to increase the level of animal production per unit area of land.

In accomplishing any or all of these objectives in these various types of veld, the cultivated pasture can, if wisely used, play a pivotal role.

1.6   CONCLUSION

Given the current state of the natural resources over much of South Africa, together with the increasing demand for protein by an ever-increasing human population, there would seem to be no alternative but to plan for a substantial increase in the amount of forage produced in South Africa in order to allow for an increase in the size of the national herd and flock. A discerning move into cultivated pastures is an absolute requirement for any major increase in forage production in this country. But the incorporation of cultivated pastures into farming systems can be economically successful only if it goes hand in hand with a high level of management and sound, objective planning. There is no doubt that there is a very great and as yet untapped potential for their use in many parts of the country. The development of this potential should make it possible to increase forage production to levels capable of supporting, at least for the foreseeable future, the necessary expansion in livestock numbers in South Africa.

The morphology and physiology of the major forage plants

2

CONTENTS

Fundamental to an understanding of the management requirements of any plant is an understanding of its morphology, i.e. its basic structure, as well as its physiology (the manner in which it functions). In this chapter we discuss important morphological and physiological characteristics of the main groups of grasses and legumes which are used as forage plants. This information should provide readers with the necessary background to enable them to understand the many management principles discussed in later chapters of this book.

2.1 GRASSES

M.M. Wolfson & N.M. Tainton

2.1.1 Morphology and development of grasses

Grass plants are made up of a grouping of units called tillers (Fig 2.1). Single-stalked grass (poaceous) plants, such as maize, have only one such tiller which lives for approximately 90 days. Annual forage grasses are made up of collections of tillers which may live for as long as 15 months. In both these examples, the tillers eventually flower and die without producing replacement tillers which survive for any length of time. With the death of the flowering tillers the whole plant dies.

In perennial forage species the individual tillers often live for no longer than

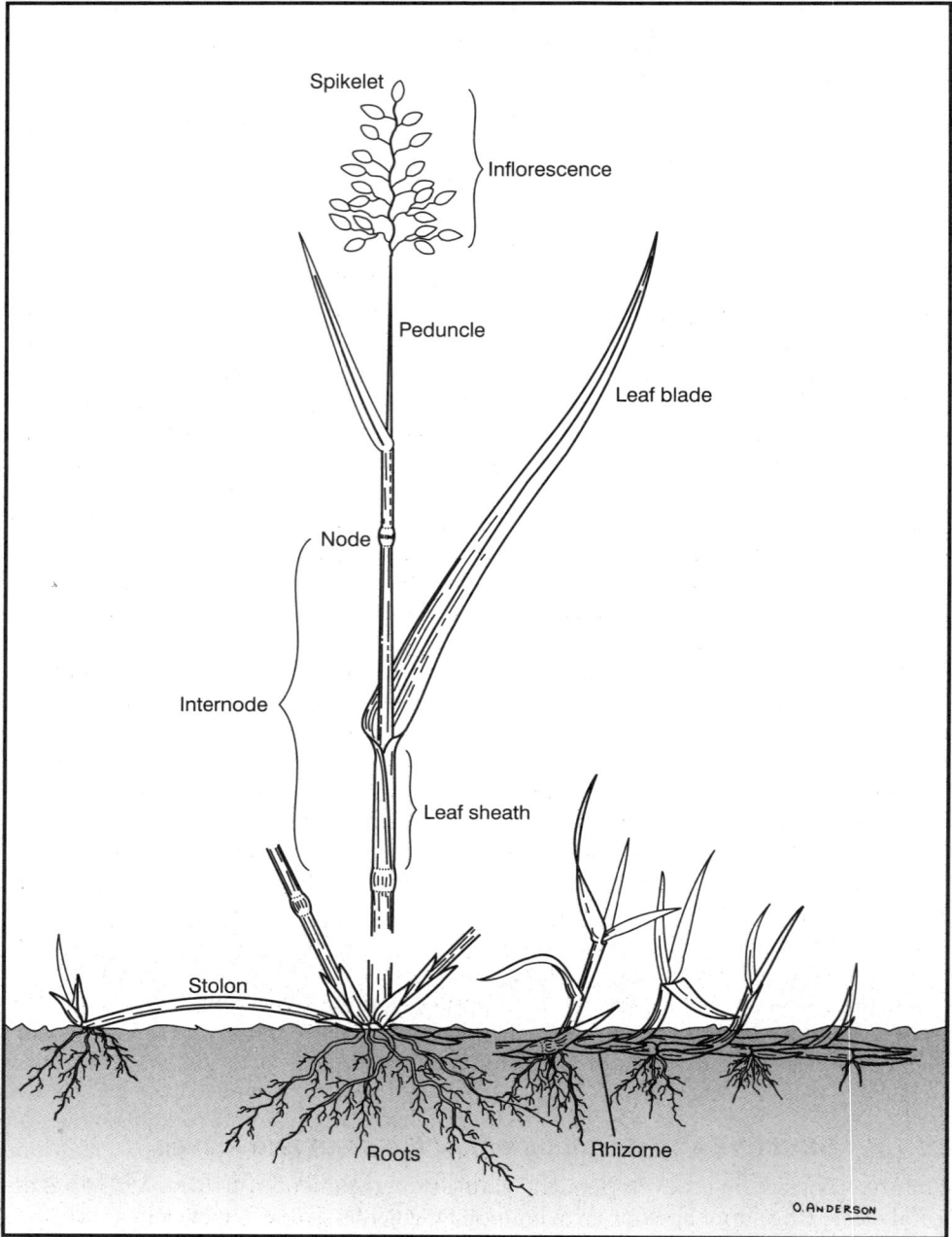

Figure 2.1   General structure of the grass plant.

those of the annual species but perenniality is conferred on them by a succession of secondary tillers which replace the original tillers. All secondary tillers are not necessarily produced at the same time so that perennial grass plants are usually made up of a collection of tillers of different ages. The tiller is the basic unit of the grass plant and an understanding of its development will provide much of the fundamental information necessary for an understanding of the growth of the plant as a whole.

2.1.1.1 Structure and germination of grass seeds

In the *Poaceae* (the grasses), the characteristic fruit is a caryopsis, an achene in which the pericarp is united with the testa. The seed is composed primarily of endosperm and an embryo, with the latter developing into the new plant. The outermost layer of the endosperm, the aleurone layer, contains lipid and protein reserves. The endosperm is a carbohydrate reserve used to sustain the embryo during germination and early seedling growth. Within the embryo are two main meristematic (growing) regions – the growing point of the stem (the plumule) enclosed in a protective sheath (the coleoptile) and of the root (the radicle), also within a sheath (the coleorhiza) (Fig. 2.2). These are attached by a short mesocotyl to a flat shield-like structure, the scutellum, which is found adjoining the endosperm. The two meristems are capable of rapid cell division and cell expansion when suitable growing conditions are provided and are responsible for most of the growth in the young embryo.

Most mature seeds contain only 5% to 10% of their total weight as water. Germination is not possible until the seed absorbs the water required for metabolic activities, such as those occurring in the aleurone layer. Aleurone cells are stimulated to secrete enzymes into the endosperm that convert insoluble starch into soluble sugars, which are then absorbed by the scutellum and passed on through the mesocotyl to the embryo. When germination occurs, the coleorhiza, which encloses the radicle or primary root, emerges first. The radicle elongates rapidly, penetrating the coleorhiza. After the emergence of the primary root, the coleoptile (the first foliar-like organ), which encloses the plumule, elongates. As the primary root elongates through continued growth at the tip, secondary meristematic areas develop on its flanks. These form secondary branches which later produce tertiary branches. In this way the seedling or seminal root system develops – a root system which supports seedling growth and later dies as it is replaced by the permanent coronal (crown) system of adventitious roots arising from the basal nodes of the main shoot and daughter tillers.

2.1.1.2 Structure and development of the stem and leaves

While the root system is developing, the young stem commences activity. At its tip is the apical dome (the shoot apex or growing point) which is a meristematically active area, variable in size and shape, but elongated in many grasses. Leaf initials (or primordia) are formed at the apical meristem in alternating order (Fig. 2.3). The cells divide and grow and the stem tip elongates. As it does so it continues to form new leaf initials nearer its tip, each of which develops a lateral tiller bud in its axil (Fig. 2.3). These buds are replicas of the apical growing point of the main stem and they may remain dormant or continue development to form new lateral (or daughter) tillers.

(a)

Scutellum

Pericarp

Scutellum vascular strand

Coleoptile

First true leaf

Plumule

Epiblast

Mesocotyl

Root vasculation

Radicle

Scutellum cleft

Coleorhiza

Hypopeltate appendix

(b)

Pericarp

Remnant of style

Scutellum

Coleoptile

Ruptured pericarp

Coleorhiza

Radicle

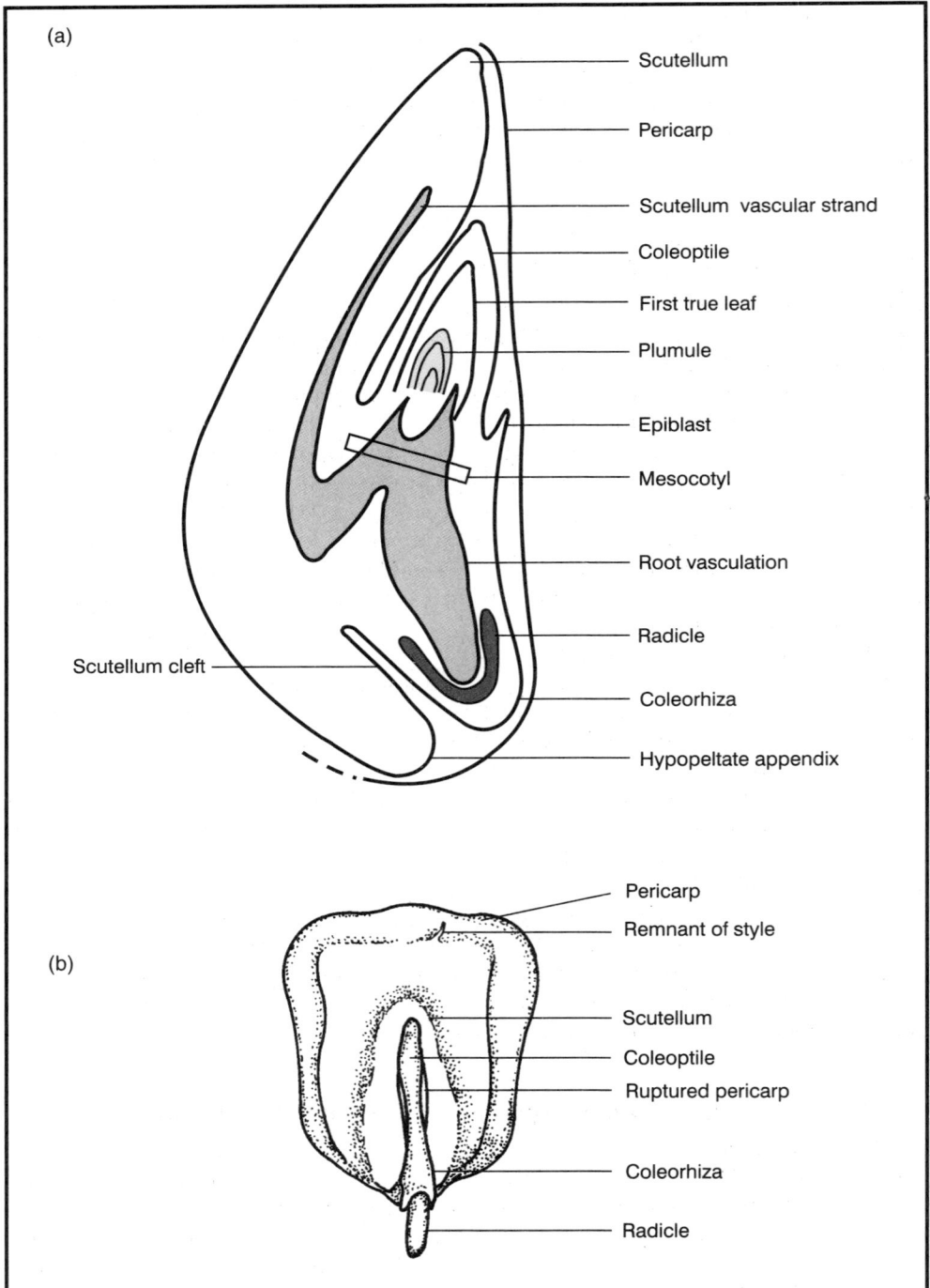

Figure 2.2   Section through a grass seed (a) and early stages of germination (b) (Chapman 1996).

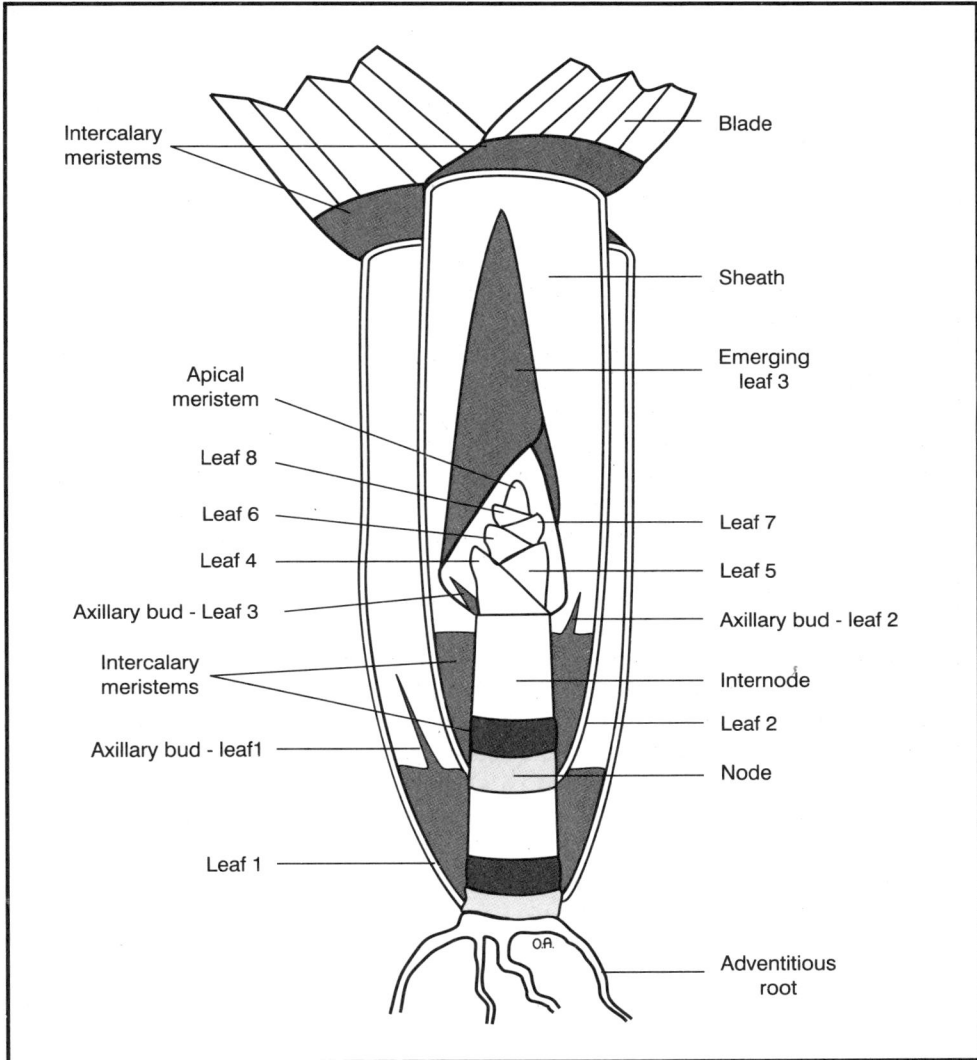

Figure 2.3 A schematic representation of a cross-section through a vegetative grass tiller.

Lateral shoots which exhibit considerable above-ground horizontal stem elongation are called stolons, while those that elongate underground are called rhizomes (Fig. 2.1) (Evans 1958; Beard 1973). Stolons have continuous growth, i.e. they continue growth along single axes and form roots and shoots at their nodes. As they grow, rhizomes may be either continuous, producing aerial shoots progressively, or terminal, when their apices turn upwards and emerge as phototrophic shoots.

As the stem tip continues to elongate, the already formed leaf initials on the flanks of the young stem grow into leaves. Leaf growth begins with rapid cell division in the outermost cell layers of the apical dome of the stem, giving rise to a microscopically visible protuberance. Further cell division changes the protuberance into a cowl which grows upwards and eventually overtops the apical dome, ultimately appearing as a leaf. The same process is repeated for as long as the shoot remains vegetative. The region on the stem where the leaf and lateral bud are attached is known as the node. Because the stem is very compact (it is often less than 10 mm long), the nodes in vegetative tillers remain at the base of the tiller where they are surrounded by leaf sheaths, and are not visible. The vegetative tillers therefore appear to be made up only of leaves. What appears to be a stem is, in effect, a collection of sheaths rolled, or folded, one inside the other, to form a pseudo-stem. During growth, young leaves are extruded in succession up the centre of the pseudo-stem so that the oldest leaf sheath is always on the outside. In most species the 'true stem', located at the base of the shoot, is concealed by encircling sheaths for as long as the shoot remains vegetative (Fig. 2.3).

At first, the entire tissue of the leaf initial is meristematic but as a leaf primordium grows, cell division and expansion are restricted to a basal meristem divided in two by a band of parenchyma cells from where the ligule grows and, in some species, two claw-like auricles develop. The upper portion of the meristem is associated with lamina (blade) growth and the lower with the growth of the sheath (Fig. 2.4). A large proportion of the leaf may be removed during grazing but as long as the meristems remain, growth continues in immature leaves. Restriction of growth to the basal meristem means that extension is complete in the part of the leaf emerging from the encircling sheaths of its predecessors. The tip of the leaf is always older than its base.

Meristematic activity in the lamina is terminated as soon as the ligule is differentiated, but the sheath continues to elongate until the ligule is exposed (Holmes 1989). The lowermost leaf matures first, followed in succession by the leaves above as they appear in strict sequence up the stem. Leaf death and senescence follow the same sequence, with the lowermost leaf dying first. This general pattern is similar in all grasses but the specific developmental pattern is, in each case, dependent on local environmental conditions (Langer 1972). Since leaves form the bulk of the herbage grazed by animals, the timing of each defoliation used to harvest grasses should relate to the rate of appearance of leaves and their longevity (Anslow 1966).

The rate of leaf appearance is not necessarily the same as the rate of leaf initiation (Anslow 1966), and it varies markedly between species and within species between seasons (Anslow 1966; Silsbury 1970). Seasonal variation appears to largely reflect changing temperatures, while illumination and defoliation may indirectly influence the rate of leaf emergence through their influence on the amount of assimilated carbon available for leaf expansion (Anslow 1966). In mid-summer, ryegrass (*Lolium perenne* L.) tillers expand new leaves every 5 to 7 days but do so at only one-tenth of that rate during mid-winter. Leaf senescence also slows, so that the mean number of mature photosynthetically active leaves remains relatively constant at about 3 per tiller in this species. The total number of leaves produced per tiller ranges from about 5 in some cereals to as many as 20 in some perennial grasses (Sharman 1947).

Once leaves have senesced, they begin to decay. Thus, in an ungrazed situation, forage can be lost through leaf decay and abscission. Hunt (1964) estimated that, on a cultivated ryegrass pasture, 8 kg/ha/day of dry matter may be lost through death and subsequent decomposition of leaves.

In annual and many perennial grasses, short shoots are produced where the stem internodes elongate very little during vegetative growth (Booysen *et al.* 1963; Jewiss 1966) so that the apex is protected during vegetative growth.

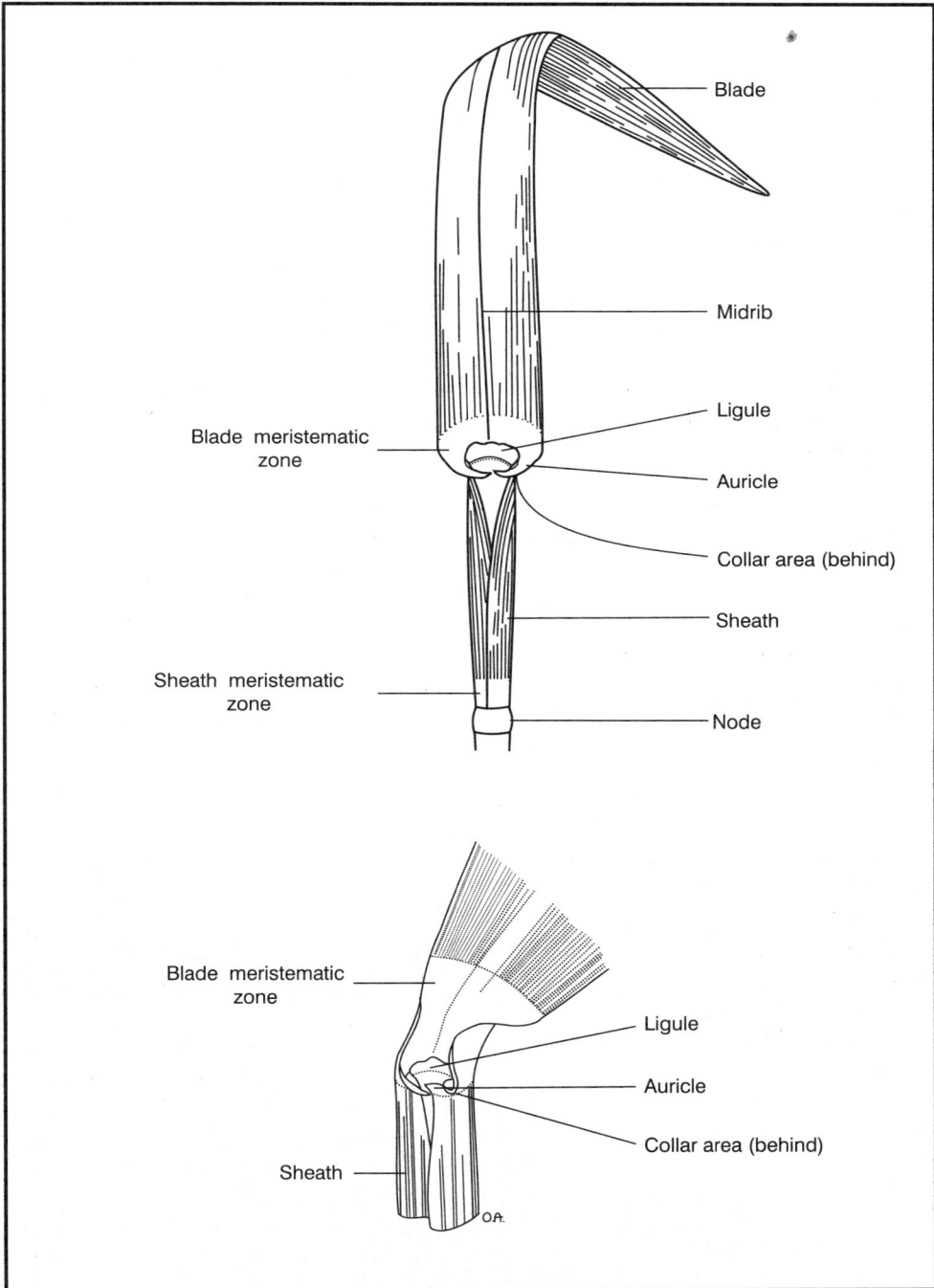

Figure 2.4 A node and its accompanying leaf showing the various leaf structures and the position of the two meristematic zones.

Vegetative growth in a grass stem may be terminated in one of two ways:

(a) the developing apex, from which all leaves of the main stem arise, may be destroyed in any one of a number of ways, particularly by being physically removed by grazing animals; or

(b) the apex may become reproductive, after which it no longer produces leaf initials. The onset of flowering is of considerable biological and economic significance because it marks the change from the continued production of vegetative organs to the ultimate production of seed (Calder 1966). Such reproductive development is characteristic of a species and cannot readily be manipulated by management. In species which are resistant to grazing, the relative proportion of stems which at any one time become reproductive and so terminate vegetative growth is relatively low. This aspect should be taken into consideration in selecting pasture species.

2.1.1.3 Flowering

Floral induction is interpreted as a biochemical process that may be genetically induced or require certain conditions such as a cold pre-treatment (vernalisation), a certain daylength or sequence of daylengths, and favourable growing conditions before it occurs (Dahl 1995).

When the flower is initiated, the apex continues to produce leaf primordia at an accelerated rate and the shoot apex lengthens. Soon thereafter bud primordia develop at the stem tip and in the axils of the older leaf primordia, giving the apex a characteristic 'double ridge' appearance (Holmes 1989). The remaining leaf initials now develop into rudimentary leaves or small scales, while the buds in their axils develop into parts of the inflorescence. This change is often accompanied by rapid growth in the lower previously formed internodes of the stem and so the characteristic flowering stems (culms) of the grass plant develop.

Although the main parent tiller is generally the most certain to produce seed in any year and is likely to produce the largest inflorescence, some of the lateral daughter tillers which begin their development soon after germination often keep pace with the main stem and flower at much the same time. Others may follow at a later stage to prolong flowering. Even in their first year of establishment, therefore, grass plants are able to develop a number of flowering stems.

2.1.1.4 Tillering

Tillering is the process by which the lateral buds which develop on the grass stem continue their development to form new tillers. These tillers may develop an upright habit typical of tufted grasses (*E. curvula*, Fig 2.5a) or they may grow laterally to produce rhizomes or stolons in the creeping grasses such as kikuyu or Rhodes grass (Fig. 2.5b). Whatever the nature of their subsequent development, the process of tillering is vital to both the production and persistence of grasses. A high tillering rate is a major requirement of almost all forage grasses, and particularly those which are used for grazing. Tillering patterns are not, however, purely a function of the particular species, but also depend on growing conditions. Wolfson (1989) demonstrated, for example, that the application of high concentrations of nitrogen (N), especially in the nitrate form, could stimulate tillering and stolon production in *Digitaria eriantha*. Defoliation, high light intensities and low temperatures will also stimulate tillering.

(a)

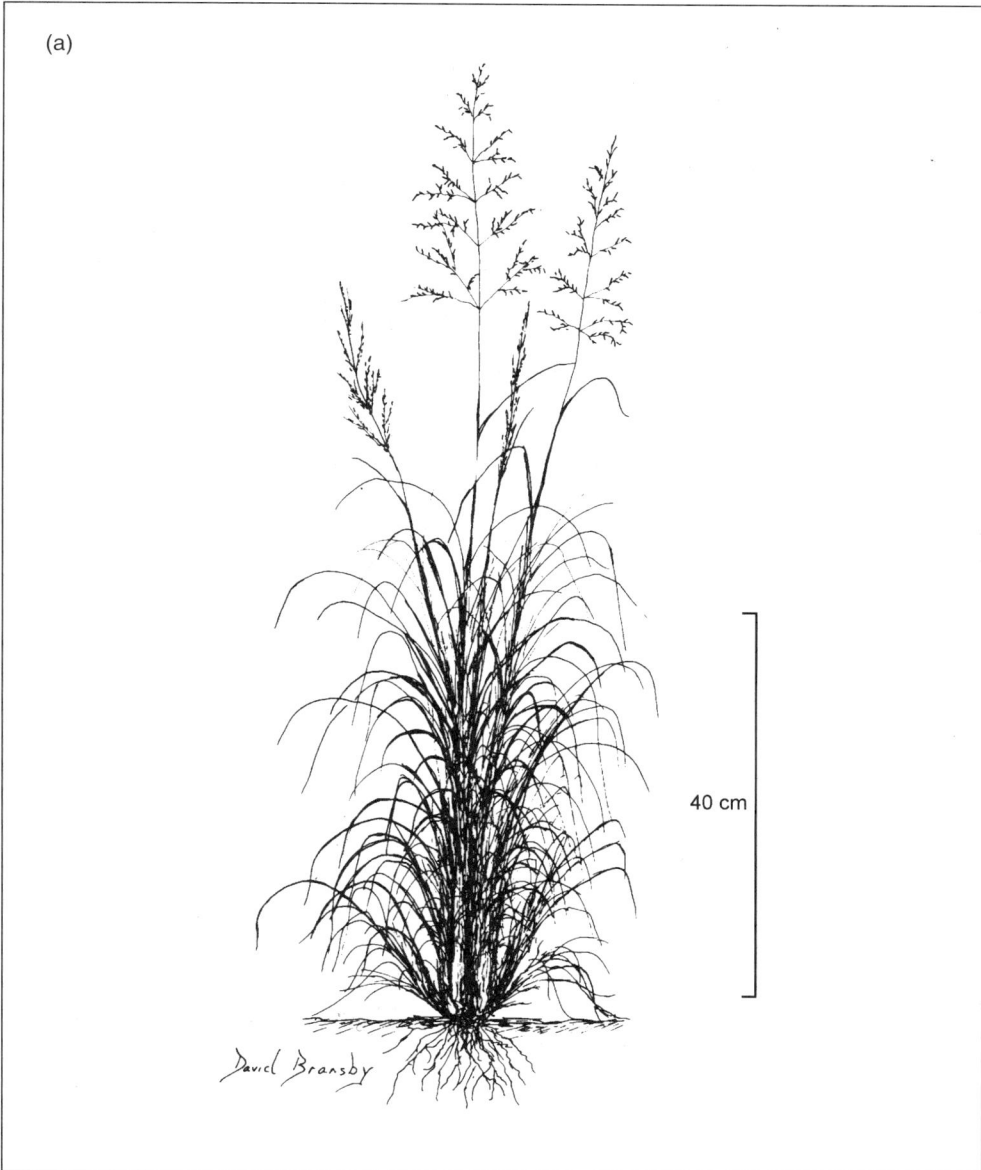

40 cm

David Bransby

Figure 2.5a An example of a tufted grass (*Eragrostis curvula*) (a) and a creeping grass (*Chloris gayana*) (b) (see next page), showing the very different patterns of tiller organisation. Examples taken from Tainton *et al.* (1990).

(b)

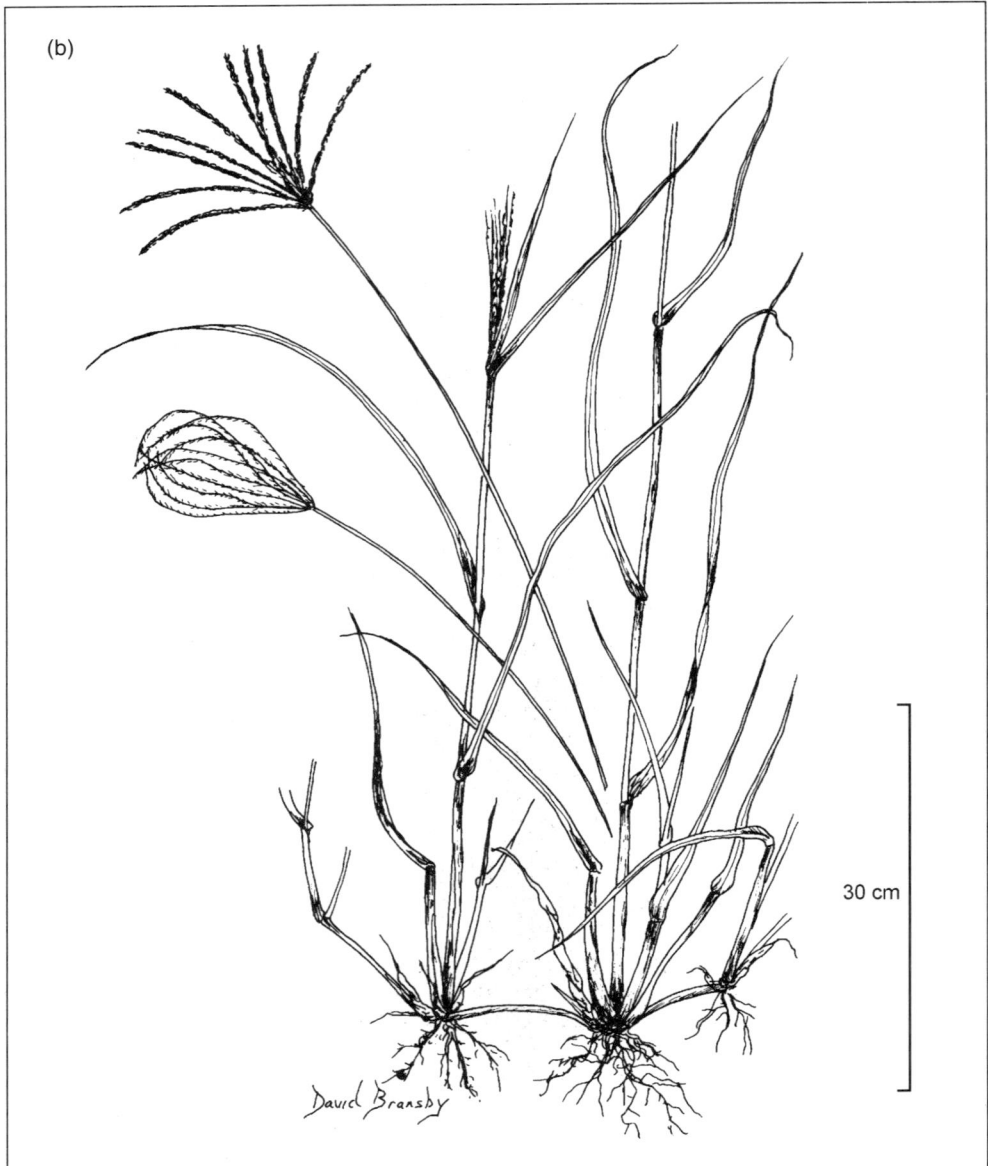

David Bransby

Figure 2.5b An example of a creeping grass (*Chloris gayana*).

Pasture production must concern itself with the provision of conditions which favour tiller development. Differences in the rate of tiller development in different pasture species may arise in a number of ways. The number of potential tillering sites may vary because of differences in the number of nodes (and therefore tiller buds) in the compact basal section of the stem. It is only those tiller initials which are situated sufficiently low down on the parent stem which are able to develop into effective tillers with their own root systems.

The subsequent behaviour of these tiller initials may vary in one of three ways:
(a) they may show no signs of dormancy and develop almost immediately into tillers;
(b) they may remain dormant for a while and develop into tillers at a later stage; or
(c) they may remain dormant and eventually degenerate.
Species vary in the proportion of tiller initials which fit into these three categories. In annuals many of the tiller buds which develop during the pre-flowering vegetative stage continue to develop almost immediately into tillers. Those that fail to do so eventually degenerate with the death of the main stem after flowering. In slow-growing perennials, on the other hand, only a small proportion of the tiller initials develop into tillers during the pre-flowering period. Some of those that do not, survive to develop after flowering. The commonly used perennial pasture species are intermediate between these two, with a proportion of the tiller initials developing into tillers during both the early vegetative and the post-flowering periods. In all grasses, whether annual or perennial, there is a suppression of tillering during the period of rapid stem growth which normally follows the inception of flowering.

Different plants therefore differ in their ability to produce tillers and in the longevity of their tillers. In annuals, most of the component tillers of the plant flower as soon as conditions become favourable for flowering and then they die, whereas in slow-growing perennials only a relatively small proportion of the tiller population flowers at any one time, after which the tillers die, except for the compact basal regions on which the basal tiller initials are situated. The tillers that remain vegetative continue to grow during the flowering and post-flowering period, when they produce the bulk of the grazeable material produced at this time. The new population of daughter tillers which develops from the basal node region of the dying flowering culms will produce little usable material at this time and, particularly in those grasses which flower from mid-summer onwards, will generally remain relatively small until the following season. Some of the tillers which remain vegetative through the flowering period will flower in the following season but many die without producing flowers. They will, nevertheless, have made a valuable contribution to the pasture by reducing the severity of the production trough which often follows flowering.

As tillers develop and grow, adventitious roots develop from the basal regions of their stems to form the coronal root system which takes over the function of the seminal root system. In this way each subsequent generation of tillers develops its own root system. During their early stages of development young developing tillers rely on the root system of their parent tiller until such time as their own root systems are sufficiently well developed to allow them to function as independent units.

Grass species vary widely in respect of the seasonal patterns of development of their constituent tillers. In some species, shoot apex elevation is rapid, while in others it may extend over many months. In some species it occurs early in the season, while in others it occurs much later. The stem apices of different species remain elevated and vulnerable for periods of varying length and during different seasons of the year, and even within the same species, stem elongation patterns vary (Booysen *et al.* 1963; Rethman 1971). Added to this variation is that induced by defoliation treatments, since the time and period during which the apex is elevated and therefore exposed to damage prior to flowering are both treatment dependent.

What is the relevance of a knowledge of tiller behaviour patterns to the formulation of management philosophy? In pastures a major phase of development typically follows soon after the initiation of the inflorescence. During this phase yield will increase rapidly, but at the expense of forage quality as stem material comes to make up an increasing proportion of the available forage. An early harvest, soon after the commencement of stem elongation, generally removes the apices of most of the developing stems and prevents any further stem development. Leafy growth normally follows but growth is likely to be slow because only a relatively small number of stem apices (i.e. those which commenced development late) is likely to have survived the defoliation treatment. The removal of a large proportion of the terminal apices will, however, normally activate some of those dormant lateral tiller initials situated in the basal stem sections. These will develop to form the nucleus of the tiller population during the post-defoliation period, but their initial development is usually slow and so there is typically a slowing down of growth immediately following a harvest treatment.

Harvests can, therefore, be timed to meet specific needs. If the need is for a large quantity of relatively low quality forage, then care needs to be taken to close the pasture before there is any sign of stem elongation. The pasture should then be allowed to grow out without interruption during the stem development phase and harvested only once stem development is complete. The same practice applies when the intention is to produce seed. Where, however, high quality forage is required, deliberate steps will need to be taken to terminate stem development by harvesting the pasture once the majority of the stem apices have elevated above harvest height. Subsequent growth will then be leafy and devoid of any appreciable quantity of stem.

2.1.1.5   *Root production*

The root systems of grasses appear to be as dynamic as the above-ground components. New roots are continuously being initiated on new tillers as those of the old tillers die and decay. Clear seasonal patterns of root initiation and growth have emerged from studies which, to date, have been undertaken largely on the temperate species. In these grasses, root initiation and growth is rapid in late winter and early spring and again in the autumn, with periods of slow root activity in summer and mid-winter. The roots may be relatively short-lived, as in perennial ryegrass where most are reported to live for only three to six months, or they may be more perennial, as has been reported for cocksfoot. It seems that there is a close relation between the longevity of a tiller and of the roots that develop from it. The average lifespan of a root of those species which have long-lived tillers would be expected to be greater than that of species with short-lived tillers.

Typically more than 85% of the root mass in grasses is to be found in the top 150 mm of the soil. Perennial ryegrass is shallow-rooted with approximately 75% of the root mass occurring in the top 50 mm of the soil (McKenzie 1996). Evidence suggests that those roots which are deeply penetrating are considerably more efficient per unit weight of root than are the surface roots, so the value of the deeper roots should not be underestimated.

The advantages of a grass such as Italian ryegrass, which is able to develop a deep root system when stressed, are therefore obvious (Steynberg *et al.* 1994).

2.1.2 The growth and physiology of grasses

2.1.2.1 Growth

The rate of growth of grass plants, and therefore the ultimate production of forage, depends on the size of the photosynthetic (leaf) area available for trapping sunlight and the efficiency with which this leaf can photosynthesise. These two aspects are closely related since leaf expansion is dependent on the availability of energy substrates, which is in turn related to photosynthetic rate. In practice, however, leaf expansion and photosynthetic rate may respond differently to changing circumstances, and particularly to different stresses.

Potential leaf area is essentially genetically controlled but the attainment of the genetic potential of the plant is influenced by the environment (Anslow 1966) and the amount of time since the plant was last defoliated. Photosynthetic efficiency is also dependent on the environment (e.g. temperature and therefore time of year) as well as on the availability of raw materials of photosynthesis (CO_2, light, water and nutrients) and the age of the plant's leaf system (Booysen 1966).

The rate of leaf extension declines more rapidly than the rate of photosynthesis when stress is experienced by a plant, e.g. N deficiency slows leaf expansion more than it does photosynthetic rate (Watson 1956; Wolfson 1989). Shortages of water also limit leaf expansion. Initially this is a result of a reduction in the rate of cell expansion, which is more sensitive to water stress than is cell division (Holmes 1989).

Light, moisture and temperature are among the most important variables affecting leaf growth. However, the temporary storage of sugars in leaf bases and other metabolites in growing leaves provides some buffer against short-term fluctuations in the supply of photosynthetic products for growth when photosynthetic rates are low (Gordon *et al.* 1977). In spaced plants grown at low light intensity, the products of photosynthesis are retained by the shoot at the expense of the root. In this way, leaf area is maximised. In a closed crop canopy, however, where light intensities experienced by individual leaves are reduced further by self shading, the greater allocation of resources to produce leaves will normally not increase light interception sufficiently to offset the effects of low light intensity. Thus total plant photosynthesis is depressed, as well as total dry matter production.

The rate of extension of a growing leaf is extremely sensitive to the current temperature, and responds to temperature changes within minutes (Holmes 1989). General temperature will also affect the final size and shape of the leaf and its rate and direction of extension. 'High temperature' leaves tend to extend more rapidly for a shorter period and to a greater final length with proportionally more lamina relative to sheath, than 'low temperature' leaves.

As a leaf ages, its photosynthetic capacity initially increases, reaches a maximum soon after it is fully expanded, and then declines (Wolfson 1989). The decline starts before there are any visible signs of senescence (Woledge & Parsons 1986). The implication of the change in photosynthetic capacity with leaf age is that, as a grass tiller or tuft ages, the average age of leaf material increases and the overall photosynthetic capacity of the tiller or tuft declines.

Thus plant or tiller age can affect vegetative growth in one of two ways:
(a) as the plants age the average age of their leaves increases and so their photosyn-
     thetic rate declines; and
(b) as the plants grow and their canopies become increasingly dense, mutual shading
     will increase in intensity, resulting in an effective decline in 'photosynthetic' leaf
     area and in photosynthetic rate.
Therefore the growth rate of individual plants and of the sward as a whole will decline.
A conceptual model of the effect of plant age on leaf area, photosynthetic capacity,
absolute growth rate and dry matter production of a grass tiller, tuft or sward, is presented in Fig. 2.6.

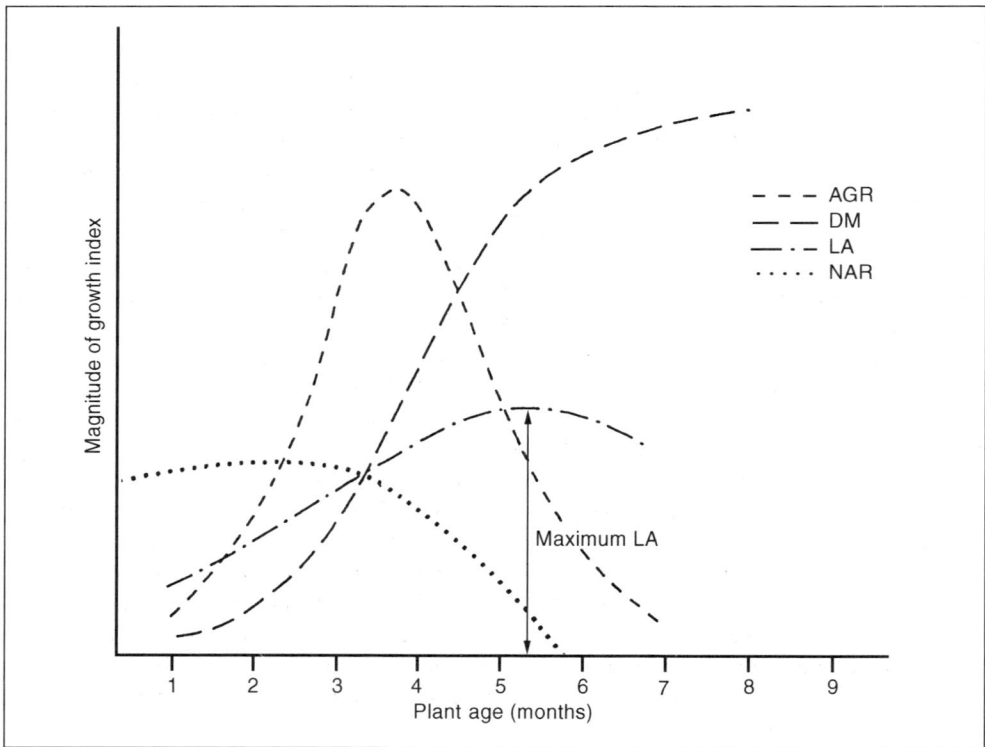

Figure 2.6   The hypothetical effect of age on leaf area (LA), photosynthetic rate (NAR), absolute
growth rate (AGR) and cumulative dry matter production (DM) of a grass tiller.

The theoretical implication of Fig. 2.6 is that there is an optimum length of successive growth cycles for maximum dry matter production. If this interval is too short, the
average growth rate will be maintained below its potential. If the interval is too long,
leaf area and photosynthetic rate will decline, in turn causing a decline in growth rate
and overall dry matter production. The optimum length of successive growth cycles has
been calculated mathematically by Maeda & Yonetani (1978) (Fig. 2.7). The length cor-
responds to a point where average growth rate is highest. It is noteworthy that the optimum length of successive growth cycles does not correspond to the time taken to

reach maximum growth rate, but to a period 1.4 times longer than this. This means that the length of time between successive defoliations should be somewhat longer than one would anticipate. The models are purely conceptual and although the trends will remain similar, the quantitative values will change from species to species and from season to season. The result of this is that the optimum defoliation frequency will vary among different species.

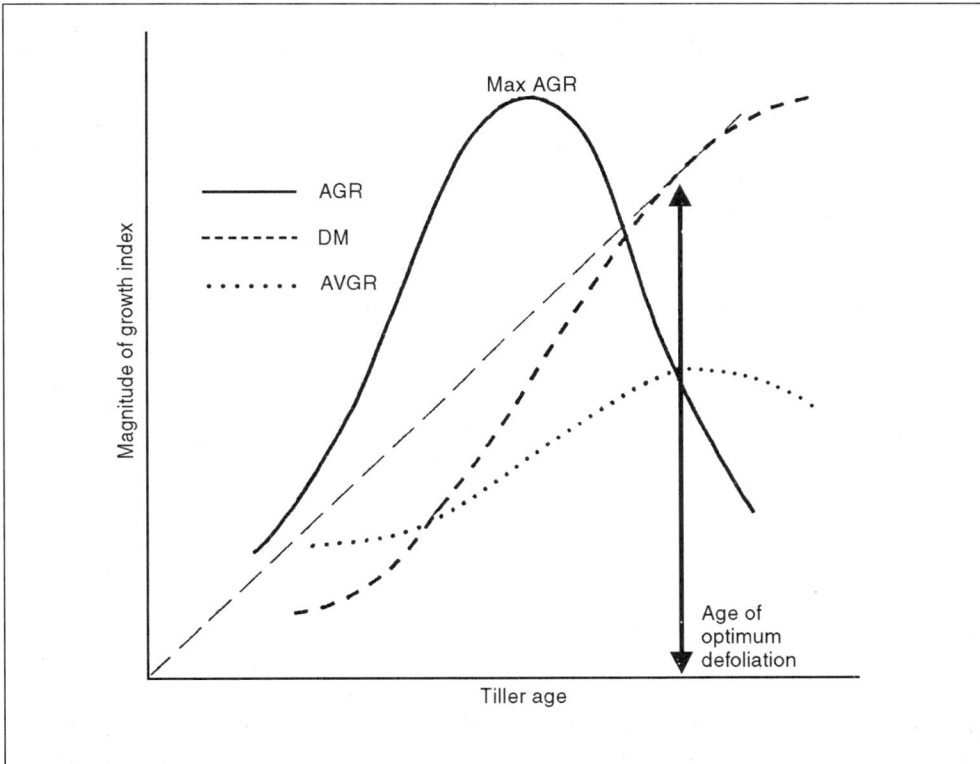

Figure 2.7 The hypothetical effect of tiller age on absolute growth rate (AGR), average growth rate (AVGR) and cumulative dry matter production (DM) of a grass tiller (after Maeda & Yonetani 1978).

Inherent differences in tillering patterns between species are further modified by environmental conditions, notably by temperature, light, moisture and nutrient supply. Their effect on tillering appears to arise chiefly through their influence on leaf area and photosynthetic capacity, which would provide energy and substrates for growth. If the plants are growing in conditions of adequate nutrients and moisture, time of year will be the major factor affecting growth since this will influence the quantitative values depicted in the models. As a result, the growth of a grass tiller, plant or sward at any particular time is difficult to predict as it is always influenced by a number of independent parameters. Thus when attempting to develop management strategies for grass swards, it should be appreciated that optimal treatments will vary continuously and, in particular, will vary from season to season and as the plant's age changes.

During early growth young tillers depend entirely on their parent tillers for energy substrates. These are made available in any quantity only when they are in excess of the immediate requirements of the growing point of the main stem. Hence, when temperature conditions are such that the growth rate of the main stem is rapid, the available energy is used largely by the parent tiller and the lateral tillers remain dormant. A decline in temperature, which reduces the growth rate of the main stem but allows photosynthesis to continue, will promote lateral tillering since their initials will then have access to a larger supply of photosynthetic products produced by the parent tiller. Optimum temperatures for tillering are therefore often lower than those for general plant growth. Thus tillering is usually more rapid in the spring and autumn than in the summer.

The effect of light intensity on tillering is likewise dependent on its effect on the accumulation of energy-containing products in the plant. Hence, high light intensities which promote high photosynthetic rates also promote tillering. In mature swards, however, the basal portions of the stem are often subjected to intense shade cast by the upper canopy. This in itself may reduce the rate of tillering from the basal nodes. Those tillers that develop at such basal nodes often die before their leaves penetrate sufficiently into the upper canopy to allow them to supply their own energy requirements through photosynthesis. In some grasses, e.g. *Eragrostis curvula*, excessive basal shading promotes tiller development from elevated nodes on the stem at the expense of those from basal nodes. Since these elevated tillers cannot develop their own root system, they soon die. Excessive basal shading of the pasture leads to low tiller densities which may result in substantially reduced production during subsequent seasons and to the invasion of weeds.

Dry conditions during the season promote tiller dormancy, as do low soil N levels. Nitrogen is often applied in the autumn specifically to increase tiller numbers, particularly in crops used for seed production.

The growth of roots is also dictated largely by the amount of carbohydrate which is available for their use and which is transported to them from the plant's leaves. The greater the amount of carbohydrate produced by the plant, and the less of it that is used for the growth of new leaves and stems, the greater will be the amount available for root growth.

Shoot:root ratio's within plants increase as the plants age. Increases in temperature, decreases in light and an increase in the N supply may also lead to increases in the shoot:root ratio.

As indicated above, root growth is rapid when environmental conditions are suited to high photosynthetic rates but do not favour the growth of leaves and stems. This often occurs in spring and autumn, when conditions are too cold for the rapid growth of above-ground components but still allow for rapid photosynthetic rates. Even during such periods, however, a number of factors may limit photosynthesis and consequently also root growth. Dry conditions and low light intensities will both reduce photosynthetic rates, as will a reduction in the size of the photosynthetic system of the plant by harvesting the leaf material. Plants which are grazed frequently or severely, therefore, normally have very much reduced root systems (Crider 1955; Tainton 1958).

2.1.2.2   *Plant growth and carbon metabolism*

2.1.2.2.1   *Photosynthesis*

Photosynthesis is the process by which solar radiation, intercepted by green leaves, provides the energy to convert carbon dioxide and water into simple carbohydrates. These carbohydrates perform a dual purpose in the plant. They may be oxidised during respiration and the energy which is released used to drive the living processes in the plant, or they may be used as the structural building blocks for the production of new tissue when they are converted into a wide variety of complex organic compounds. The leaf blades of grasses are more ideally structured to intercept light and capture carbon dioxide from the air than are the leaf sheaths or stems, and so it is these blades which produce the greatest proportion of the carbohydrates formed within a grass plant. These blades also largely constitute the harvestable product of pastures. Thus competition exists for the leaves as a source of energy for the plants themselves, and as a source of feed for animals.

The initial photosynthetic product formed when carbon dioxide (CO_2) is fixed during the Calvin Cycle of photosynthesis, is 3-phosphoglyceric acid (PGA) (Gutierrez *et al.* 1974). Plants exhibiting this as the primary carboxylation reaction are called C_3 photosynthetic pathway plants. In other plants, the initial carboxylation products are four carbon acids such as malate and aspartate rather that PGA. Such plants have been classified as C_4 photosynthetic pathway plants.

The leaf structure of C_3 and C_4 plants clearly shows anatomical and ultrastructural differences (Fig. 2.8). In grasses with the C_3 pathway the vascular tissue is surrounded by two bundle sheaths. The inner, called the mestome sheath, has few or no chloroplasts and has thickened cell walls. This sheath is encircled by a second, the parenchyma sheath, containing only a few chloroplasts. Its cells are sometimes smaller than those of the mesophyll. The bundle sheath cells play no significant role in photosynthetic CO_2 assimilation and metabolism. CO_2 is fixed in mesophyll cells and the mesophyll chloroplasts form starch.

In plants with the C_4 pathway, the leaves have a 'Kranz' anatomy, where the vascular bundles are surrounded by two concentric chlorophyllous layers (Fig. 2.8), an inner bundle sheath layer consisting of large cells with fairly thick walls and an outer mesophyll layer. The chlorenchyma cells of the mesophyll are arranged more or less radially around each vascular bundle and each cell is in direct contact with, or is separated from, the Kranz sheath cell by not more than one other chlorenchymatous cell (Hattersley & Watson 1976). Species of some genera, e.g. *Aristida*, have a double bundle sheath. The differentiation of biochemical activity between the mesophyll and sheath cells in C_4 plants appears to be the main functional difference between the C_3 and C_4 species. Nevertheless the chloroplasts of the sheath and mesophyll cells are functionally closely associated with each other by means of numerous pits and plasmodesmata (Ellis 1977).

Extensive work done by Hatch & Slack and co-workers (Hatch & Slack 1966; Hatch *et al.* 1967; Johnson & Hatch 1969) confirmed the C_4 pathway and showed that atmospheric CO_2 is assimilated in the mesophyll cells into oxaloacetic acid, which is rapidly converted to malate (Fig. 2.9) and/or aspartate. These early products of photosynthesis are transferred to the bundle sheath cells where they act as a source of carbon for the sugar phosphates, via a decarboxylation step. The CO_2 that is released is re-fixed into

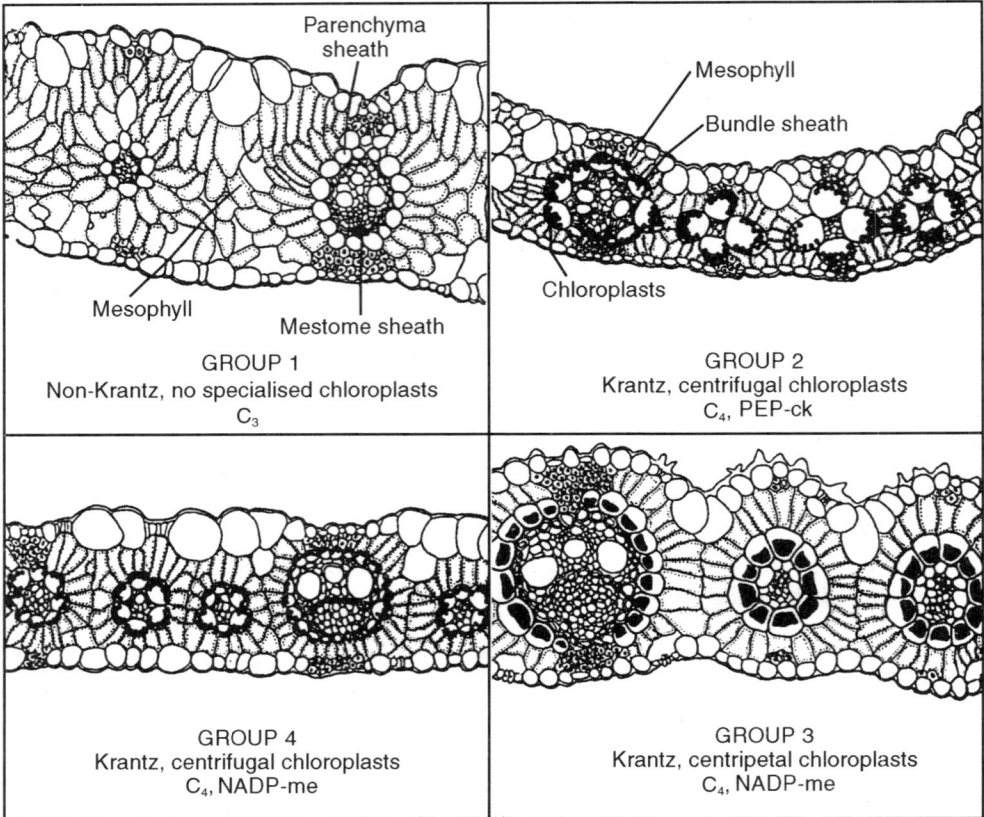

Figure 2.8 Transverse sections of leaf blades of *Panicum*, showing the different anatomical types associated with C_3 and C_4 photosynthetic pathways (Ellis *et al.* 1980).

3-PGA during normal photosynthetic carbon reduction in the chloroplasts of the bundle sheath cells.

Grass plants which exhibit the C_4 dicarboxylic acid pathway are characterised physiologically by low CO_2 compensation points varying from 0 mℓ/ℓ to 10 mℓ/ℓ. The CO_2 compensation point is the steady state CO_2 concentration attained in a lighted, closed system containing a plant. At this point of equilibrium the CO_2 fixed by photosynthesis is exactly balanced by the CO_2 produced by all respiratory processes (Coombs 1985). C_4 plants also exhibit high photosynthetic rates varying from 19 mmol/m²/s to 65 mmol/m²/s of CO_2 (Coyne *et al.* 1995) and high temperature optima for photosynthesis usually in the range of 30°C to 45°C (Bjorkman 1973). The temperature optima for photosynthesis in *D. eriantha* and *Eragrostis pallens* at 1 400 mmol photosynthetic photon flux density (PPFD)/m²/s¹ was 30°C and decreased with decreasing radiant flux density (Cresswell *et al.* 1982). C_3 plants usually have CO_2 compensation points ranging from 30 mℓ/ℓ to 70 mℓ/ℓ, lower photosynthetic rates ranging from 6 mmol/m²/s to 22 mmol/m²/s of CO_2 and temperature optima for photosynthesis ranging from 15°C to 30°C (Coyne *et al.* 1995). C_3 species saturate at about 50% of full sunlight (Coombs 1985).

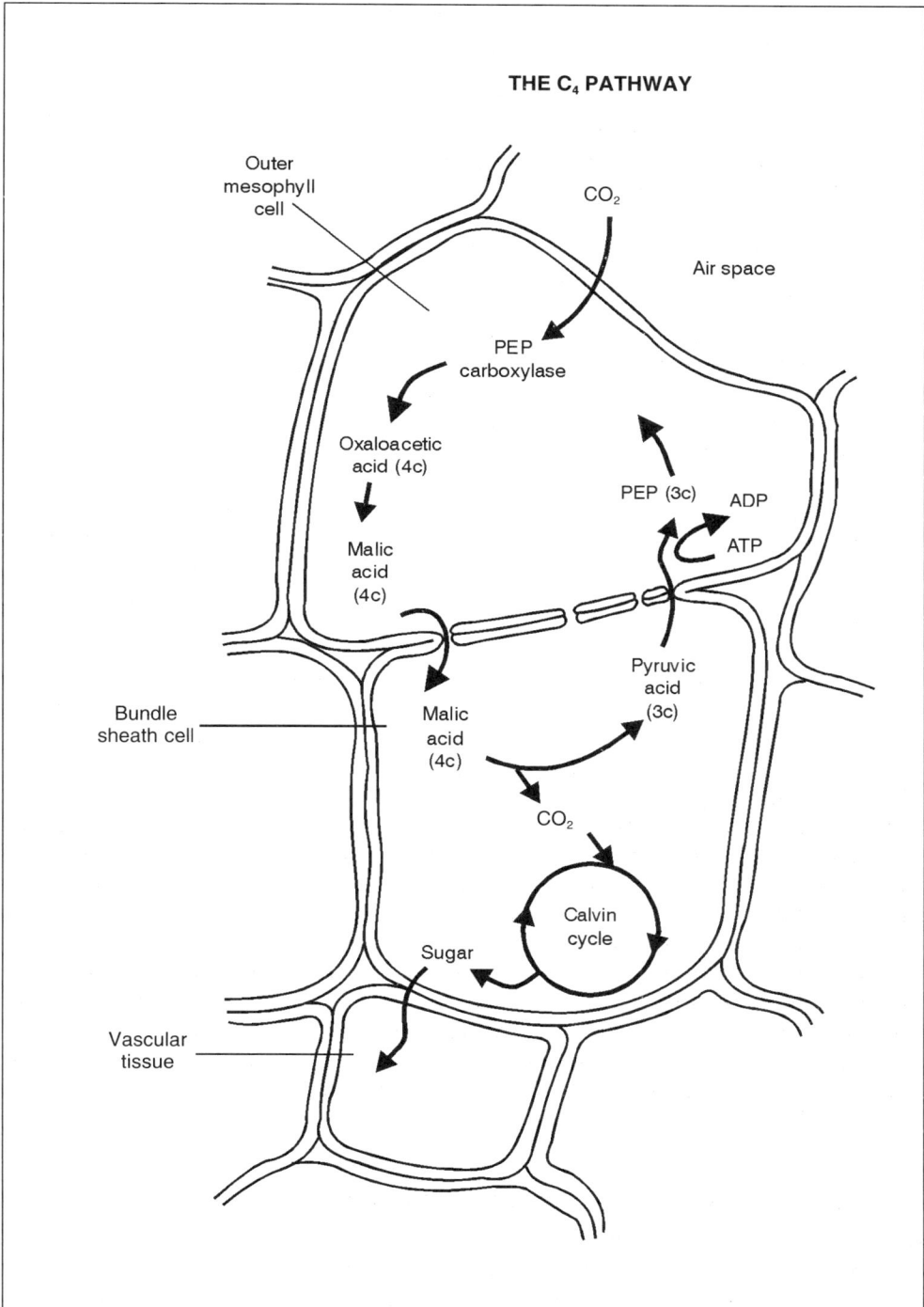

Figure 2.9 The C_4 pathway of photosynthesis. PEP = Phosphoenolpyruvate; ADP = Adenosine diphosphate; ATP = Adenosine triphosphate.

The global CO_2 concentration is increasing at an unprecedented rate (Keeling *et al.* 1982). Owing to their specialised photosynthetic mechanism it would be expected that C_4 plants would generally exhibit a smaller photosynthetic and growth stimulation in elevated CO_2 than C_3 species (Arp *et al.* 1993). The general hypothesis is that increasing CO_2 levels will shift the competitive balance in favour of C_3 species (Johnson *et al.* 1990). However, interactions between elevated CO_2 and other environmental factors such as water stress may alter the CO_2 response (Owensby *et al.* 1993). Wand *et al.* (1996) showed that net CO_2 assimilation rates in *Themeda triandra*, a C_4 species, were unaffected by elevated CO_2, but stomatal conductances and foliar N levels decreased and water-use efficiency increased. The biomass of all vegetative fractions and, in particular, leaf sheaths and leaf blade lengths, increased when CO_2 levels were enhanced. However, tiller numbers declined, possibly because the greater leaf area caused increased shading of the basal nodes.

The compensation point of C_4 plants is unaffected by oxygen concentration whereas C_3 plants are sensitive to increased oxygen concentrations. This is considered to be linked to photorespiratory activity (i.e. light induced respiration). Recent evidence appears to suggest that some photorespiratory activity also occurs in C_4 plants.

In spite of the apparent advantages of the C_4 over the C_3 pathway, the former does not necessarily lead to higher growth rates and dry matter yields. There are examples where C_4 plants are inferior to C_3 plants in both these characteristics, although their rates of photosynthesis were higher (Ludlow 1976). The C_4 pathway confers the potential for high productivity but this is achieved only in the absence of biological limitations and in a favourable environment.

The primary advantage of the C_4 photosynthetic pathway to plants in arid and semi-arid environments is that it permits efficient photosynthesis while simultaneously restricting water loss due to partial closure of stomata (Ludlow 1976, 1980; Waller & Lewis 1979; Osmond *et al.* 1980; Jones 1985; Pearcy *et al.* 1987). These species are well adapted to high temperatures but are unable to tolerate low temperatures during the growing season (Caldwell *et al.* 1977). In southern Africa, therefore, the C_4 grass species predominate within the natural subtropical grasslands (Vogel *et al.* 1978). These exclude the grasslands of the highest parts of the Drakensberg mountains, the winter rainfall areas of the Western and Northern Cape and the outliers of the mountains of the Cape folded belt (Ellis *et al.* 1980). Here the C_4 species are largely replaced by C_3 species better adapted to these colder conditions. The latter species are also planted extensively throughout the country for forage largely because they grow more actively during the winter season when water is available and because of the much higher quality of the forage which these species provide (refer to Chapter 5). A summary of the main photosynthetic features and characteristics associated with the different pathways is presented in Table 2.1.

2.1.2.2.2   *The use and storage of carbohydrates*

The main sites for the use of carbohydrates in a growing plant are the meristematic regions – the stem apex, growing leaves and root tips. These organs, with the exception of growing leaves in the later stages of their development, must be supplied with carbohydrates which are transported from the photosynthesising parts of the plant or from stored reserves. Transport to the meristematic regions appears to be well ordered, with

Table 2.1 Generalised comparison of plant characteristics associated with C_3 and C_4 photosynthetic carbon reduction pathways (developed from information in Downtown 1971; Fitter & Hay 1981; Devlin & Witham 1983; Salisbury & Ross 1985; Farquhar *et al.* 1989; Hay & Walker 1989).

Characteristic	C_3	C_4
Leaf anatomy	Diffuse mesophyll; no distinct bundle sheath	Radially arranged mesophyll surrounding distinct bundle sheath ('Kranz')
Initial product of carboxylation	3-carbon acid	4-carbon acid
Primary carboxylation enzyme	Ribulose-1,5-biphosphate carboxylase/oxygenase	Phosphoenolpyruvate carboxylase
Maximum photosynthetic rates ($\mu mol/m^2/s$)	6–22	19–65
O_2 inhibition of photosynthesis	Yes	No
Photorespiration	High	Low (bundle sheath only)
CO_2 compensation point ($\mu \ell / \ell$)	30–70	1–10
CO_2 saturation point ($\mu \ell / \ell$)	500 or higher	200–250
Discrimination against ^{13}C	High	Low
Chloroplasts	One structure	Two structures
Stomatal conductance	High	Low
Mesophyll conductance	Low	High
Light saturation level	Low	High
Temperature optimum	Low	High
Transpiration	High	Low
Water-use efficiency	Low	High
Nitrogen-use efficiency	Low	High
Productivity	Low to high	High

leaves in different stages of development or at different positions on the stem contributing carbohydrates to different parts of the plant.

The stem apex and the unexpanded leaves arising immediately below it on the stem import all their energy requirements from the more mature expanded leaves lower down the same stem. As the young leaves expand and are exposed to light, they begin to photosynthesise themselves and are able to supply at least a part of their own energy needs. However, they do not become self-sufficient until their growth has been completed, when they reach maximum photosynthetic efficiency and their own demand for energy declines. It is only at this stage that they begin to export carbohydrates to the stem apex. As they mature further the role of supplying the apex is taken over by more recently developed leaves above them on the stem. Their own photosynthetic products are then diverted downwards for use by the root systems and lateral tiller buds. This occurs only once the demands of the apex have been met. Having satisfied the requirements of the root systems and the apices of any young lateral tillers that may be developing, excess carbohydrates are accumulated in the stem bases, rhizomes, runners and roots. These accumulated non-structural carbohydrates are commonly known as the 'carbohydrate reserves' and include the monosaccharides glucose and fructose, and the more complex oligosaccharides, sucrose and fructosan. Starch is also a major reserve carbohydrate in some perennial grasses, particularly in those from the tropics and subtropics, while those of temperate origin accumulate fructosans.

Non-structural carbohydrates are traditionally accepted as the primary source of carbon for initial growth of new tillers or regrowth of old tillers after defoliation (Trlica 1977; Briske & Woie 1984). Regrowth of the plant must depend on these stored supplies of carbohydrates when the leaf material has been harvested and the plant can no longer produce sufficient carbohydrate to meet its immediate needs. However, these stored carbohydrates form only a small proportion of total assimilated carbon relative to that used for respiration or that metabolised into structural carbohydrate and protein (Gordon *et al.* 1977). The stem bases are the most important storage organs in many of the tufted grasses and are sometimes modified in such a way that they can perform this role very effectively, e.g. the swollen bases of such species as *Phalaris tuberosa*. In the rhizomatous grasses in particular, the underground stems serve as storage organs and contribute to the ability of these grasses to recover after heavy use.

Storage becomes particularly important in pasture species which are harvested frequently and in perennial species which must recover following periods of drought or cold which completely destroy their leaf systems. The plants then draw on the stored reserves to produce new leaf material and in so doing deplete the storage organs. It is only when sufficient leaf has been produced to provide the plant with energy compounds to divert into growth, that growth can again proceed rapidly, and the storage organs can be replenished. As previously indicated, storage occurs only when energy compounds are produced in excess of the plant's immediate requirements for growth. Hence, storage may be delayed when conditions promote rapid growth following defoliation. Until such time as reserves have been replenished, the plant is particularly vulnerable. If the young developing leaf is removed, the plant has neither sufficient reserves nor sufficient leaf to provide it with the carbohydrates it needs to recover, even at a moderate rate. Recovery growth is therefore likely to be slow even under ideal growing conditions and the root systems in particular will suffer from the lack of available carbon.

Considerable research aimed at determining seasonal and post-defoliation trends in the concentration of total non-structural carbohydrates in various plant tissues (Weinmann 1940a, 1940b, 1943, 1944; Weinmann & Reinhold 1946; Daitz 1954; Hyder & Sneva 1959; Bartholomew & Booysen 1969; Nursey 1971; Trlica & Cook 1971, 1972; Gifford & Marshall 1973; Steinke 1975; Buwai & Trlica 1977; Daer & Willard 1981; Menke & Trlica 1981) has produced generalised patterns for reserve carbohydrate accumulation and depletion in grasses which can be summarised as follows: Initially, a new tiller imports substrates from its established parent tiller for a period of time. It should eventually develop sufficient leaf area to become photosynthetically self-sufficient. It may, nevertheless, remain dependent on the primary tiller for water and nutrients. As the tiller approaches maturity, sugars and polymers are photosynthesised by the leaves at a rate exceeding the requirements of aerial growth, and the excess is translocated via the phloem to the roots, stem bases and other storage organs. The plant then draws on the reserves whenever the demands of new growth for energy exceed its ability to supply such energy from current photosynthesis. This occurs not only with the emergence of new tillers but whenever new growth is produced after cutting or grazing.

In addition, Briske & Woie (1984) suggest that a physiologically independent tiller can itself revert to being an importer of carbon should it enter a period of negative carbon balance. Carbon transport may in fact occur continually among tillers regardless of the stage of vegetative development and this transport can apparently increase several fold within minutes of either shading or partial defoliation.

Work by Sagar & Marshall (1966) and Marshall & Sagar (1968) suggested that the grass tuft is a highly integrated organism. Instead of becoming dominant, intact shoots supply damaged shoots with assimilates, so that each tiller tends towards an evenness of size. In contrast, Langer (1972), in a review on growth of grasses, suggested that carbohydrate movement between older tillers appears to be restricted. Unpublished results of research at the Motopos Research Station in Zimbabwe likewise suggested that reserve re-allocation among established tillers may well have been minimal, even after partial defoliation of tufts (Barnes pers. comm.). Recent data produced at Ukulinga in KwaZulu-Natal (Daphne 1993) support the contention that there is little within-tuft transfer of carbon between non-defoliated and defoliated tillers, at least in the veld grass *T. triandra*.

Danckwerts & Gordon (1987) monitored the flux of ^{14}C from a fed tiller to the rest of the plant in a temperate *(L. perenne)* and a tropical (*T. triandra*) grass species. In both species the amount of ^{14}C recovered was greatest in secondary tillers subtended by the fed tiller. However, some ^{14}C was also translocated to other tillers not subtended by the fed tiller, implying that tillers do not act as independent units even where there is no defoliation. After uniform defoliation, the recovery of ^{14}C in the regrowth of the various tiller categories was approximately proportional to the occurrence of ^{14}C in the same categories before clipping. It thus seems that there is little change in the pattern of redistribution of reserve carbon after uniform clipping of the whole plant (Danckwerts & Gordon 1987).

In the summer-growing species which characterise the grasslands of most of southern Africa, storage reserve levels may be expected to be relatively high in grass plants which have retained a reasonable amount of active leaf through the autumn period. A small proportion of these reserves will be used during the winter period to sustain the life processes in the plant. In spring, with the onset of growth, the reserve levels will decline rapidly in the storage organs as heavy demands are made on them by the growing regions of the plant (Weinmann 1955). These reserves will be replenished only once carbohydrates are produced in excess of the current demand by leaf systems, so that reserve levels will remain low unless an adequate leaf system is allowed to develop. Recent work in the Highland Sourveld of KwaZulu-Natal suggests that storage may not commence in any meaningful way until mid-summer and that reserve levels will remain low until late January (Peddie *et al.* 1996).

Even if reserve recovery commences early in the season, recovery through the summer may be expected to be checked at flowering time by the heavy demands of the rapidly developing stem and inflorescence. Work with radioactive carbon has suggested that this may largely be supplied by the flag (uppermost) leaf and other leaves occurring towards the top of the tillers, so that reserve withdrawal may in fact be minimal (Langer 1972). On completion of flowering, reserves can again accumulate, particularly in autumn when growth rates are slow but conditions remain suitable for continued photosynthesis. The critical seasonal periods in the plant therefore seem to be the early spring (as demonstrated by Weinmann 1940b), during the flowering period and during any period when growth rates are high following reserve withdrawal. The autumn period assumes importance (Weinmann 1940b) because this is when reserves which ensure winter survival and spring recovery are accumulated. By way of example, root reserves in *T. triandra* have indeed been shown to reach a minimum in summer (8.4% of dry matter) and a maximum in winter (18–19%) (Daitz 1954).

2.1.2.2.3 Plant growth and nitrogen

Of all the plant nutrients required for metabolism and growth, N seems to be one of the most limiting (Barneix *et al.* 1984). Not surprisingly, more N is applied to grasslands than any other nutrient (Brown & Ashley 1974).

The main sources of N in the soil are nitrate and ammonium ions. The physiological response to these ions is very different and the ability of plants to absorb and/or metabolise them varies greatly (Hewitt 1975). The rate of dry matter production in *L. perenne*, for example, diminishes at very low nitrate levels (Alberda 1968) and the root:shoot ratio of *Dactylis glomerata* is strongly dependent on the nitrate concentration of its growth medium (Caloin *et al.* 1980). In contrast, calcifuge (calcium-loving) plants growing naturally in acidic soils where there is little nitrification, seem to be adapted to use ammonium in preference to nitrate.

For the majority of species, however, it seems that a mixture of ammonium and nitrate N produces the greatest growth and protein content, although the optimum ratio of ammonium and nitrate probably differs for different species and may change with the age of the plant and both the temperature and the pH of the growth medium (Haynes & Goh 1978).

In essence, the N economy of plants can be considered as a balance between the need to produce vegetative and reproductive growth (Millard 1988). Leaf senescence is often associated with N deficiency as a consequence of the remobilisation of the N they contain for reproductive growth. Nitrogen may also alter the balance between carbon storage and carbon utilisation for growth. Nitrogen deficiency, which reduces the amount of N available for cell division and/or cell expansion in meristematic tissue, for example, would reduce the overall assimilate demand of these sinks and so lead to assimilate storage in a plant (Oparka *et al.* 1986).

Evidence for the involvement of N availability in carbon accumulation and, through this in the production of tannins, is that low tannin levels have been measured when soil N levels are high and reserve carbon levels are low because of their use to promote growth (Du Toit 1992) (Table 2.2). Nitrogen deprived plants of *Panicum maximum* seemed to be unable to mobilise starch accumulated during the photosynthetic period to the same extent as N supplied plants (Ariovich & Cresswell 1983). Wilson (1975) suggested that N stimulated the translocation of assimilates from the leaves as well as their utilisation at a separate site, with a corresponding reduction in the level of carbohydrates accumulating in the leaves.

Energy from photosynthesis is employed for N acquisition and reduction and a plant's photosynthetic capacity is strongly coupled to leaf N content. Nitrogen fertilization is frequently accompanied by a stimulation of photosynthesis, as well as by an increase in leaf area. Robson & Parsons (1977) found that plants receiving low concentrations of N have a smaller leaf area than plants receiving high N levels. This difference was due much more to the greater allocation of new growth to leaf growth in high N plants than to their slightly higher relative growth rate. Shortages of N may also lead to chlorosis (yellowing), which may reduce photosynthetic rates.

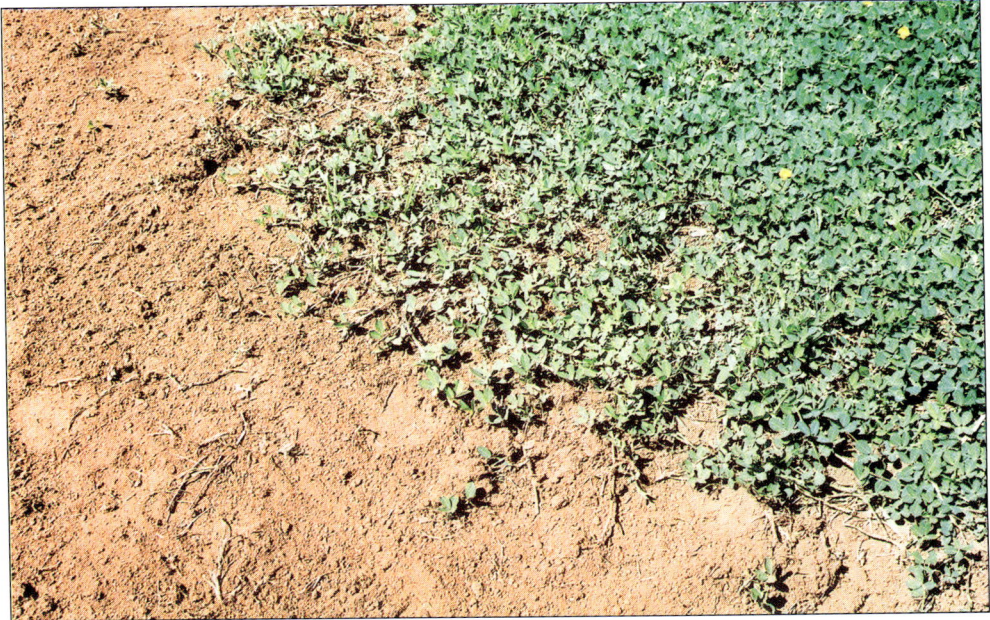

Plate 1   The prostrate growth form of the forage legume represented by the forage groundnut (*Arachis* spp.) (refer to section 2.2.1).

Plate 2   The tufted growth form of the forage legume, represented by lespedeza (*Lespedeza cuneata*) (refer to section 2.2.1).

Plate 3   The rambling form of the forage legume, represented here by silverleaf desmodium (*Desmodium uncinatum*) (refer to section 2.2.1).

Plate 4   The shrub form of the forage legume, represented by leucaena (*Leucaena leucocephala*) (refer to section 2.2.1).

Table 2.2　Tannin concentration of *Eulalia villosa* treated with no nitrogen (control), 50 mg nitrogen/dm³ (low nitrogen treatment) and 200 mg nitrogen/dm³ (high nitrogen treatment). The results are expressed in mg/g dry mass. Values are means ±SD (Du Toit 1992).

Treatment	Total phenols	Condensed tannin	Proanthocyanidin	Protein precipitation
Control	36.0 ± 1.2	114.0 ± 3.7	39.1 ± 1.2	11.7 ± 0.8
Low nitrogen	34.5 ± 1.1	106.4 ± 6.8	42.4 ± 1.7	13.2 ± 0.9
High nitrogen	22.7 ± 1.8	65.4 ± 2.4	29.3 ± 1.1	8.7 ± 0.9

Although N fertilizer increases the protein content of plants, it usually leads to a decline in the amount of carbohydrate reserves (Brown & Ashley 1974), but the effects of N fertilizers on carbohydrate reserves are complex and variable. Increased protein is associated with the production of new tissue and thus increases may be expected in respiration, because of the increased amount of tissue which needs to be supported, and/or because respiration provides the energy needed for protein synthesis, cell division and cell expansion.

2.1.2.2.4　Growth and moisture stress

The effects of moisture stress on the growth and development of grasses will vary among different plant species according to the growth stage of the plant, the duration of the moisture stress period and management prior to and during the stress period. Seedlings are generally more susceptible to water deficit mortality than mature plants primarily because of their less extensive root systems (Johnson 1980).

Generally, the first response of plants to a diminishing water supply is reduced cell enlargement, followed by a decrease in the rate of cell division (Slayter 1967). Results of studies made by Turner & Begg (1978) indicate that the rate of photosynthesis is less sensitive to moisture stress than is leaf expansion in the species they studied. The first visible response by the plant to an increased intensity or a lengthening of the period of a water deficit was a slowing of leaf growth and shoot development (Brown 1995). Leaf expansion has been found to cease after 40% soil water depletion, with significant leaf senescence only after 70% soil water depletion (Danckwerts 1988). Maximum growth rates and net assimilation rates reached by the plants were reduced under conditions of moisture stress (Danckwerts 1988), and even partial stress adversely affects plant growth.

In a study on the impact of moisture supply on the year-to-year variation in kikuyu yields and the response of this grass to fertilizer, Miles (1997) found that at Cedara, in the relatively dry 1982/83 season when the October to March rainfall was 66% of the long-term mean, the efficiency of response to N was only about 50–65% of that of the 1983/84 season when rainfall was 101% of the long-term mean.

Relatively short moisture stress periods during any phenological phase have been shown to cause significant increases in the root mass of some grasses, as well as significant increases in the total available carbohydrate content of roots and stubble (Opperman & Human 1976). This suggests decreased utilisation of carbohydrates during such periods and a greater effect of moisture stress on growth processes than on photosynthesis and therefore also on stomatal action (Brown & Blaser 1965). In *P. maximum*, water stress delays stem elongation and flowering and may also delay the normal ontogenetical changes in the leaf which may have implications for herbage quality (Wilson & Ng 1975).

2.2 PASTURE LEGUMES   *V.D. Wassermann, M.B. Hardy &*
                                                          R.J. Eckard

2.2.1 General morphology

The legume family *Fabaceae* comprises a large number of species within which there is a broad morphological variation. This family is now commonly subdivided further into the three sub-families: the *Papilionoideae, Mimosoideae* and *Caesalpinioideae*. Most of the agriculturally important legumes, including most pasture legumes, belong to the *Papilionoideae* sub-family (Allen & Allen 1981; Kretschemer 1989).

Due to the large variation in soil and climatic conditions in South Africa, as many as 50 annual and perennial pasture legumes are used to a greater or lesser extent. Although mostly herbaceous, the growth form of these legumes ranges from the twining (such as the *Desmodium* spp.) and creeping, prostrate types (such as white clover – *Trifolium repens*), to the more upright, tufted (e.g. lucerne – *Medicago sativa*) and shrubby (e.g. Leucaena – *Leucaena leucocephala*, a member of the *Mimosoideae*) growth types (Plates 1–4). Both tropical and temperate legumes are used in pastures, with a predominance of temperate legumes in the cooler, higher rainfall regions.

It is not possible, nor appropriate, in this text to provide a detailed morphological description of all the legume species used in South African pastures. Only a general outline of the morphology of legumes will be provided. For greater detail, the reader is referred to Sharman (1947), Duke (1981) and Taylor (1985).

2.2.1.1 *Vegetative structure*

(a) The stem

The stem morphology (including woodiness) of pasture legumes varies greatly among different species, but has one common feature which differentiates them from the grasses and results in a substantially different response to defoliation. While both grasses and the legumes have their primary growing points terminal to the stem, in legumes this terminal growing point is, except in some of the prostrate legumes, completely exposed to removal by even light grazing, whereas in grasses, as previously discussed (section 2.1.1.2), the apex remains low to the ground during vegetative growth in most species. Here it readily escapes defoliation damage.

Stems of annual legumes are relatively unlignified compared with those of perennials, and particularly of the shrub legumes such as Leucaena. The upright types usually have secondary stems branching from a recognisable primary stem. Certain perennial legumes such as lucerne (*M. sativa*), the trefoils (*Lotus* spp.), sainfoin (*Onobrychis viciaefolia*) and joint vetch (*Aeschynomene falcata*) develop secondary stems from the axillary buds of the lower leaves on the original primary stem. This process is repeated as tertiary stems develop on secondary stems, and so on. As the stems of these upright, perennial legumes mature their bases become thickened and woody, forming the crown which is comprised of a collection of thick, short lengths of basal stem material. The crown is usually situated at or below the soil surface and from here further stems develop. Such development often takes place in spring or after grazing or cutting of top-growth. Once harvested, these plants no longer possess a main stem.

Some of the more prostrate, creeping legumes possess rhizomes (underground stems) from which aerial tillers develop, thus allowing the plant to spread vegetatively. Examples of rhizomatous legumes used in South Africa include crown vetch (*Coronilla varia*) and greater lotus/big trefoil (*Lotus uliginosus*, formerly *Lotus pedunculatus*). Yet other perennial prostrate legumes spread by developing stolons from the base of their primary stems. Examples include the widely used white clover (*T. repens*) and strawberry clover (*Trifolium fragiferum*).

Stoloniferous and rhizomatous legumes tend to be reasonably tolerant of heavy grazing, as their apical growth points (developing from the nodes) usually remain close to, or below, the soil surface. Certain annual forage legumes, like subterranean clover (*Trifolium subterraneum*), also develop prostrate stems, but these stems do not produce roots at their nodes. The annual temperate legumes do not produce underground stems which would allow them to escape defoliation.

Another group of herbaceous, subtropical forage legumes develops robust runners which either remain close to the soil surface, or creep along available supports such as grass tufts and shrubs. In perennial species such as siratro (*Macroptilium atropurpureum*) and *Desmodium* species, lateral roots may develop where runners come into contact with the soil. These roots do not normally form on the runners of annual species.

(b) The leaf

Apart from the leguminous shrubs, most of which have microphyllous leaves, forage legumes tend to be characterised by trifoliate leaves. These are usually borne alternately on stalks (petioles) which arise from node positions along the stem or stolon. Individual leaflets of the trifoliate leaf are themselves normally joined by short stalks, with the stalk to the terminal leaflet longer than those to the two lateral leaflets in some species (e.g. *Glycine, Desmodium, Medicago, Melilotus*). Annual *Medicago* species (the medics) have the typical trifoliate compound leaves of the perennial clover species, but in contrast to the clover leaflets, the central leaflet of the medic leaf has a distinct stalk (which is a useful diagnostic tool to distinguish between medic and clover seedlings). The leaves of annual legume species such as serradella (*Ornithopus* spp.) and vetch (*Vicia* spp.) are pinnate and 'feather-like' with 9 to 18 and 4 to 16 pairs of leaflets respectively (see Wasserman 1982; Frame *et al.* 1998). The leaflets of individual species vary considerably in size, shape, hairiness, edge serration and leaf pattern. Such variation is used extensively in species identification.

(c) The root system

Legumes are dicotyledonous plants with a typical taproot system which branches to a greater or lesser extent. In some species (e.g. lucerne), the tap root can be several meters long, while in others (e.g. white clover) it may be no longer than 40 cm (Smith 1988). The perennial legumes which produce stolons or runners from the primary stem (e.g. *T. repens, T. fragiferum* and *Desmodium* spp.) develop secondary roots at the stem nodes. The primary root eventually dies, leaving an independent system of secondary roots to nutritionally support the plant. In some plants these nodal root systems remain fibrous, whereas in others they develop a tap root (Thomas 1987). Annual legumes (e.g. *T. subterraneum* and *Medicago truncatula*) also develop a shallow branching tap root system. Unlike the perennial clover species, the prostrate to semi-upright stems produced from the primary stems do not root from the nodes to develop a secondary root system. Thus the plant dies when the primary root system dies at the onset of the dry season.

With few exceptions, legumes develop root nodules. These result from a symbiotic relation between *Rhizobium* bacteria in the soil and the roots of the leguminous host. Most nodules develop on the secondary, fibrous roots (Allen & Allen 1981). The symbiotic benefit to the legume host is access to the N fixed from the atmosphere by the *Rhizobium* bacteria. This process is discussed further in section 2.2.2.3.

2.2.1.2 *Reproductive structures*

(a) The inflorescence
The inflorescences of pasture legumes develop either at the end of the main stem, at the ends of the branches of the main stem (both terminal) or, more commonly, in the axils of leaves along the stem (axillary). Annual forage legumes tend to be self-pollinating, while perennial legumes cross-pollinate. Certain species are self-sterile and rely heavily on insect (mainly bee) pollination for seed production. Other legume species, although able to self-pollinate, require cross-pollination for successful seed-set. Red clover (*Trifolium pratense*) produces little seed in South Africa as it requires a specific insect, the bumble bee, for successful pollination.

Seed production in the annual temperate legumes which are widely used in the Western Cape is particularly important since, as with all annuals, survival depends on regular seed production. Inflorescences are produced as day length increases from the end of winter and during the early spring months. Time of flowering varies among species and cultivars within species. It is essential therefore that time of flowering is taken into account when selecting an annual temperate legume pasture to ensure that flowering and seed-set are synchronised with local climatic conditions.

(b) The fruit
The fruits of forage legumes are usually long, roughly cylindrical or flattened pods. In a number of species, like *Medicago*, the pod develops into a spiral. Pods vary greatly in size and may contain from one to many seeds. They may also be covered with spiny burrs, as in Barrel medic (*M. truncatula*) and Burr medic (*Medicago polymorpha*), to assist in seed distribution. While the fruiting bodies of most pasture legumes are borne and remain above ground, those of *T. subterraneum* are buried. After fertilization the florets reflex and the peduncle turns towards the soil surface and the developing pods are pressed into the soil surface – hence the common name of this species (subterranean clover) (Wasserman 1982; Frame *et al.* 1998). Thus the fruiting bodies of this species can escape grazing and maintain soil seed banks – an important survival strategy for this annual plant.

(c) The seed
Legumes are dicotyledonous, with seeds varying greatly in size, shape and colour. The seeds are encased in an extremely hard seed coat, which may be variably im-

pervious to moisture. As the seed coat degenerates, often over a number of years, the restriction on moisture imbibition is reduced, allowing the seeds to germinate. 'Hardseededness' is important in limiting germination to only a portion of the seed population on any particular occasion, thereby ensuring adequate seed reserves in the soil for subsequent opportunistic germinations (Jones & Carter 1989). This is particularly important in the annual leguminous species commonly used in pasture systems in the Western Cape, although it does vary quite considerably among the different annual pasture legumes used. In this area the plants normally die at the start of the summer dry season and a sufficient number of seeds must be able to survive the hot, dry summer months. Medic species are particularly well adapted to this survival strategy as they may produce as much as 400 kg/ha of seed in a season, only a portion of which will germinate at the start of the next season. This means that a high proportion of potentially viable seed remains in the soil for several years. Seradella also exhibits a considerable degree of hardseededness but this is not true of *T. subterraneum* (subterranean clover) and *Trifolium balansae* (Balanse clover). In these two species unseasonal rains may induce germination, thus diminishing soil seed banks and the ability of the pasture to re-establish.

Because of hard seededness, scarification is normally required to induce uniform germination when establishing a leguminous pasture (refer to Chapter 10).

2.2.2   General physiology

2.2.2.1   *Light*

(a) Plant growth

Legumes photosynthesise via the C_3 pathway, so that their leaves light-saturate at intensities well below full sunlight (Table 2.1). When exposed to full sunlight these leaves cannot match the photosynthetic levels achieved by C_4 plants. However, due to the growth habit of legumes in a grass/legume pasture, most legume leaves are either partially shaded, or not inclined perpendicularly to incoming radiation. Reasonably dense grass/legume swards are, therefore, still capable of making full use of available sunlight. If the pasture canopy becomes so dense that new leaf growth (at the top of the canopy) equals premature senescence of older shaded leaves, light-use efficiency is greatly reduced.

As with all pastures, legume yields will be maximised when the leaf area index (LAI; leaf area per unit area of pasture) is such that all leaves make a positive contribution to energy capture, with little incident light reaching the soil surface (Donald & Black 1958; Donald 1963). Brougham (1958, 1960) established optimal LAI values of three to four for white clover (*T. repens*) with its horizontally orientated leaves. This compares with values of between seven and eight in plants with more vertically oriented leaves, such as ryegrass (*L. perenne*). In general, the taller the plant the higher its optimal LAI (Brown & Blaser 1968).

(b) Plant structure

Increased light intensity results in a higher leaf appearance rate which is, in turn, responsible for a higher rate of secondary branch development. This ultimately results in a greater leaf area available to capture incident sunlight, a very important factor influencing productivity.

Under conditions of diffuse light, as in a dense canopy, plants are able to partially compensate for the lower light intensities by producing thinner leaves, thus maximising the leaf area per unit energy expended on leaf production (Blackman 1956). Stems developing under low light intensities tend to grow taller than normal, thereby distributing their leaves over a greater vertical distance (Buxton 1989). This reduces self-shading. However, this increased production of top growth is often at the expense of root development (Black 1957; Buxton 1989). This reduces the tolerance of the plants to harsh conditions.

(c) Photoperiodism

Plants are usually adapted to flower when climatic conditions, in their region of origin, are optimal for flowering and seed production, with the largest response being that to changing daylength (Chang 1968; Aitken 1974). This response is known as photoperiodism. Plants are broadly classified into three categories of photoperiod response:

1) short-day plants which flower only, or most rapidly, when exposed to less than a certain species-specific daily photoperiod;

2) long-day plants (mostly temperate plants) which flower only, or most rapidly, when exposed to more than a certain species-specific daily photoperiod; and

3) daylength neutral or indifferent plants (mostly tropical and subtropical plants) in which the flowering response is independent of photoperiod.

Long-day plants may also require exposure to a specific (low) temperature before flowering. In both short- and long-day plants, the response to changing daylength may either be absolute, i.e. they will not flower unless their daylength requirements are met, or quantitative, i.e. short or long days may promote flowering, but are not mandatory.

Plants grown in latitudes different from those of their regions of origin may never flower. A species response to photoperiod therefore needs to be understood if seed production is an important part of the production system, as for example for 'breeder seed' production, in self-seeding annual legumes, or when the seed provides an important source of dry-season animal protein (as with the burrs of the annual medics (*Medicago* spp.)).

2.2.2.2   Temperature

The optimum and minimum temperature requirements for growth varies markedly among species (Chang 1968; Buxton 1989). A typical temperate species has a minimum temperature requirement of between 0°C and 5° C for growth, a maximum tolerance to between 31°C and 37° C, and grows optimally between 25° and 31° C. The annual temperate legumes commonly used in the winter rainfall regions of the Western Cape are adapted to mean daily temperatures in the range 5°C to 20°C. Day/night temperatures of 25/15°C appear optimum for the germination of seed of these legumes, e.g. subterranean clover (Frame *et al.* 1998). Germination rate is high at soil temperatures above 15°C but temperature induced dormancy occurs when soil temperatures are much higher than this.

Subtropical legumes have a minimum temperature requirement of between 15° and 18° C, a maximum tolerance to between 44° and 50° C and grow optimally between 31° and 37° C. These temperature ranges also apply to germination. Temperatures below 15° C may severely restrict nodulation of legumes (Buxton 1989).

As mentioned previously, long-day plants often require low winter temperatures, in addition to a change in photoperiod, before they will flower. The required degree of exposure to cold (vernalisation) varies with species, while some species opportunistically respond to either short days or vernalisation.

2.2.2.3 Biological nitrogen fixation

Legumes are unique in their ability to utilise the limitless supply of chemically inactive atmospheric N to satisfy their N requirements. They achieve this through a symbiotic relation with N-fixing *Rhizobium* bacteria which form nodules on their roots. By supplying the bacterium with carbohydrate, the host plant becomes independent of soil N. Rhizobia are aerobic (oxygen demanding), non-spore forming, gram-negative, rod-like soil bacteria, or more specifically rhizosphere bacteria, which require root exudates for multiplication (Rovira 1961; Van Egeraad 1975).

Although indigenous strains of rhizobia are present in most soils, most legume species require a highly specific species or strain of *Rhizobium* for effective nodulation and N fixation. Species of *Rhizobium* bacteria are, therefore, designated by their host range or group (Table 2.3). But even within these host groups, particularly with *Rhizobium japonicum*, there are strains which vary in their ability to effectively nodulate a specific legume host – even different cultivars within a legume species may require different strains of rhizobia for effective nodulation.

Table 2.3 Legume host groups for specific *Rhizobium* species (after Date 1970; Allen & Allen 1981).

Host group	Rhizobium species	Host affinities	Plant
1	R. meliloti	Medicago, Melilotis	lucerne, medics
2	R. leguminosarum	Pisum, Lathyrus, Vicia, Lens, Cicer	peas and vetches
3	R. trifolii	Trifolium	clovers
4	R. lupini	Lupinus, Ornithopis	lupins and seradellas
5	R. japonicum	Glycine, Arachis, Vigna, Stylosanthes, Puereria, Macroptilium atropurpureum, Desmodium, Lotus and many others	soya beans, groundnuts, various beans, stylo, kudzu, siratro, trefoils

When a *Rhizobium* bacterium comes into contact with the roots of an appropriate host seedling, it multiplies rapidly and enters the root through the root hairs. These hairs then curl noticeably. Once inside the root hair, rapid multiplication continues and an infection thread develops which enters the root cortex. The presence of the rhizobia stimulates cell division in the root, forming the nodule. It normally takes three to four weeks before the nodule is recognisable (Allen & Allen 1981).

The nodules vary greatly in size among the different host legumes. The interiors of active nodules (those fixing N) are pink to deep red in colour, changing to green as the nodule decays. Ineffective nodules appear translucent or white inside.

3

The response of forage plants to defoliation

CONTENTS

3.1 GRASSES *M.M. Wolfson*

Consideration will be given in this chapter to the influence of defoliation on the growth and development of grass and legume plants since it is their responses to this treatment that largely determines their forage production under practical field conditions. A good understanding of such responses will allow managers to manipulate grazing practices so as to maximise the forage yields of their pastures.

3.1.1 Effects of defoliation on individual plants

3.1.1.1 *The effect of leaf removal*

The effect of leaf removal on plant growth is dependent on whether the whole or only part of the leaf is removed, the stage of development of the leaf which is removed and the extent to which the leaf area of the plant as a whole is reduced. The apical meristem, along whose flank new leaves are initiated (see Fig. 2.3), is usually situated at the base of the tiller while the meristematic regions in young leaves are also situated at their base. Therefore, in immature leaves the removal of only part of the leaf will not affect its subsequent growth provided the basal meristems remain intact, although the final size of the leaf and its ability to produce photosynthates will be reduced. The removal of either the sheath or blade meristem, however, will terminate growth in that leaf organ.

In older leaves in which the lamina has stopped expanding but the sheath is still active, growth of the sheath will continue provided its basal meristem is not removed. In yet older leaves which have emerged ligules and have reached their full size, growth cannot be resumed following defoliation of any part of the leaf. The removal of part or the whole of a mature photosynthetically active leaf will therefore have no direct effect other than to reduce the size of the photosynthetic system of the plant. A small reduction in the available photosynthetic area may result in a compensatory increase in the photosynthetic efficiency of the remaining leaves. However, if any quantity of leaf is removed, the effect will be to reduce the amount of carbohydrate produced and available for export to the growing or storage organs (Fig. 3.1). This in turn will lead either to a reduction in growth rate, with its associated effects which may include a change in the growth pattern of the stem and delayed stem development and flowering, or to a reduction of the plant's ability to recover following subsequent harvests.

The often, only temporary, decline in photosynthetic rate (Briske & Richards 1995) which follows partial defoliation (as in normal harvesting operations which remove only the upper part of the canopy) is not necessarily proportional to leaf area or biomass removed. In this respect there are a number of possibilities. When previously shaded leaves remain on the plant following partial defoliation, canopy photosynthesis may be reduced to a greater extent than the proportion of the leaf area removed because of the low photosynthetic capacity of the remaining basal leaves (Ludlow & Charles-Edwards 1980; Gold & Caldwell 1989). Alternatively, leaves in partially defoliated plants may exhibit higher photosynthetic rates than comparable leaves on undefoliated plants, either because the normal decline in photosynthetic capacity associated with ageing is inhibited (Gifford & Marshall 1973) or because of the exposure of leaves remaining on the plants to greater light intensity and quality (Wallace 1990; Senock *et al.* 1991). Compensatory photosynthesis has been documented in a large number of species, including crested wheatgrass (Caldwell *et al.* 1981), Italian ryegrass (Gifford & Marshall 1973) and perennial ryegrass (Woledge 1977). This may result in a lower reduction in post-defoliation photosynthetic rates than the proportion of leaf material removed.

Where a high proportion of relatively young leaves remains on the plant following partial defoliation, whole-plant photosynthesis may be very closely related to the amount of leaf area removed (Briske & Richards 1995), but repeated severe defoliation may reduce photosynthetic rates even in very grazing-tolerant species (Wallace 1981; Wallace *et al.* 1985).

The removal of old leaves which have become photosynthetically inefficient should not materially affect the subsequent growth of the plant. These leaves produce little or no carbohydrates for export, nor do they import from other leaves.

Complete removal of green leaves will make initial regrowth dependent on carbohydrate reserves, leading to a reduced growth rate after clipping. However, different species may respond differently to different intensities of defoliation at different times of the year. An example of this in grasses is that of *Themeda triandra* growing in the semi-arid savanna of the Eastern Cape which seems sensitive to intense and frequent defoliation during the growing season, particularly during spring, and the green leaf remaining after defoliation plays an important role in regrowth. In contrast, *Sporobolus fimbriatus* seems well adapted to severe defoliation during the growing season, when it

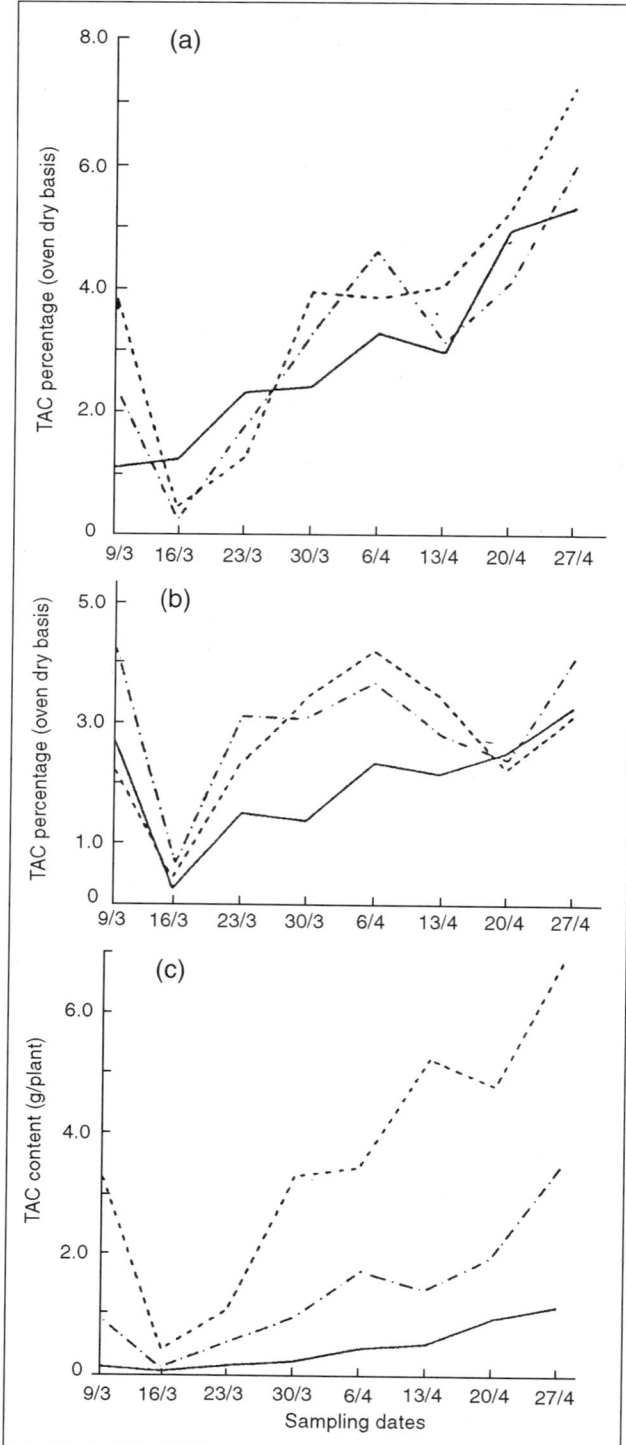

Figure 3.1

Changes in percentage Total Available Carbohydrates (TAC) in the (a) crowns and (b) root systems and (c) the mean TAC content of both the crown and the root systems of *Eragrostis curvula* during the recovery period following defoliation (after Steinke & Booysen 1968). Treatment cutting intervals: 2 weeks, ———; 4 weeks, —•—; 12 weeks, - - - - - -

is less adversely affected by frequent defoliation than is *T. triandra* (Danckwerts 1984). These results indicate that, even though the two species grow side by side, they prefer very different defoliation regimes.

The traditionally accepted role of storage carbohydrates in plant recovery following any defoliation treatment has often been questioned (May 1960; Ward & Blazer 1961; Ryle & Powell 1974; Caldwell *et al.* 1981; Atkinson & Farrar 1983; Richards & Caldwell 1985). The basis of the dispute has generally been a lack of correlation between non-structural carbohydrate concentrations in storage organs and the rate of regrowth after defoliation (Fig. 3.2). Richards & Caldwell (1985) suggest three possible reasons for this:

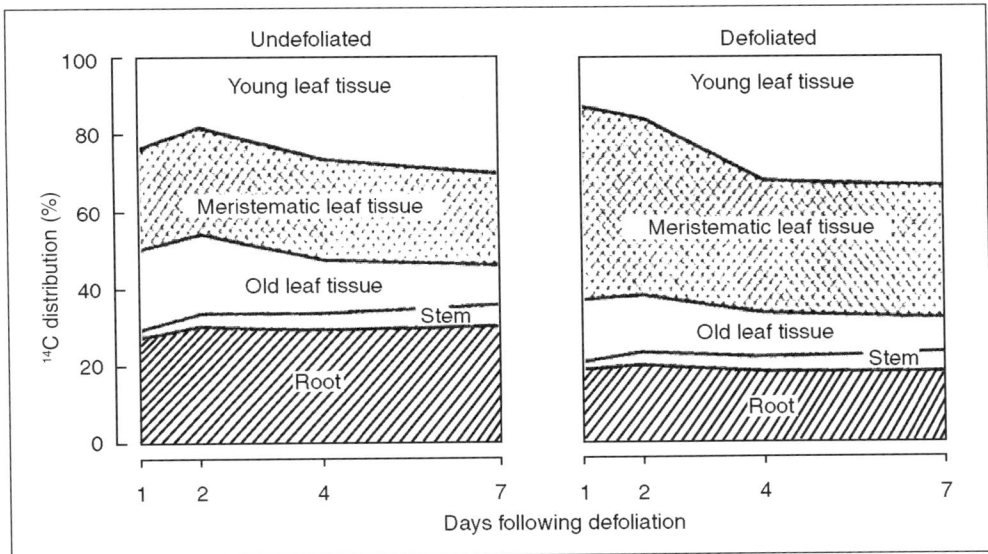

Figure 3.2   Relative allocation of ^{14}C to various components of defoliated and undefoliated barley plants. Plants were labelled for 25 minutes and harvested at 1, 2, 4 and 7 days. Defoliated plants were cut at the ligule of the third leaf at the beginning of day 1. Allocation to meristematic leaf tissue (apical meristem, leaf primordia, and unexpanded leaves) increased and allocation to roots decreased at day 4 and 7 of regrowth (from Ryle & Powell 1975).

(a) the contribution of concurrent photosynthesis to regrowth is large relative to that of the reserve substrates;

(b) non-structural carbohydrate reserves, as assessed by traditional procedures, do not adequately represent available substrates for regrowth; and

(c) meristematic restrictions may limit regrowth more than the insufficiency of energy substrates.

An additional possible reason is:

(d) the compensatory increase in photosynthetic rate of the remaining photosynthetic surface.

Data collected by Zarrough *et al.* (1984) indicate that the quantitative contribution of carbohydrate reserves to plant growth following defoliation is much smaller than was previously assumed. The non-structural carbohydrates represent less than 10% of the stored carbon used for regrowth (Richards & Caldwell 1985). These authors came to this conclusion after studying etiolated regrowth of *Agropyron* species, where changes in the total available carbohydrate pool could explain only 52% of the change in crown biomass. Therefore soluble compounds other than carbohydrates must have been utilised as a source of carbon. Similar results have been reported with other pasture species by Davidson & Milthorpe (1966), Chung & Trlica (1980) and Dewald & Sims (1981). However, the identity of such soluble compounds is not clear.

Studies of carbon balance and of the redistribution of ^{14}C-labelled reserves following defoliation have, in a number of instances, shown little or no mobilisation of root or crown reserves for regrowth (Marshall & Sagar 1965; Davidson & Milthorpe 1966; Smith & Marten 1970; Chung & Trlica 1980). Smith & Marten (1970) suggested that the contribution of concurrent photosynthesis varies considerably between species and phenological state, thus obscuring correlations between regrowth and reserve carbohydrates. Only after very severe defoliation did the contribution of reserve carbohydrate exceed that supplied by concurrent photosynthesis, and then only for a few days after defoliation (Richards & Caldwell 1985). Therefore the rapid production of a new photosynthetic canopy immediately following defoliation is critical for the response of species which are tolerant of herbivory, especially when defoliated plants must compete with undefoliated neighbours (Richards *et al.* 1988).

Caldwell *et al.* (1981) showed that culms and sheaths themselves can be important photosynthetic surfaces after defoliation and may therefore contribute substantially to the available carbon for regrowth following partial defoliation. However, in situations where photosynthetic areas are eliminated by severe defoliation or drought, or a combination of the two, regrowth must inevitably be triggered by reserve labile carbon, provided meristematic sites for regrowth are available. Under these circumstances reserve carbon is essential, even if only for a short period. Studies by Danckwerts (1993) showed that, although the contribution of reserve carbon may be relatively small, it nevertheless plays an important role in regrowth after the removal of all leaf material. Although its contribution may exceed concurrent photosynthesis for only a short while, the indirect effect of the initial rapid production of green leaf material, and therefore of the photosynthetic surface, is manifested for considerably longer. In addition, different species differ in their ability to remain independent of carbon reserves after partial defoliation. For instance, *Lolium perenne* remains independent of carbon reserves if only about 10% of the green leaf area remains after defoliation, but *T. triandra* utilises carbon reserves even when leniently defoliated (Danckwerts & Gordon 1987).

Mineral nutrient status may also be important in affecting responses to defoliation. The application of N will, for example, bring about a decline in the concentration of water soluble carbohydrates, but the N content will increase, i.e. there is a negative correlation between tiller N and carbohydrate content. This can be seen as evidence for the use of carbohydrate reserves in early tiller regrowth (Bahrani *et al.* 1983).

Nitrogen pools in the roots of grasses can be mobilised to support shoot growth following defoliation (Millard *et al.* 1990; Ourry *et al.* 1990). Nitrogen has been shown by Welker *et al.* (1987, 1991) to be preferentially concentrated in rapidly growing, defoliated daughter tillers which were attached to non-defoliated parental tillers that had been labelled with ^{15}N prior to defoliation. This preferential allocation continued until the defoliated daughter tillers had re-established a substantial amount of the leaf area removed by defoliation.

The allocation of carbon, N and other resources from undefoliated to defoliated tillers within a plant may provide a potential mechanism of herbivory tolerance through facilitating tiller survival and growth following defoliation. Growth of defoliated tall fescue (*Festuca arundinacea*) tillers increased progressively when either a greater percentage of tillers remained undefoliated or when the defoliation intensity (of the partially defoliated tillers) was reduced (Matches 1966; Watson & Ward 1970). A portion of this growth response was attributable to resource allocation from non-defoliated to defoliated tillers. Any carbon allocation between connected shoots or branches within a plant is apparently rapidly modified when a portion of the shoot system remains undefoliated or is defoliated less severely than the remainder of the plant (Marshall & Sagar 1965, 1968; Gifford & Marshall 1973; Welker *et al.* 1985). Also, carbon import from undefoliated, attached parental tillers can be maintained for a longer period or harnessed to a greater extent when the importing tiller is repeatedly defoliated (Gifford & Marshall 1973).

The pattern of inter-tiller resource allocation is dependent on the relative leaf areas of the undefoliated parental tillers and the attached daughter tillers remaining after defoliation (Gifford & Marshall 1973), as well as on the number of actively growing meristems following defoliation. When actively growing shoot sinks are absent or limited, available carbon is allocated to alternative sinks including roots (Richards 1984) and storage sites in the sheath and stem bases (Bucher *et al.* 1987a, 1987b).

Root respiration and nutrient acquisition are also reduced following defoliation but to a lesser extent than is root growth (Davidson & Milthorpe 1966; Chapin & Slack 1979; MacDuff *et al.* 1989). The rapid decline in soluble carbohydrates in roots often observed following defoliation (Deregibus *et al.* 1982) results from a reduction in the amount of photosynthetic carbon translocated from the shoot system and a continuation of carbohydrate utilisation by root respiration. Quantitative carbon balance studies have shown that the root system continues to function as a net sink for carbon immediately following defoliation (Richards & Caldwell 1985; Danckwerts & Gordon 1987). Therefore it is unlikely that soluble carbohydrates in the root systems are remobilised to meet the demands of the shoot system during regrowth.

Leaf removal results in either a slowing down or a complete stoppage of root growth, depending on the severity of the removal treatment. This is well illustrated by the classic work of Crider (1955) with Rhodes grass *(Chloris gayana)*. He showed that at least a proportion of the roots stopped growing when 50% or more of the top-growth was harvested. Root growth stopped completely when 80% or 90% of the top-growth was removed and the root systems remained inactive for 12 days and 17 days respectively (Fig. 3.3). At lighter intensities of defoliation not all the roots stopped growing, but there was a pronounced trend towards a greater stoppage of root growth as the harvest intensity increased above 40%. As expected, the total number of roots produced by the plants also varied widely according to the severity of leaf removal. The plants from which 90% of the top-growth had been removed produced 32 main roots in the 33 days following treatment, as against 132 produced by plants from which 10% of the top-growth had been removed. This effect was shown in a number of different species (Table 3.1). Richards (1984) pointed out that reduced root growth following defoliation is an effective mechanism to aid re-establishment of the photosynthetic canopy and the root:shoot balance. As such it contributes to both herbivory tolerance and the maintenance of competitive ability (Dahl 1995).

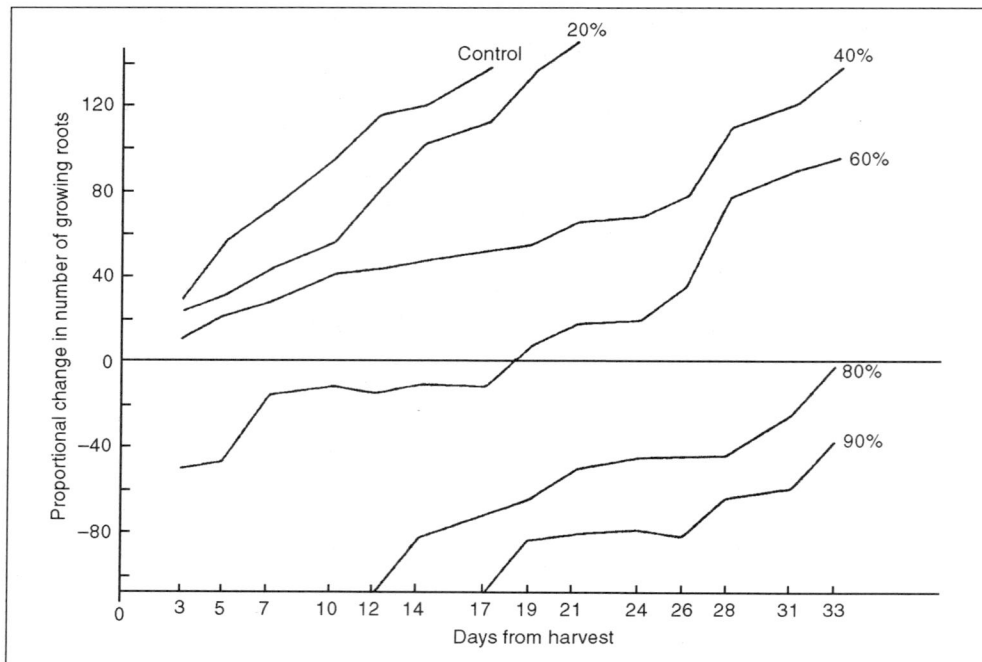

Figure 3.3 Effect on *Chloris gayana* (Rhodes grass) of single clippings involving the removal of different percentages of top-growth (after Crider 1955).

Similar responses were observed by Tainton (1958). The activity of the root systems increased as cutting height increased and the cutting frequency decreased (Table 3.2). This, as would be expected, affected root depth (Table 3.3), particularly since repeated severe cutting led to the death of root tips and a slow die-back of the roots. These effects are not, however, equally pronounced in all grass species. Species with a low growth habit, and particularly those which have underground stems which act as storage organs able to provide carbohydrates for root growth, are less affected than are tall tufted species. One would therefore expect that root growth in such species as *Pennisetum clandestinum* or species of *Cynodon* would be less affected by heavy and frequent grazing or cutting than the more upright species like *Eragrostis curvula*, and therefore the former species would be more adaptable to all but the most lenient grazing treatments.

High grazing frequencies in *L. perenne* have been shown to reduce root production more than low grazing frequencies, but grazing intensity has no effect on root production. These results suggest that grazing frequency may be more important than grazing intensity in influencing root production in perennial ryegrass grown under subtropical conditions (McKenzie & Tainton 1996).

The third cause of a lack of correlation between regrowth and reserve carbohydrates which has been suggested is the question of meristematic limitation. Removal of apical meristems of two *Agropyron* species affected etiolated regrowth considerably more than did the concentration of carbohydrate reserves. There were also considerable differences in the etiolated regrowth of the two species even though large amounts of stored carbohydrate were present in both (Richards & Caldwell 1985). This difference

Table 3.1 Effect of severe top reduction on root growth of a number of pasture species (Crider 1955).

Species	Date cut	Ht. cut (cm)	Duration of root growth stoppage (days)	Final root mass (g)	
				Clipped	Unclipped
Bromus inermis	Apr 28	6.5	12		
	May 20	6.9	17		
	July 7	7.5	8	4.3	61.0
Festuca arundinacea	Apr 28	6.5	13		
	May 20	6.9	12		
	July 7	7.5	7	6.2	20.8
Dactylis glomerata	Apr 28	6.5	0		
	May 20	6.9	18		
	July 7	7.5	7	4.6	26.2
Paspalum floridanum	July 15	5.0	11		
	July 29	5.0	6		
	Aug 4	5.0	10		
	Sep 9	5.0	18	1.0	9.6
Andropogon ischaemum	July 22	5.0	8		
	Aug 12	5.0	7		
	Sep 9	5.0	18	1.5	17.6
Panicum virgatum	July 22	5.0	18		
	Aug 30	5.0	11	1.4	17.4
Bouteloua gracilis	July 29	5.0	17		
	Aug 23	5.0	13	0.6	3.8
Cynodon dactylon	May 20	2.5	9		
	June 21	2.5	6		
	Sep 28	2.5	16		

Table 3.2 Root growth of *Themeda triandra* plants harvested at different heights and different frequencies, expressed as (a) the product of the number of growing roots and the number of days on which growth was recorded and (b) the number of days on which no root growth was recorded (Tainton 1958).

Frequency of harvest (weeks)	Height of harvest (mm)						Mean	
	62		187		312			
	a	b	a	b	a	b	a	b
2	453	31	835	24	1418	11	902	22
3	551	31	1396	24	3023	9	1657	21
5	819	14	1332	19	1563	4	1238	12
Mean	607	25	1188	22	2001	8	1265	18

was attributed to the inability of *Agropyron spicatum* to produce secondary tillers as readily as the more rapidly responding *Agropyron desertorum* after the apical meristems of both species had been removed (Olson & Richards 1988a, 1988b).

Morphological characteristics of the plant may therefore sometimes limit regrowth to a greater extent than availability of carbon resources. Such a meristematic limitation may be particularly relevant in semi-arid regions, where shoot apices may be destroyed by both defoliation and drought. Regrowth following drought could therefore be particularly slow even in the presence of adequate labile carbon reserves. However, while

Table 3.3 The effect of different intensities of harvesting *Themeda triandra* on the average number of roots which penetrated to a depth of 60 cm. Data expressed as a percentage of the number recorded on uncut plants (Tainton 1958).

Frequency of harvest (weeks)	Height of harvest (mm)			
	62	*187*	*312*	*Mean*
2	11.1	22.2	61.1	31.5
3	11.1	30.6	141.7	61.1
5	33.3	97.2	61.1	63.9
Mean	18.5	50.0	88.0	52.2

not disputing the findings of Richards & Caldwell (1985), it would be unwise to ignore the importance of energy reserves in triggering the immediate development of new photosynthetic surfaces.

In the longer term, grasses adapt to defoliation by producing lateral tillers. When tillers are decapitated by defoliation, they are usually readily replaced by new tillers growing from lateral buds produced lower down the stem. The ability to produce lateral tillers varies considerably from species to species. Although it is genetically controlled, attainment of genetic potential is influenced considerably by environmental conditions. Furthermore, plants that are not defoliated become moribund and may cease to produce lateral tillers.

The production of lateral tillers is often stimulated by defoliation, whether it involves growing point decapitation or not, provided defoliation is not so frequent and intense that it diminishes the vigour of the plant. But, the impact of defoliation on tillering may vary considerably between species and will also vary according to the time of year it is applied and to the stage of the plant's physiognomic development.

The overall effect of leaf removal may therefore be twofold. Firstly, the reduced photosynthetic capacity of the plant will invariably affect vigour and in particular the growth rates of the roots and lateral daughter tillers, both of which are much more sensitive to a deficiency of energy-rich substrates than is the main stem. Frequent leaf removal will therefore reduce the capacity of plants to produce tillers and will reduce the size and depth of penetration of the root systems. Secondly, in dense swards, the increased light penetration to the base of the sward resulting from leaf removal may stimulate the development of daughter tillers from the basal nodes in many species. Alternatively, the growth of young tillers which have already begun development and which may not have survived in the dense canopy may be stimulated. The overall effect of defoliation in dense swards is often, therefore, to increase tiller density, provided sufficient energy is available for the development of these tillers.

3.1.1.2 *The effect of removing the stem apex*

Stem apex removal will affect vegetative growth only if it occurs before the apex becomes reproductive. As far as we know, this seldom if ever occurs in pasture species used in South Africa. Apex removal will, however, affect seed production since the inflorescence of the grass plant develops directly from the apex of the stem. The most obvious effect of removing the apex is that of terminating stem growth and preventing flowering and therefore seed production. However, where secondary tillers which develop from the elevated nodes on the stem are able to develop almost immediately into secondary inflorescences, the removal of the apex reduces the size of the inflorescence but does not eliminate stem production or flowering. The seed-producing capacity of such secondary inflorescences is, however, often extremely poor.

As an alternative, treatments may be specifically designed to either promote stem production in crops managed to produce large yields of relatively low quality stemmy material or they may be managed to promote flowering and seed production. When neither of these apply there may be some merit in designing grazing management systems which will prevent stem development and maintain the plant in a vegetative condition when high quality forage is required. This will apply particularly to species in which the ratio of flowering (reproductive) to non-flowering (vegetative) shoots is high (Booysen *et al.* 1963). Continued leaf development can be at least partly promoted by heavy utilisation at the time of early stem elongation. Such treatment usually induces the development of new daughter tillers which may produce leaf material through the remainder of the season.

An important secondary effect of the removal of the stem apex is the associated removal of the inhibiting effect of the apex on the development of lateral tillers lower down on the stem. This inhibition may partly result from the secretion of growth regulators by the apex. It seems probable that the sudden availability at the tillering sites (basal nodes) of carbohydrates which would otherwise have been used by the main apex, may be an important contributory factor to the stimulation of lateral tillering. In any event, the removal of the apex leads to renewed growth of previously dormant tiller initials at the basal nodes of the stem, resulting in an increased tiller density. This effect would be expected to be most pronounced if the apex were removed early in stem development before the energy reserves of the plant had been depleted during the period of rapid stem growth and flowering. However, by preventing stems from developing, the overall yield of the plant may be reduced quite considerably. Nevertheless, the improved quality of the predominantly leafy material which is subsequently produced may more than compensate for this decline in yield. In addition to the above effects, the removal of the stem apex will reduce the eventual number of leaves where elevation takes place in the vegetative stage. The subsequent photosynthetic capacity and herbage yield of such plants will therefore be reduced.

The complex physiological mechanisms regulating axillary bud growth and a large number of other factors, including environmental variables, the state of phenological development and the frequency and intensity of defoliation, contribute to an inconsistent tillering response to defoliation (Briske & Richards 1995). Nonetheless, apical meristem removal is consistently associated with increased tillering in several tropical grasses. It would appear that physiological processes and environmental variables, in addition to direct regulation by apical meristems, play a role in the regulation of tillering in perennial grasses (Tainton & Booysen 1965; Richards *et al.* 1988; Murphy & Briske 1992).

In practice, the method, timing and intensity of defoliation will determine the pattern of removal of leaf and stem material and therefore the effect which such defoliation treatments have on the individual plant and on the grass sward as a whole. The effect is, therefore, very management dependent.

3.2 PASTURE LEGUMES   *R.J. Eckard & V.D. Wassermann*

The response of pasture legumes to defoliation varies with the growth habit of the plant, the amount of carbohydrate reserve stored in the undefoliated portions and the residual leaf area after defoliation. The intensity and selectivity of defoliation, in relation to the positioning of the developing apical buds, remain crucial to the legumes' ability to withstand defoliation (Curll & Jones 1989).

3.2.1 Growth habit

Only three broad growth-habit classes (prostrate, upright and twining) of herbaceous legumes will be considered in this chapter.

In the prostrate legume types, like white clover (*Trifolium repens*), the apical growth points usually remain close to, or under, the soil surface. Here they are partially or completely protected from the grazing animal (Tainton 1979; Thomas 1987; Forde *et al.* 1989). This very important pasture legume does, however, have a rather peculiar growth pattern. The main deep tap root system develops from the seed and provides major support to the plant for approximately 18 months from the time of seed germination. The lateral runners, as they develop on the mother seedling plant, produce a large number of roots from their nodes but these roots remain relatively shallow. It is this shallow root system which supports the runners once the main tap root has died, but because of their shallowness they are considerably less effective than the original tap root in supporting the plant during dry periods. Not only are such plants extremely susceptible to direct drought stress, but they are also affected indirectly through a reduction in mineral uptake under dry conditions. This is shown, for example, by the very positive response of such plants to high levels of phosphatic fertilizers.

The upright tufted types (such as *Trifolium pratense* and *Medicago sativa*) and the twining types (such as *Desmodium uncinatum* and *Neonotonia wightii*) are generally less resistant than the prostrate types to heavy defoliation, although there are some considerable differences in the levels of resistance among these types. Lucerne, for example, is a great deal hardier than red clover because of differences in the general structure and perenniality of the crown area of the plant. In these species axillary buds develop either from a crown region situated above the soil surface, or from nodes along the twining stems (Curll & Jones 1989). Under conditions of heavy defoliation, particularly by selective grazers, new stem apices and trailing stems are vulnerable to removal. Recovery growth could, therefore, be inhibited by a lack of residual growth sites.

3.2.2 Time of year and physiological growth stage

The response of prostrate legume types to defoliation at various times of the year depends largely on the relative growth rates of any associated grasses and of the legume in question, as well as on the legumes' ability to capitalise on any reduction in competition which may come about following any grazing of the grass component (Curll & Jones 1989). In white clover, a reduction in grass competition in mid-spring and autumn, induced by grazing at these times, would result in a competitive advantage to the clover (Tainton 1979). However, in mid-summer and winter, temperature constraints might limit the rate of clover growth and therefore the ability of the clover to compete with a grass.

The upright tufted legume types, like red clover and lucerne, are more tolerant of defoliation before flowering than after flowering (Curll & Jones 1989). In these types a new population of stems normally develops from the crown during the early stages of flowering, and these may be destroyed during grazing, leading to slow recovery. In both these species specific types have been selected on the basis of their response to winter temperatures and therefore the time at which they will make maximum growth. Red clover is well known for its early and late season types, while a considerable amount of work has been undertaken to select a range of dormancy types in lucerne.

The damaging effect of severe defoliation of the twining legumes is particularly severe in spring when large numbers of new stems develop. These legumes should be either rested or very leniently defoliated during this spring growth phase. The twining legumes tend to require longer intervals between defoliations than prostrate types (Skerman 1977; Curll & Jones 1989).

Leucaena is the only leguminous shrub currently used to any extent in South African pastures. Not surprisingly, this shrub appears to react to being browsed in much the same way as do the indigenous leguminous shrubs in that frequent and severe defoliation leads to a gradual decline in the vigour and production of the shrub. Pastures composed of this plant should therefore be browsed rotationally (for more information on the use of shrub legumes, refer to Stuart-Hill (1999)).

3.2.3   Carbohydrate reserves

Regrowth, particularly in the early stages following defoliation, is generally considered to be very much dependent on the mobilisation of carbohydrate reserves stored mainly in the roots and other subterranean organs of the plant. However, the stems and crown regions of the plant may also contain carbohydrate reserves vital to plant regrowth. The removal of these plant parts would deplete plant reserves as well as remove new growth points, thereby resulting in slower regrowth or even plant death.

Studies of the role of carbohydrate reserves in the promotion of regrowth have demonstrated a cyclic decline and accumulation of reserve carbohydrates following defoliation (Smith 1962; Nelson & Smith 1968; Etzel *et al.* 1988). In spite of such evidence, however, some researchers have questioned the importance of carbohydrate reserves in regrowth (May 1960; Volenec 1985). Volenec (1985), for example, reported a low correlation between root carbohydrate levels and regrowth rate following the defoliation of several lucerne types. Chatterton *et al.* (1977) suggested a more indirect relationship whereby plants with high reserves may better withstand the 'shock' of severe defoliation than those with low reserves. They argue that plants with a high reserve status are likely to survive for longer than those not given the opportunity to accumulate reserves.

3.2.4   Residual leaf area

The size of the residual leaf area retained by the plant has been shown to greatly affect the rate of recovery following defoliation (Etzel *et al.* 1988). If ample above-ground tissue remains after defoliation and this tissue has a high photosynthetic potential, then levels of carbohydrate reserves may be of little relevance to regrowth.

It would appear, therefore, that the response of legumes to defoliation depends on the physiological stage of growth of the plant at the time it is defoliated and on the quantity of important plant parts retained by the plant, i.e. plant reserves, active stem apices and active photosynthetic material.

4

The animal factor in pasture management

N. Owen-Smith

CONTENTS

Feeding behaviour forms the link between the animal and the vegetation and influences the impact of grazing and browsing on plants. In many respects such behaviour can be flexible in that animals can respond to changing conditions but there are, nonetheless, intrinsic anatomical, physiological and behavioural differences among animal species which have a fundamental influence on feeding patterns.

The aim of this chapter is to document aspects of foraging behaviour that may influence the impact that herbivores have on their food resource. The approach will largely be a comparative one, involving considerations of similarities and differences between different types of domestic stock and among some of the wild game species. Although the latter obtain their food largely from veld rather than from pastures, an understanding of their behaviour provides a useful insight into patterns of animal behaviour pertinent to the pasture farmer. We make no apology, therefore, for including some information on game species in this chapter.

4.1 ANATOMICAL AND PHYSIOLOGICAL DIFFERENCES AMONG HERBIVORES

4.1.1 Body size

Size has a fundamental influence on an animal's metabolism and hence on its food (and energy) needs (Fig. 4.1). Basal metabolic rate, which determines an animal's energy requirements, varies among animal species according to their body mass raised to the power three-quarters. This means that the specific metabolic rate per unit body mass decreases as mass increases ($M^{0.75}/M^{1.0} = M^{-0.25}$, where M = body mass). Therefore, a unit mass of a 50 kg animal metabolises almost twice as much energy as a unit mass of a 500 kg animal.

The standard formula for calculating basal metabolic rate (BMR) is given by Kleiber (1975) as

BMR = 293 $M^{0.75}$ kJ/day

with body livemass, M measured in kg.

Strictly, the basal metabolic rate applies to animals that are immobile, have empty guts and are not growing. Resting (non-fasting) metabolic rate is typically about 30% higher than the basal rate. Normal daily activity increases the rate so that the daily metabolic requirement of normally active animals can be assumed to be about 1.5 to 2 times the basal rate, the multiplier depending on the animal's general activity level. Weather conditions will also affect the metabolic rate, as will the reproductive state of the animal, with energy needs increasing approximately 1.5 fold during late pregnancy and early lactation. Factors such as body composition and body insulation may also influence the standard metabolic rate co-efficient of 293 commonly used. Nonetheless, in spite of these deviations from the standard Kleiber regression, the nutritional needs of species of different size are conveniently compared using units of 'metabolic mass equivalent' calculated according to the basal metabolic rate. Using this rate, the metabolic livemass equivalent of a standard 454 kg animal unit (AU) is 98 kg, while that of a 250 kg 'tropical livestock unit' is 63 kg.

Sheep typically have a metabolic rate co-efficient about 30% lower than that of cattle, possibly due to the insulating effect of their wool and perhaps also to differences in patterns of fat deposition. Among animals of the same species, small, growing animals metabolise more energy per unit body livemass than would be predicted by the standard regression (Meissner 1982; Hudson & Christofferson 1985).

Requirements for protein (nitrogen) and other nutrients are closely related to energy metabolism, although daily protein requirements increase somewhat more during the critical stages, such as pregnancy and lactation, than do energy needs (Moen 1973).

Because the metabolic rate per unit of body livemass decreases as body mass increases (Fig. 4.1), daily food intake as a fraction of body livemass declines with increasing body size. So, for example, a cow will typically eat between about 1.5–2.5% of its body livemass per day (expressed as the dry weight of the food), sheep will eat 3–4% and large herbivores (like elephant) 1–1.5% or less of their body livemass (Fig. 4.2). The rate of this decline with increasing size is, however, less sharp than the three-quarters power formula ($M^{0.75}$) would predict because larger species can tolerate a lower quality (i.e. a higher fibre content) diet than can the smaller species.

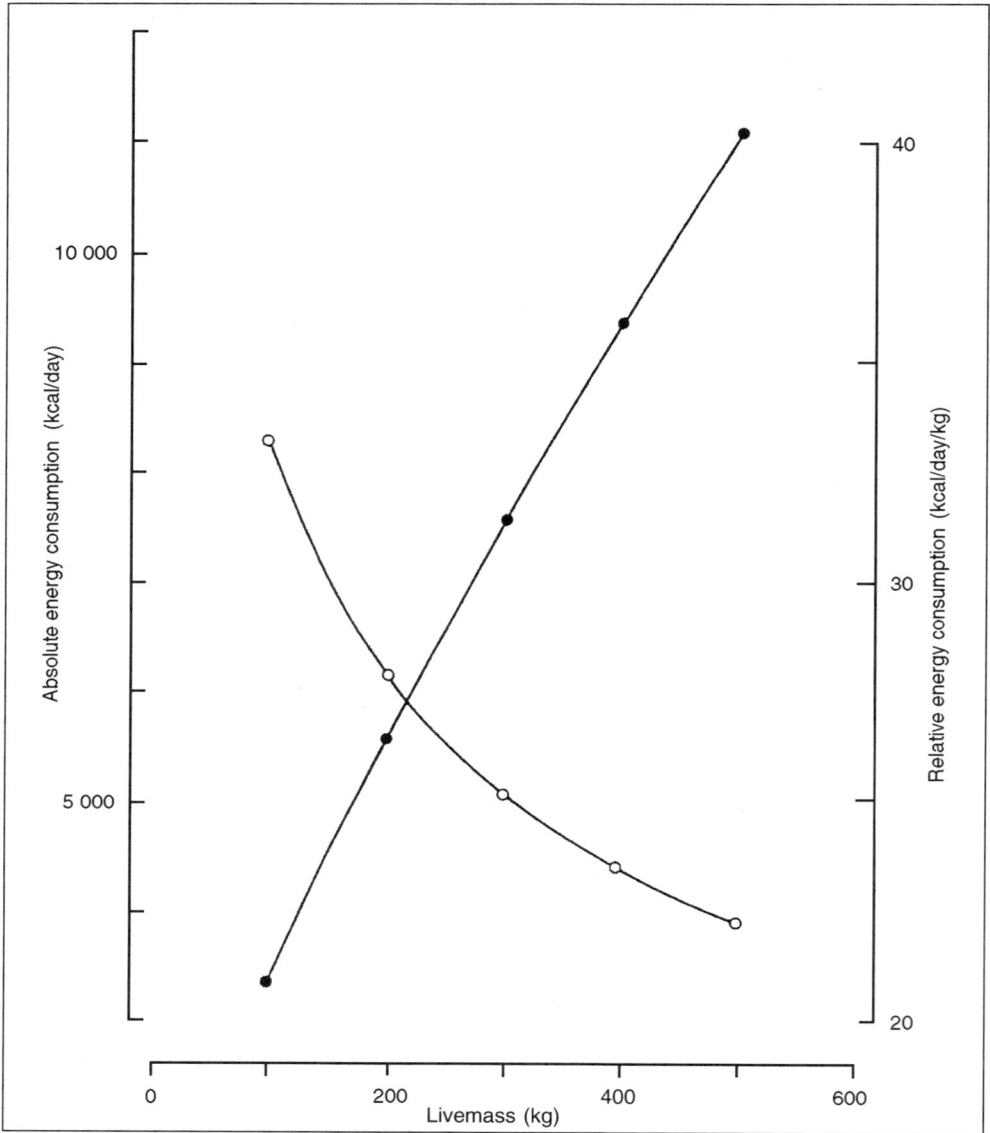

Figure 4.1   The relation between livemass and absolute (•——•) and relative (○——○) energy
consumption in mammals.

   To meet its high relative requirement for food, a small animal may
(a)   have a relatively high intake;
(b)   select a diet high in nutritive value; or
(c)   have a high digestive efficiency.
Strategy (c) can apparently be ruled out (Blaxter 1962; Arman & Hopcraft 1975; Vorster
1976; Hoppe 1977). Some small animals such as the dikdik have almost no ability to
digest fibre (Hoppe 1977) and cannot be said to have a high digestive efficiency prob-

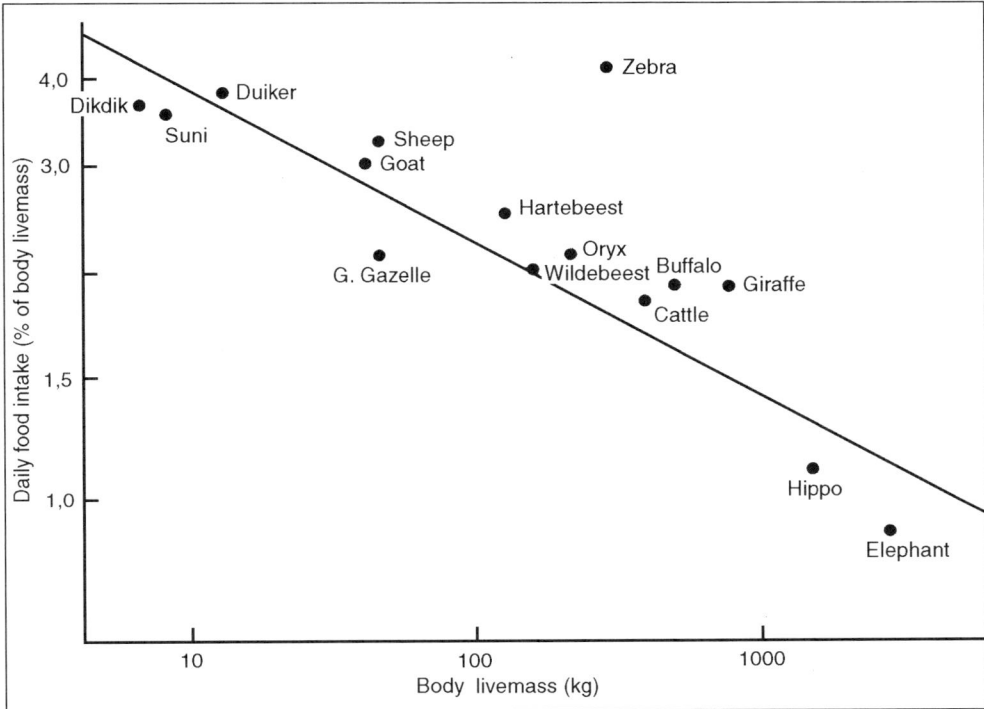

Figure 4.2   Daily food intake in relation to body mass. Regression line: daily food intake (dry mass as % of livemass) = 6.0 M$^{-0.191}$ (adapted from Owen-Smith 1988).

ably because of the reduced food retention time. Cattle and their bovine relatives (buffalo and bison), on the other hand, are extremely efficient in digesting grass cell wall material as Fig 4.3 shows.

It is unlikely that strategy (a) can be used by small animals to fully meet their high nutrient needs since stomach capacity is related directly to body size rather than to metabolic body mass. This suggests that small ruminants must select concentrated foods, i.e. strategy (b) before. Such selective behaviour is possible for two physical reasons. Firstly, by virtue of their low absolute requirement, small ruminants have relatively more foraging time than do large ruminants per unit amount of material consumed. Secondly, small ruminants have small mouths and are therefore able to select small items of concentrated foods (e.g. young newly emerging grass leaves). Generally, the proportion of such high quality items in the diet increases with a decrease in the size of the animal (Bell 1970; Blankenship & Qvortrup 1974; Hoppe *et al.* 1977; Kautz & Van Dyne 1978) (Tables 4.1 & 4.2). As expected, the proportion of stem in the diet increases with increasing body size (Fig 4.4).

Because large animals individually consume a greater total amount of food per day than small animals, they cannot afford to be as selective for plant parts as can smaller animals. They nonetheless still prefer the best quality material that is available to them. White rhino, for example, commonly graze grass just as short and as nutritious as sheep, but they also have the ability to subsist on dry grass that would be nutritionally inadequate for a smaller ungulate.

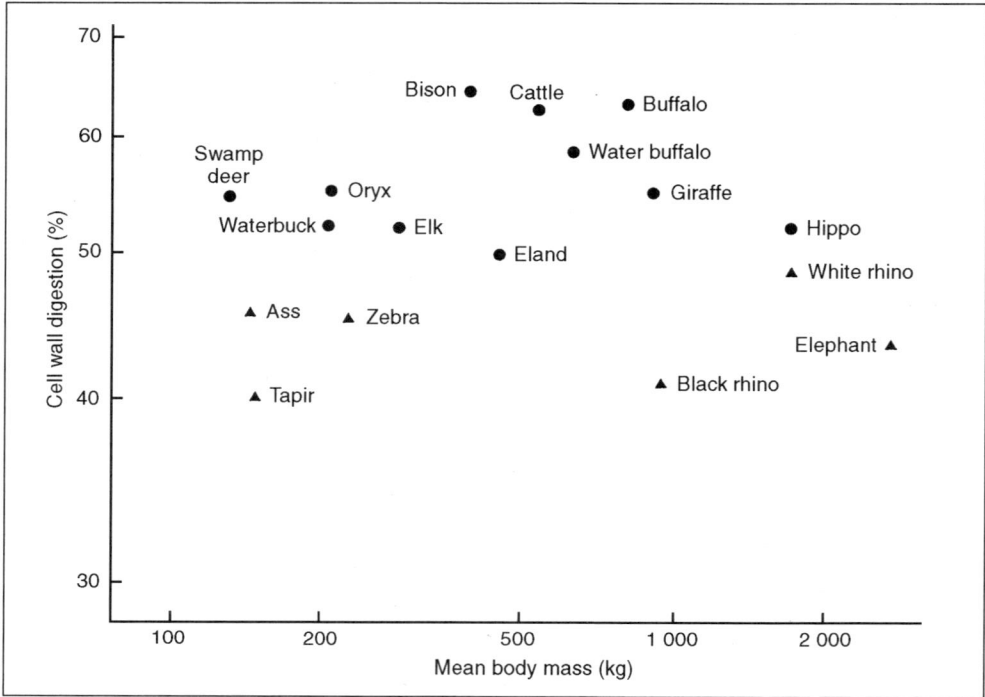

Figure 4.3   The extent of digestion of the cell wall component of grass hay in relation to body mass, distinguishing foregut fermenters (including ruminants – circles) from hindgut fermenters (triangles) (adapted from Owen-Smith 1988).

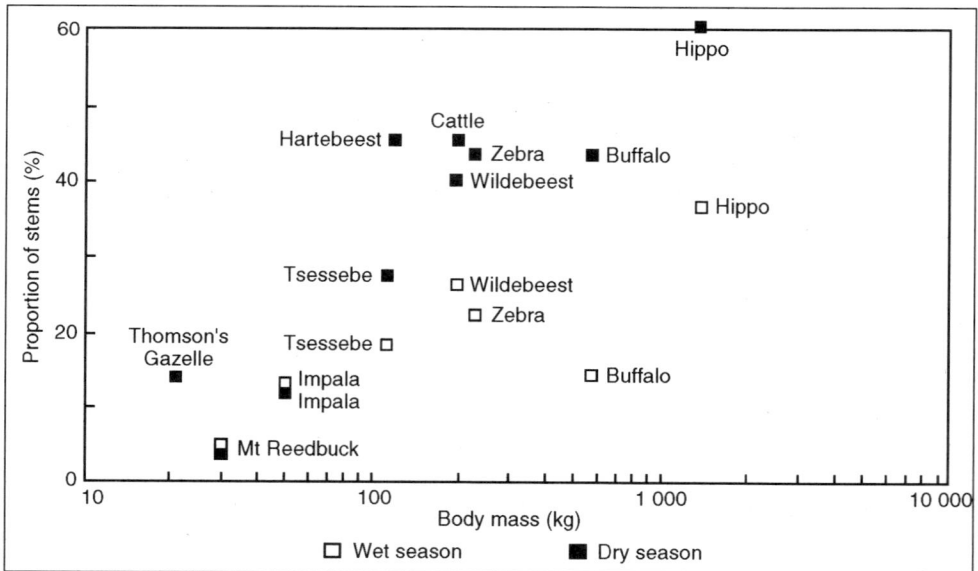

Figure 4.4   Proportion of grass stem in the rumen or stomach contents of various grazing ungulates in relation to body mass (from Owen-Smith & Cumming 1993).

Table 4.1 Mean livemass and percentage composition of the rumen contents of antelope and cattle (Hoppe *et al.* 1977).

Species	Livemass (kg)	Grass leaves and sheaths	Grass stems	Forbs and dicots.	Green grass %*
Cattle (*Bos indicus*)	197	53	40	7	31
Wildebeest (*Connochaetes taurinus*)	200	57	37	6	49
Haartebeest (*Alcelaphus buselaphus*)	120	51	45	5	56
Topi (*Damaliscus lunatus*)	114	81	15	3	56

* = Green grass as a percentage of total grass

Table 4.2 Percentage chemical composition of rumen contents of cattle and antelope (Hoppe *et al.* 1977).

Species	Crude protein	Crude fibre	Ether extract	N-free extract	Ash
Cattle	3.3	51.0	3.0	35.8	6.9
Wildebeest	4.0	51.0	3.1	36.0	5.9
Haartebeest	3.9	49.5	3.3	36.6	6.7
Topi	6.2	44.1	2.9	41.5	5.3

4.1.2 Dentition and feeding mechanism

Grazing ungulates have high-crowned cheek teeth with a finely ridged surface which facilitates grinding the fine, fibrous leaves of grasses. Browsers, in contrast, have prominent cusps on their molar surfaces which facilitate the maceration of leaves of trees, bushes and dicotyledonous herbs (forbs).

Grazing and browsing ruminants also differ in muzzle width relative to body size and in the angle of insertion of their lower incisors. Animals which have comparatively wide muzzles favour short creeping grass genera over the more upright genera. Browsers have narrow muzzles suited to plucking leaves from branch tips or from between thorns and spines. Grazing antelope, on the other hand, have protruding incisors that aid in plucking grass leaves, which are gripped between the incisors and a pad on the upper palate, while allowing the more resistant stems to slip through. Cattle and other bovines feed in a somewhat different way in that they use their tongues to sweep grass over their teeth for plucking although cattle can switch to grasping grass with the lips when it is short, but at the expense of bite size (Beekman & Prins 1989) and therefore of intake. Consequently they do not feed effectively on short grass.

There are also differences in the relative muzzle width among grazing ruminants. Species preferring short grass have relatively wide muzzles compared with those favouring somewhat taller grass. In tall grass swards a narrow muzzle is beneficial in that it allows the animal to pluck only green leaves from within the tall sward. Small ungulates like sheep can feed effectively on both tall and short grass, but tend to prefer the latter (Owen-Smith 1985, 1988).

4.1.3   Digestive system

A major anatomical difference among herbivores is that between ruminants (foregut fermenters) and non-ruminants (hindgut fermenters). Species of ruminants (cattle, sheep, antelope) have a fermentation chamber (the rumino-reticulum) anterior to the acidic stomach abomasum (Fig. 4.5). In hindgut fermenters the caecum (a blind sac at the junction of the small and large intestine) is the site of cellulose fermentation. Hindgut fermenters include equines (horses and zebra), rhino and elephant, as well as hares and herbivorous rodents. The hippo has foregut fermentation but does not ruminate (chew the cud).

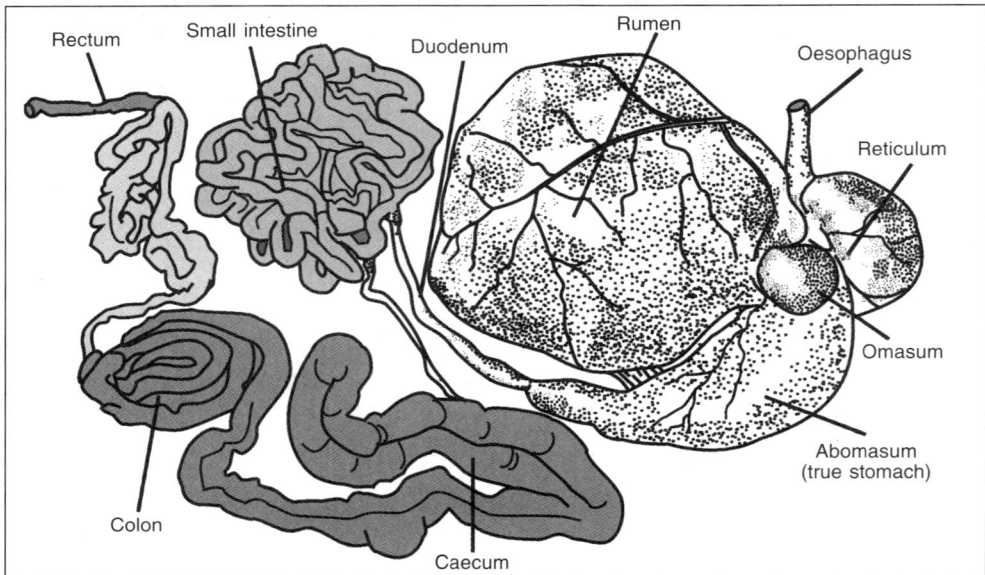

Figure 4.5   Digestive tract of the ruminant or foregut fermenter showing the evaginations (rumen, reticulum and omasum) which precede the abomasum or true stomach where nutrients are fermented prior to digestion in the lower intestinal tract. With hindgut fermenters the oesophagus goes directly to the stomach (abomasum) and the caecum and colon are significantly larger to facilitate digestion of roughages.

The advantage of foregut fermentation is that it permits regurgitation of the ingesta for further chewing. This increases the surface area of the ingesta for microbial fermentation, leading to a more efficient breakdown of cell wall constituents into volatile fatty acids. In addition, the food is retained in the rumen until its particle size has been reduced sufficiently for fermentation to be almost complete.

Non-ruminants are less efficient digestors of cell wall material than ruminants, although not due to any inherent limitations in microbiota. To compensate for this, they have a high throughput rate of digesta. Equines like asses and zebra may digest hay only two-thirds as well as do cattle (Fig. 4.3) but are able to consume as much as 50% more food per day relative to body mass. In this way they can achieve a similar energy gain. However, in order to do so, these animals must spend a longer time feeding each day.

There are interesting differences in the anatomy and physiology of the digestive tracts of grazers and browsers (Hofmann & Stewart 1972). Grazers have capacious rumens which enable them to retain the slow-fermenting leaves of grasses long enough to allow them to extract most of the digestible energy they contain. The leaves of trees and forbs ferment somewhat faster than those of grasses. Hence, browsers have relatively smaller rumens but they have larger openings between the rumino-reticulum and the following two compartments (the omasum and abomasum). Increased papillation within the rumen of browsers allows the latter to rapidly absorb fatty acids, preventing excessive acidity from developing in the rumen fluid. The rumen of browsers is generally not filled to capacity, and is also relatively free of adhesions to the body wall. Hence browsers are better able to expel the gasses generated by rapid fermentation in the rumen than is a large grazer like a cow. They are therefore not subject to bloat, even when feeding on high quality green herbs.

Mixed feeders like goats have the digestive capacity of a grazer but other adaptations similar to those of browsers (Hofmann & Stewart 1972). Their rumen papillation changes seasonally, depending on the ratio of grass and browse consumed. The diet selected by browsers tends to be somewhat higher in protein, and lower in total fibre content (although not necessarily in digestibility), than that selected by grazers of similar body size (Owen-Smith 1988).

Many browsers have large salivary glands, and in certain species these have been shown to secrete proteins that neutralise the digestibility-reducing effects of tannins in tree leaves. Browsers have relatively larger livers than grazers, evidently an adaptation for detoxifying the secondary chemicals prevalent in dicotyledonous plants (Hofmann 1989). Hence grazers are far more susceptible to being poisoned by toxic plants. So for example, cattle and goats are highly susceptible to being poisoned by gifblaar (*Dichapetalum cymosum*), a toxic plant, but indigenous ungulates have some degree of tolerance.

4.1.4 Water dependency

The degree of water dependence is influenced by the ability of an animal to absorb water from faecal material during its passage through the large intestine, and by mechanisms of thermoregulation.

Obligatory water loss through the skin, respiratory tract, urine and faeces also varies among different animal species. It has, for example, been reported to amount to 4% of body mass per day in sheep, compared to 12% for cattle (MacFarlane & Howard 1970, 1972). Because of this, sheep do not need to drink in cool conditions if grass has a 30–50% water content. Zebu cattle can get by with half the water intake of Hereford cattle (Taylor 1968). Cattle with dark coat colours absorb more solar radiation, and hence drink more water per day, than do those with light coats (Finch & Western 1977). The latter have been shown to survive droughts relatively better than those with dark coats at low altitudes in Kenya, but at high altitudes where there is less heat stress, this pattern was reversed, apparently because of the energy saved by heat absorption. Zebu breeds of cattle can be moved to water only every third day, with little apparent effect on their productivity (Jewell & Nicholson 1989).

The distances moved by animals in search of water will also, of course, depend on the moisture content of the ingested forage and is also dependent on breed type, but this should seldom, if ever, be a problem in cultivated pastures because of the relatively small areas usually involved.

However, where the provision of drinking water is difficult it is possible, within reason, to select animal types which can tolerate conditions in which water is poorly distributed or where the overall supply of water is restricted.

4.1.5   Patterns of diet selection

Food preferences are influenced by a number of factors. These include:
(a)   the species of animal and its basic anatomic and physiological adaptations, as discussed previously;
(b)   the relative palatability of the plant species and plant parts available (refer to Chapter 5);
(c)   the physical accessibility of the plant material;
(d)   the surrounding conditions;
(e)   the physiological condition of the animal; and
(f)   its previous history.

Cattle expand the range of species grazed as the period of occupation of an area increases, with the pattern being influenced by stocking rate and season (Daines 1980; O'Reagain & Mentis 1989a; Stoltz & Danckwerts 1990). In mixed pastures cattle will initially concentrate their feeding on certain highly preferred species. Use of species of intermediate preference increases as ungrazed tufts of the preferred species become less available. Tufts of preferred species are re-grazed before intermediate species become well grazed, but some intermediate species may eventually be heavily grazed. Some unacceptable species remain ungrazed even with heavy stocking.

Within patches, herbivores select for tufts of particular grass species or, in the case of browsers, for particular tree or shrub species or size classes. Animals also temporarily avoid faeces contaminated patches, particularly patches contaminated by their own faeces. At yet a finer level, animals bite off particular plant parts, e.g. sheep and goats select the seed-heads of grasses at certain times of the year. Flowers or fruits of trees may be plucked by browsers.

In general, different animal types may complement each other by selecting different plant species or species groups, or by grazing different vegetation components or particular categories of plants. It remains important to emphasise, however, that animals tend to select the most nutritious food available to them, bearing in mind their particular abilities to harvest that food.

4.1.5.1   Standing biomass as a factor in selection

As indicated previously, different animals have different abilities, largely by virtue of variations in the structure of their mouth-parts, to graze grass plants of different length.

They can, however, compensate to a limited extent for the influence of changing grass height on intake rate. Sheep do so by taking more bites per minute from short grass than from long grass swards and may also increase their daily grazing time. The limit to their compensatory ability is reached, at least in ryegrass pastures, when grass height drops below 60 mm, causing daily food intake to decline (Hodgson & Grant 1985; Penning 1986). The rate of food intake of cattle drops when grass height falls below about 100 mm (although animal production may not show a corresponding decline because short grass is often more nutritious than longer material, depending on the circumstances). Nonetheless, variations in grass height can be an important factor in changing preferences for particular grass species.

However, the opposite can also occur. In some situations the highest rate of intake is achieved from quite short grass because the biomass concentration or bulk density (mass of leaves per unit of sward volume) is highest in this state. Of interest in this respect is that aggregation by grazing ungulates in herds can promote the development of grazing lawns, which increase feeding efficiency (McNaughton 1984). For medium-large browsers, the rates of intake obtained from browse may be inadequate for their needs, so that animals must supplement their diets. When browsing, goats preferentially select the shoots offering the greatest rate of intake (Teague 1989). Grazers like cattle with large mouths browse rather clumsily and their rate of intake is normally rather low.

4.1.5.2 Effect of environmental conditions and management on selection

Wind, temperature and rain influence the direction in which animals travel when feeding, whether they take shelter or not, and where they take shelter when they do. These weather factors cause localised concentrations of animals, with corresponding localised defoliation, trampling and deposition of dung and urine. Sheep and cattle often congregate on knolls to avoid the cold air which sinks into hollows on calm nights. The location of drinking water and feed supplements will also cause localised areas of animal concentration.

4.1.5.3 Implications of selective patterns of grazing

Much heterogeneity exists in planted pastures (Gordon & Lascano 1993). In mixed legume–grass swards, clover or other legumes may be concentrated in patches. The heavy feeding pressure and consequent addition of dung and urine to these patches tends to enhance the contrast. Sheep are able to exploit these patches more effectively than cattle although ungulates generally avoid faeces-contaminated patches of pasture. Horses have the habit of concentrating their droppings in particular localities in a camp and ungrazed tall grass often develops in these localities. Since horses readily eat grass cut and removed from these patches, the aversion is to smell of the faeces rather than any change in the palatability of the grass. Cattle readily graze where horses and sheep have defecated (Odberg & Francis-Smith 1977).

It follows that the type of animal and the proportional mix, if there is more than one animal type, are important considerations in determining defoliation patterns. Among grazing animals the need for managerial skill increases as the animals become smaller. This is necessary to compensate for the increasing degree of selection in small animals. Problems raised by selection can, however, be at least partially overcome by mixed grazing/browsing.

4.2 TIME SPENT FEEDING

Livestock usually feed less at night than during the day but the amount of night-time feeding varies widely among species and with conditions (temperatures and the amount of light available to assist in foraging). Most ruminants forage for 4 to 7 hours during daylight (Owen-Smith 1988), but horses feed for 7 to 8.5 hours (Duncan 1991). Free-ranging cattle typically feed for 9 to 10 hours per 24-hour cycle. Their daily feeding time tends to decline when the maximum temperature exceeds about 26°C in humid climates, but temperature has less effect when the air is dry. Cattle and sheep can, however, forage for up to 13 or 14 hours per day when pressed (Smith 1959; Arnold 1981). What sets this upper limit is unclear, since 25% of the time is spent idle (Zemo & Klemmedson 1970). Sleeping occupies no more than 1 to 2 hours per day in most ruminants.

Most animals have a feeding rhythm, with most feeding done in the early morning and the late afternoon to early evening, although there is commonly a feeding spell around midnight, presumably to top up the rumen. Resting and ruminating are the predominant activities over midday and at night, although cattle may sometimes graze predominantly at night under tropical conditions (Payne *et al.* 1951).

Grazing ruminants may reduce their daily grazing time in the dry season when food quality is poorest because of the effects of high fibre content on the rate of passage of food through the digestive tract.

The type of forage selected may vary with time of day and may not be the same at night as it is during the day. As their appetite becomes satiated during daylight grazing, animals can presumably afford to graze more slowly, and hence more selectively. So, for example, the crude protein content of the diet selected by sheep has been found to peak in the afternoon (Langlands 1965). Cattle have been recorded to eat mainly grasses during the morning but an increasing amount of forbs during the afternoon (Obioha *et al.* 1970).

Rumination time varies between 5 and 9 hours per day for cattle, but is somewhat less in sheep (Arnold 1981).

4.3   PRODUCTION IN RELATION TO FEEDING STRATEGY

Data on the productivity of a variety of mammalian herbivores is presented in Table 4.3. Efficiency per animal increases with a decrease in animal size, cattle being an exception with respect to their high growth per unit of food consumed. It appears that this difference results largely from the greater ability of small animals to select concentrated foods of high digestibility and nutritive value. However, where the quality of pastures is low, as in many which comprise tropical/subtropical species, large animals are efficient producers per unit mass of primary production or per unit area in spite of being poorer producers per unit mass of intake because of their ability to use low quality forage. In contrast, small animals may be expected to exhibit high production per animal, but low production per unit area.

4.4   OTHER EFFECTS OF ANIMALS ON THE VEGETATION

4.4.1   Physical damage to plants

In addition to defoliating plants, animals physically damage plants by cutting, bruising, breaking and debarking them. Whole plants may be dislodged and uprooted, particularly among plants whose leaves have a high tensile strength. Uprooting by those concentrate feeders that feed particularly close to the ground may be markedly severe.

Table 4.3 Herbivore efficiency ratios in kcal/m^2 year, compared to mean livemass (from Mentis 1977).

Efficiency	Elephant	Cow	Moose	Blesbok	Deer	Mouse
Mean livemass (kg)	2 000	460	300	55	50	0.1
Growth/unit standing crop	0.059	0.115	0.25	0.29	0.5	2.5
Assimilation/unit standing crop	3.38	—	—	20.6	33.9	87.5
Food consumed/unit standing crop	10.38	1.9	—	29.5	41.4	131.6
Growth/unit food consumed	0.0057	0.060	—	0.0097	0.012	0.020
Assimilation/unit food consumed	0.326	—	—	0.7	0.80	0.70
Maintenance/unit food consumed	0.32	—	—	0.62–0.69	0.75	0.68
Growth/unit assimilation	0.017	—	—	0.014	0.016	0.029

Sheep, for example, often uproot large numbers of Rhodes grass stolons when stocking rates are high.

4.4.2 Soil disturbance by animals

Soil moved by walking animals may cover short plants, and animals may trample litter (and so promote decomposition) and bury seeds (and so promote germination). The dust they raise may coat plants and, in so doing, reduce acceptability. Also, animals may alter the structure of soils by chipping or loosening the soil surface or they may deform or compact the soil, depending on the type of soil and its moisture content. Without disturbance, the soil surface may seal off (become capped) and, in at least some soils, animals' hooves may break up this seal, particularly when soils are dry and, in this way, promote infiltration. This loosening may also lead to increased soil loss associated with either wind or water erosion.

When soils are moderately wet there is a tendency for hoof action to compact rather than chip its surface, particularly where its clay content is high. This will cause a loss of soil structure, increased bulk density and reduced pore space, which in turn will result in reduced infiltration, aeration and water holding capacity. General conditions for plant growth will become less favourable. Many studies have correlated the above effects with high grazing pressures (Heady 1975).

Therefore it would appear that while the trampling effects of animals are neither universally good nor bad, based on our knowledge of the compacting effects of animal hooves and the detrimental effects of such compaction on plant growth, it is reasonable to assume that excessive trampling resulting from high grazing pressures will have a detrimental long-term effect on the pasture.

4.4.3 Distribution of faeces and urine

Since animals generally spend considerably more time in selected areas than elsewhere within a paddock, there is a redistribution of nutrients within paddocks through the uneven distribution of faeces and urine. The overall effect is for animals to collect nutrients from a large area and concentrate them on relatively small areas.

Cattle typically defecate about once every two hours and urinate once every three hours. Over the course of a year at a stocking rate of one animal per hectare, 20% to 40% of the pasture will be affected by urine and 1.5% to 4% by dung (depending on overlap). Phosphorus is eliminated primarily in the faeces and so accumulates where dung is concentrated (Wilkinson 1973).

Daily defecation and urination rates for cattle, daily outputs of excreta and the extent of areas covered are presented in Table 4.4. While the nutrients contained in urine are almost immediately available to plants (except for the N, some of which may be volatilised), that contained in faeces may be released slowly, depending on the rate of decomposition of the material. Only about 20% of the faecal N is returned to the soil unless coprid beetles, of which there are upwards of 2 000 species in Africa, are active (Gillard 1967; Bornemissza & Williams 1970; Waterhouse 1974). When dung beetles are active, dung pads may be disintegrated within a few hours, and Gillard (1967) estimated that these beetles buried 90% of the faecal N in the soil. In a pot trial, Bornemissza & Williams (1970) measured higher plant yield and nutrient uptake where dung was mixed with soil by dung beetles than in the control treatment. Norman & Green (1958) measured an 11% higher plant crude protein content and a 57% higher dry matter production near dung pads than away from them. In the vicinity of urine patches, plants had a 24% higher crude protein content and a 5% higher dry matter production than elsewhere.

Table 4.4   Daily defecation and urine rates for cattle, and daily outputs of excreta and the extent of areas covered and affected per day and per year.

	Times/day	Daily output (kg)	Area covered/day (m²)	Area covered/year (ha)	Area affected/day (m²)	Area affected/year (ha)
Defecation	12	25	1.1	0.04	4.4	0.16
Urination	8	9	2.2	0.08	2.2	0.08
Total excreta	20	34	3.3	0.12	6.6	0.24

The accumulation of N from urine and faeces around water points can lead to N losses of up to 50% through volatilisation as ammonia (Henzell & Ross 1973).

4.4.4   Export of nutrients in animal products

Most of the plant nutrients ingested by animals are returned to the soil, with only a small proportion being retained by the animal and exported from the system. Wilkinson & Lowry (1973) estimate that 3.3 kg of P/ha, 1.1 kg of K/ha and 5.6 kg of Ca/ha are exported annually from a typical pasture system. Nitrogen losses are likely to be similar to the Ca losses. The potentially high losses in dairy systems associated with the daily removal of milk from the system are likely, in most situations, to be balanced and in many situations more than balanced, by the input of nutrients derived from concentrate feeds. However, where the animals graze on the pasture for only short periods each day and spend much of their ruminating and resting time off the pasture, little of the urine and faeces excreted by the animals will be returned to the pasture and this may lead to a depletion of nutrients from the pasture soils.

4.5 COMPETITION AND FACILITATION

Herbivore species, with the exception of the mixed feeders, partially separate them-selves ecologically in that some prefer grasses (the grazers) and others prefer dicotyledonous plants (the browsers). Within these categories animals of all species generally prefer leafy plant species offering relatively high nutrient contents and low contents of fibre, potential toxins and digestibility-reducing compounds. Even within a uniform sward type, cattle and sheep will tend to select different types of material to the extent that sheep individually perform better when stocked with cattle than when stocked alone. This situation represents one of facilitation rather than competition.

The opposite may also, however, be true where small or medium-sized species, able to crop the grass close to the ground, may hold the sward in a state that is too short for species preferring taller grass (Illius & Gordon 1987). On theoretical grounds it might be expected that the minimum sward height on which a grazer can survive in-creases with the size of the animal. Observations do not, in general, conflict with this statement (Page & Walker 1978; Scotcher 1979). Hence high stocking levels of sheep are detrimental to cattle. However, severe grazing pressure by short grass grazers is usually restricted to localised patches. This reduces competitive effects.

4.6 INTRASPECIFIC EFFECTS

The most important of the intraspecific effects is the inevitable competition for re-sources, particularly for food. The extent to which this takes place is, of course, stocking rate dependent. Other important effects are behavioural in nature. Cattle, sheep and goats are gregarious animals. Within groups, pecking-order or social hierarchies de-velop so that frequent exchanges of animals between groups has an unsettling effect. The social hierarchy has to be re-ordered and this reduces animal performance. Among the hoofed animals a strong mother–offspring bond develops shortly after birth. Dis-turbance or separation during the first few hours *post partum* should therefore be avoided if the intention is to keep the two together since it causes stress to both parties. This will impact on their performance. Dams often also tend to hide their offspring dur-ing the first few days *post partum,* with only occasional visits made for suckling and grooming. In rotational grazing programmes, therefore, hidden offspring may be inad-vertently separated from their mothers when the latter are moved to a new paddock.

4.7 LIMITATIONS IMPOSED ON GRAZING MANAGEMENT BY ANIMAL BEHAVIOURAL CHARACTERISTICS

Animals need to be given an opportunity to adapt physiologically, anatomically and in symbiotic rumen flora to specific diets. Because of this, animal performance can be ex-pected to decline when there are changes in diet and this needs to be taken into ac-count in management planning.

5 Forage quality (feed value)

H.H. Meissner, P.J.K. Zacharias & P.J. O'Reagain

CONTENTS

5.1 DEFINITION OF FORAGE QUALITY OR FEED VALUE

Animal performance depends on an inter-relation between a number of factors, both internal and external to the animal itself. These relations are shown in Fig. 5.1. Of interest in this chapter are those factors associated specifically with forages and the influence that these factors have on animal performance.

As shown in Fig. 5.1, nutritive value (chemical or nutrient composition), digestibility and feed intake are the main factors which determine animal performance, recognising that these characters are in turn influenced by a number of other factors related to both the animal and the forage, and that they interact among themselves. Taken together, these three main factors define what is commonly referred to as feed value.

Chemical composition and digestibility are often linked to the term nutritive value, which describes the amount and types of nutrients that the animal can derive from the feed. In this chapter chemical composition and digestibility will be discussed separately since they may be influenced by different plant and environmental factors.

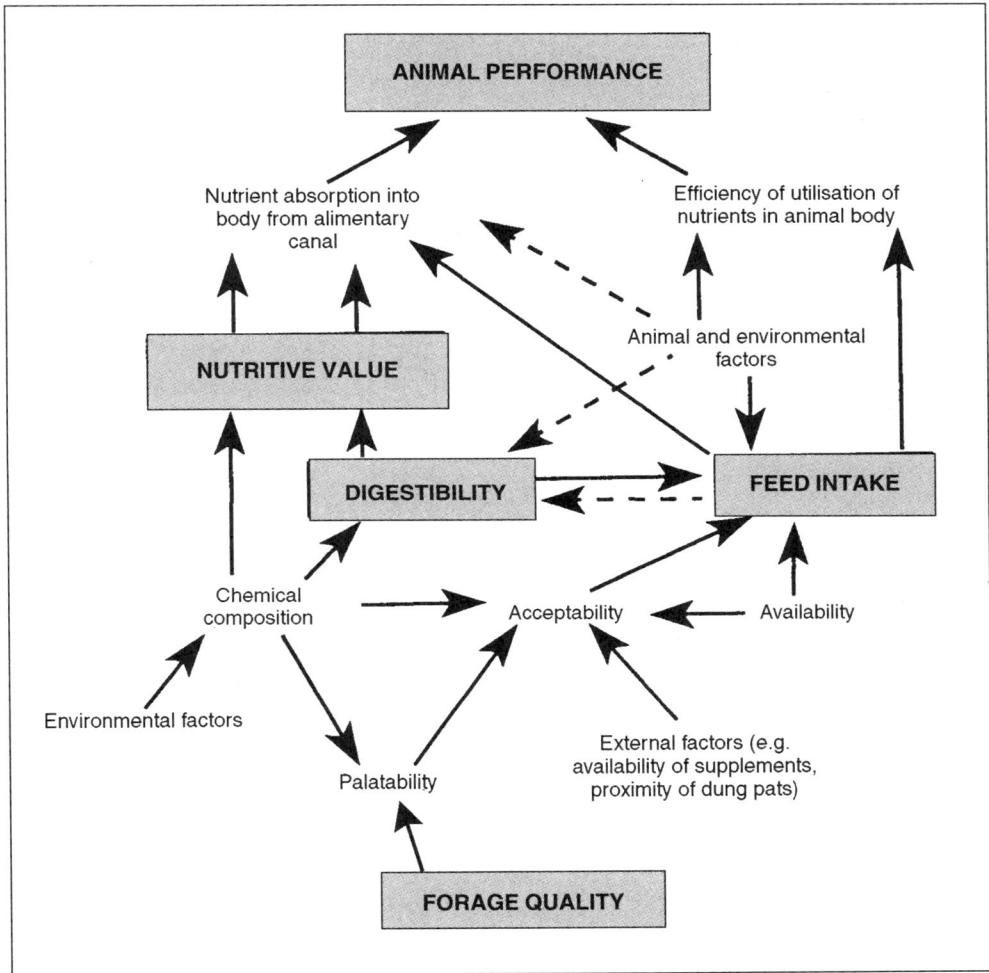

Figure 5.1 The relation between different aspects of nutritive value which influence animal performance (modified from Bransby 1981).

5.2 CHEMICAL COMPOSITION

Herbages contain a variety of chemical constituents which serve as nutrients for herbivores. Some nutrients are sources of energy while others satisfy a specific requirement in the animal's body. These chemical components can be divided into cell wall constituents (CWC) and cell contents (CC), and into digestible and indigestible or poorly digestible fractions (Van Soest 1967). In general, the cell contents are highly digestible (Table 5.1) and the cell wall constituents (commonly referred to as fibre) are either indigestible or poorly digestible. The former are soluble in neutral detergent while the latter are only partially soluble in acid detergent (Van Soest 1982).

Associations of soluble carbohydrates, starch, organic acids, cellulose and hemicel-

Table 5.1 Chemical constituents of roughages, their separation based on the Van Soest (1984) method, their real digestibility and factors which limit their use by animals.

Plant constituent	Analytical constituent	Chemical constituent	True digestibility	Limiting factor
Cell contents (CC)	Soluble in neutral detergent	Soluble carbohydrate	100	Intake
		Starch	90+	Retention time in digestive tract
		Organic acids	100	Intake
		Pectin constituents	95+	Intake
		Real protein	90+	Fermentation and accumulation of indigestible material
Cell wall component (CWC)	Insoluble in neutral detergent	Cellulose	0–100	Lignification, cutin formation, silica
		Hemicellulose	0–80+	formation, retention time in digestive tract
		Lignin	0	These chemical
		Cutin	0	constituents restrict the
		Silica	0	digestibility of cellulose
		Tannins, polyphenols	0	and other potentially digestible fractions

lulose, together with lipids (fats), contribute to the energy content of forages. Proteins, vitamins and minerals provide essential components of the animal's diet and are required in an appropriate balance if animals are to perform adequately. Forage plants may also contain anti-quality factors such as tannins, or even poisonous constituents, which may affect animal performance.

The major chemical constituents found in forages are the proteins, minerals, the structural components of the plant cells, the vitamins and anti-quality and toxic substances. These are discussed separately below.

5.2.1 Proteins

Protein is often the constituent which most limits the performance of animals on pasture. Crude protein comprises natural proteins as well as non-protein constituents (NPN) and is estimated by multiplying the N content of the forage by a factor of 6.25. This provides only a gross measure which does not distinguish between the protein needs of the microflora in the rumen and the protein available for absorption in the lower digestive tract. Nor does it take into account the quality or origin of the protein. It nonetheless gives a useful indication of the need for protein supplementation.

The protein requirements of an animal depend on its species, age, the physiological functions being undertaken (such as growth or lactation) and its level of production. In general, a minimum of between 7% and 8% crude protein is required by ruminants to meet their needs, but high producing animals require levels approaching 13% to 14%. Some livestock breeds appear to have a lower requirement than these levels.

Where crude protein levels in forages are insufficient to meet the animal's requirements, protein needs to be supplemented. Animals may also respond to protein supplementation even if crude protein levels are well above the levels shown previously when a large proportion of the crude protein is in the form of NPN.

The crude protein contents vary widely among forage plants but in all species and in all seasons, it declines with increasing age of the forage. The application of N fertilizer will normally increase the crude protein concentration in a forage, but much of this may be in the form of NPN which is of little value to the ruminant and may in fact be harmful and cause nitrate poisoning. Animals should therefore be kept off N fertilized areas for about three weeks following fertilizer application. This will allow the concentration of NPN in the forage to decline to safe levels as it is converted to protein.

5.2.2 Minerals

A detailed discussion of the mineral requirements of livestock will not be entered into here, and interested readers are referred to the tables presented by the NRC (1984, 1985) or, for South African data, to Van der Merwe & Smith (1991) and Boyazoglu (1973). Phosphorus (P) is invariably in short supply in pastures throughout most of South Africa (Du Toit *et al.* 1940). Supplementation with P is therefore often recommended throughout the year on many types of pasture, although opinions vary as to the need for this. Other minerals which are important in livestock feeding are sodium, calcium, potassium, magnesium, sulphur, zinc, cobalt, copper, manganese, molybdenum, iodine and selenium.

Both inadequate levels and imbalances of minerals in forages may lead to physiological disorders and suppress animal performance. For example, a deficiency of P will reduce conception rates while a deficiency of magnesium in cultivated pastures often gives rise to the well known hypomagnesaemia (grass staggers). It is important to note that both the level and the balance of minerals in the forage are important to livestock.

The concentration of minerals in forages is determined to a large extent by the maturity of the material. Mineral concentration declines with age, but the rate and extent of this decline cannot be predicted because it varies with time of year and is also influenced by soil type, soil nutrient levels and seasonal conditions. The best that can be done here is to give a range of expected mineral concentrations (Table 5.2).

5.2.3 Structural (cell wall) constituents (fibre)

The structural constituents of plant material include polysaccharides, lignin and some proteins, and can be divided into matrix polysaccharides (hemicellulose and pectic substances) and fibre polysaccharides (cellulose, lignin and proteins) (Jones & Wilson 1987). These have traditionally been termed fibre and may be incompletely or variably digested by the animal.

The stems of most forages have a larger proportion of structural polysaccharides and lignin than the leaves. This proportion increases with maturity in both temperate and tropical species. Tropical species appear to have a greater cellulose content, and a higher hemicellulose:cellulose ratio than temperate species due to the nature of the vascular tissue in C_4-type tropical plants.

Variations in the total content of structural constituents or their components are apparently less significant in herbivore nutrition than the interactions between the constituents, i.e. once lignin has been removed the polysaccharides of the cell wall become much more digestible (Jones & Wilson 1987). The lignin in plant fibre, however, resists microbial enzyme attack (Harkin 1973) and reduces digestibility through its linkage to specific points on polysaccharide chains and it prevents the physical attachment of rumen bacteria to plant cell walls.

Table 5.2 Approximate chemical composition (DM basis) of some common pastures (Bredon *et al.* 1987).

Pasture	DM range %	Crude protein g/kg	Crude fibre g/kg	CWC[1] g/kg	DOM[2] %	ME[3] MJ/kg	Calcium g/kg	Phosphorus g/kg
Ryegrass (annual)								
Autumn	16–20	231	206	—	64.6	10.10	4.0	3.0
Winter	18–22	200 (211)	220	330	63.7 (79.2)	9.90	3.5	2.7
Spring	20–24	174 (202)	246	507	62.0 (70.6)	9.60	3.5	2.5
Ryegrass (perennial)	20–25	195 (213)	200	472	62.0 (71.2)	9.60	3.0	2.5
Grass/clover (irrigated)								
More clover than grass	17–22	175	225	—	62.4	9.67	15.0	4.0
Grass dominant	18–25	135	250	—	63.7	9.90	8.0	3.5
Mid-summer	18–22	184	230	—	62.9	9.75	5.0	2.7
Autumn	18–25	146	265	—	60.2	9.30	5.5	2.5
Bermuda	25–40	120 (194)	235	645	56.0 (63.3)	8.10	2.3	2.5
Eragrostis curvula (mean)	30–40	110 (144)	380	605	58.0 62.2	8.85	4.0	1.4
Eragrostis/lucerne (rows)								
Late spring	—	— (318)	—	333	— (72.5)	—	—	—
Late summer	—	— (263)	—	365	— (67.2)	—	—	—
Kikuyu								
Standing hay	—	— (134)	—	692	— (55.7)	—	—	—
Spring	15–22	180 (185)	290	673	58.7 (61.9)	9.15	3.2	3.5
Summer	18–25	150 (232)	300	646	55.9 (65.7)	8.70	2.2	3.5
Autumn	20–35	115 (249)	320	671	54.9 (66.2)	8.55	1.8	3.0
Bana (mean)	—	— (153)	—	681	— (64.6)	—	—	—
Lucerne								
Before flowering	15–22	225 (307)	258	344	62.0 (68.7)	9.40	23.0	3.1
Early flowering	20–25	204	260	—	59.3 (63.6)	9.15	23.0	3.0
Smuts' fingergrass								
Standing hay	—	— (116)	—	740	— (51.3)	—	—	—
Spring	—	— (156)	—	693	— (59.1)	—	—	—
Summer	—	— (157)	—	600	— (68.7)	—	—	—
Guinea grass (winter)	—	— (134)	—	—	— (61.3)	—	—	—

Table 5.2 *Continued*

Pasture	DM range %	Crude protein g/kg	Crude fibre g/kg	CWC[1] g/kg	DOM[2] %		ME[3] MJ/kg	Calcium g/kg	Phosphorus g/kg
Blue buffalo grass									
Young	25–30	117	354	—	57.1		8.77	4.0	2.0
Mature	35–45	78	300	—	54.4		8.33	3.5	1.1
Rhodes grass (winter)	—	(66.2)	—	—	—	(45.4)	—	—	—
Anthephora	—	(118)	—	—	—	(50.7)	—	—	—
Cocksfoot									
Spring	18–27	219 (192)	220	522	63.0	(67.4)	10.00	3.5	3.5
Summer	25–35	161	300	—	59.0		9.30	3.5	3.0
Autumn	30–40	132	330	—	57.0		8.93	3.0	2.6

[1] Cell wall components determined as the insoluble fraction in neutral detergent (Van Soest 1982). Values in brackets are derived from animals fistulated at the oesophagus and recorded from the available South African literature. The data show that animals select a better quality diet than is suggested by samples harvested by hand.
[2] Digestible organic matter.
[3] Metabolisable energy.

The nature of lignin and its association with polysaccharides in plants has been called 'Nature's most closely locked secret' (Harkin 1973). It is basically a phenyl-type polymer which interconnects in various proportions and sequences such that it cannot be described exactly, and is not broken down by hydrolysis.

5.2.4   Vitamins

Vitamins are another group of essential chemical constituents, but are normally required in only small amounts. The most important of these is vitamin A, which is usually well provided for by green roughages and leafy lucerne hay. Cows usually need to be injected only once during a breeding season with vitamins A, D and E to prevent any possible deficiencies of these vitamins which can be stored in the body. For further information on vitamin requirements of livestock, refer to Van der Merwe & Smith (1991).

5.2.5   Anti-quality and toxic substances

The final group of chemicals which needs to be considered are the toxic substances. Certain legumes contain substances which cause bloat. Others contain tannins which reduce the digestibility of the forage. Some well fertilized pastures accumulate nitrate in toxic concentrations (refer to Chapter 11). The tree legume *Leucaena leucocephala* contains mimosine, which causes hair loss and liver damage in livestock (Minson 1977). Tall fescue causes 'Fescue foot' in certain regions in South Africa. Alkaloids can cause gangrene in the hooves of animals, and prussic acid poisoning is a problem associated with some sorghum varieties under certain conditions.

5.3   PALATABILITY AND ACCEPTABILITY

Definitions of palatability range from that provided by the Society of Range Management as the 'relish with which a particular species or plant part is consumed by the animal' (Jacoby 1989), which is rather broad and non-specific, to that suggested by Mentis (1981) which relates palatability to 'those factors of the feed itself which determine the absolute attractiveness of the feed to the animal'. The latter definition is widely accepted in southern Africa (Trollope *et al.* 1990).

While Jacoby (1989) does not define acceptability, Trollope *et al.* (1990) use the following definition: 'It is the attractiveness of feed to animals, as determined by factors of the forage and the environment.' It is therefore a relative term, and depends on the circumstances under which the forage is presented to the animal.

The distinction between the terms palatability and acceptability is not always clear, and Mentis (1981) uses the following example to explain the difference between them. Mature grass in the Highland Sourveld in winter is both unpalatable and unacceptable. However, if urea lick is provided, the material becomes acceptable to livestock, even though its chemical and physical properties are not altered and so neither is its palatability. This apparently arises from the improved digestibility of the material when fed together with a N source. Another example is that of the avoidance by animals of grass in close proximity to dung or urine patches. This material is unacceptable *in situ* but, if removed from the soiled area, it may readily be acceptable to animals, even though its palatability has not altered. Clearly, then, potentially palatable feed can be unacceptable, and unpalatable feed can be made acceptable. Yet another example is the use of molasses to improve the acceptability of low quality unpalatable hay or soiled pasture (Marten 1978).

It was suggested earlier that palatability is an absolute term, but that it cannot be measured because of the many factors, such as rumen fill and the physiological status of the animal, which will affect the extent to which a feed is consumed. This means that only relative palatability, as between two feeds which are on offer simultaneously, can be measured (Marten 1978). In effect, therefore, palatability, like acceptability, becomes a relative term.

5.3.1 Preferred and principal foods

A major difficulty in predicting the value of a feed is the difficulty in measuring its acceptability and palatability in any absolute way. Of interest here to the pasture manager is the choice the animal will make between various feeds on offer. This has important consequences for grazing management in mixed swards because it controls selective grazing. So, while the palatability of the feed is difficult to quantify in absolute terms, the order of choice by the animal of a number of species can be quantified. This raises the issue of preferred and principal feeds.

Preference is exhibited when an animal consumes a particular food in a larger proportion than that in which it is presented to that animal (Mentis 1981), irrespective of the extent to which that food contributes to the total diet of that animal. This defines a preferred food (Table 5.3). A principal food, on the other hand, is one that makes up a large proportion of the animal's intake, irrespective of its preference relative to other foods on offer. Principal foods are often, therefore, determined by circumstance. If, for example, an animal is presented with forage from only a single species then, assuming that it is prepared to consume that food, that species will automatically serve as the principal food species. In a mixed sward, the principal food is a function of the number of species and the relative acceptability and abundance (equitability) of each.

Table 5.3 Hypothetical species composition of pasture (forage on offer) and diet (forage selected) of grazing animals used to illustrate the concepts of preferred and principal foods.

Species	Species composition (%)		Preference ratio (Log scale)	Comment
	Pasture	Diet		
Digitaria eriantha	20	35	1.75	preferred
Chloris gayana	60	60	1.00	neutral (principal)
Eragrostis curvula	20	5	−0.60	unpreferred
Total	100	100		

5.3.2 Characteristics determining acceptability

Acceptability has generally been found to be positively correlated with the concentration of protein, energy, minerals, anthocyanins, ether extract and water content, and negatively correlated with fibre and lignin content (Heady 1964; Theron & Booysen

1966). However, it cannot be assumed that animals select directly for or against these constituents. Firstly, most are not constituents at the molecular level but simply arbitrary groupings for the convenience of forage chemists. Since only simple molecules can act on receptor sites on the tongue of the animal, these constituents cannot activate the sensory system and consequently are largely meaningless entities to the animal (Arnold & Hill 1972). Secondly, constituents which are positively associated with acceptability tend to be part of the cell content, which are in turn negatively correlated with cell wall constituents such as fibre and lignin. Hence, selection for positive factors could in fact constitute selection against cell wall constituents, and vice versa.

Acceptability is also strongly influenced by the physical properties and structure of the plant (Arnold 1964). In grasses, for example, selection by both cattle and sheep has been found to be negatively correlated with leaf strength (Theron & Booysen 1966; O'Reagain & Mentis 1989a). Plant structure may influence acceptability by affecting the accessibility of leaf to the grazing animal. The acceptability of *Hyparrhenia hirta*, for example, declines rapidly as it becomes increasingly stemmy over the growing season (O'Reagain & Mentis 1989a). The phenomenon is also likely to occur in *Digitaria* or *Panicum* pastures. Thorns and spines may reduce the acceptability of certain woody browse species below levels which would be expected from their leaf chemistry (Cooper & Owen-Smith 1986). Acceptability may also be reduced by the presence of awns, hairs or stickiness (Mentis 1981), as well as by the coarseness or harshness of the leaves (Shewmaker *et al.* 1989).

Plant secondary metabolites, such as tannins and alkaloids, are common amongst woody browse species and may significantly depress their acceptability to browsers such as goats (Cooper & Owen-Smith 1985). While these chemicals are relatively scarce among grasses, when encountered they affect acceptability, as is the case in many tropical legumes.

Acceptability is therefore quite clearly determined by a number of factors. Animals are apparently able to balance the favourable and unfavourable factors of a species and select accordingly (Arnold 1964; Field 1976; Cooper *et al.* 1988).

Aside from the characteristics of the plant species itself, acceptability is also strongly influenced by the situation in which the plant grows. Neighbouring plants of other species may modify a plant's acceptability by masking its chemical cues, by discouraging animals from grazing in their vicinity because of their smell, or by physically reducing access to the plant (Crawley 1983). Acceptability is also influenced more generally by the relative abundance and associated preference of other species growing in the same area. Thus the acceptability of a species may increase, for example, as the relative availability of other more acceptable species declines during grazing (O'Reagain & Mentis 1989b).

5.3.3 The determination of acceptability and palatability

It has become clear that a number of plant and animal factors interact in determining the extent to which a particular food item will be selected and that the relative importance of the different factors will vary according to the circumstances prevailing at any time. Therefore, the determinants of acceptability and palatability are poorly understood and their determination difficult. Because of these difficulties we can do little more than classify species according to the order in which they are selected by grazing animals (Daines 1980; Danckwerts *et al.* 1983; O'Reagain & Mentis 1989a). Any such approach must of necessity be regionally based and specifically related to the animal type used, and therefore have limited general application. For this reason most workers do no more than assign species to palatability classes. In practice, however, such a classification is often based on no more than a subjective judgement (Grunow 1980; Barnes *et al.* 1984).

5.4 INTAKE

5.4.1 Regulation of intake

Intake is the most important factor influencing the feeding value of roughages and therefore in determining the performance of grazing livestock on pasture (Waldo & Jorgensen 1981). The equations of Heany (1970), which link intake and digestibility to digestible energy intake in order to index the quality of a feed, suggest that intake is more than twice as important as digestibility in determining animal performance. Therefore, an understanding of the factors which affect intake on pastures is extremely important and so some knowledge of the digestion process in ruminants is needed.

When a ruminant swallows, the bolus enters the rumen. This organ is so structured that the outflow of food particles is restricted. This provides the opportunity for fermentation by micro-organisms, but it also restricts further intake. Further intake is possible only when the forage particles disintegrate sufficiently, as a result of rumination and fermentation, to allow them to pass through the valve of the reticulo-omasum. The rate of food-particle disintegration and its passage out of the rumen therefore regulates intake and largely determines the differences in intake between different species and forages.

The rate of forage disintegration in the rumen is closely related to the abundance and nature of the cell wall constituents in the forage since these constituents depress fermentation and outflow. In effect, the rate of fermentation and disintegration is closely related to digestibility, although digestibility accounts for only about 30% of the variation in intake (Van Soest 1984). The reason for this is that the ratio between cell content and cell wall components is itself highly influential in determining the rate of fermentation and outflow of indigestible residues. The greater the proportion of cell wall constituents, the slower are these two processes, and so the lower is the rate of intake. In general, therefore, cell wall constituents have a greater impact on intake by livestock on pastures and other roughages (>50%) than does digestibility, with subtropical crops being the worst in this respect (Rohweder *et al.* 1978) (Table 5.2).

The rate of cell wall fermentation generally imposes little limitation on intake at digestibilities above 70% and for cell wall contents below 35%. In pasture species with these characteristics intake is limited rather by the poor availability of forage, as well as by palatability, moisture content, grazing management, and other factors such as excretal contamination of the feed. These factors are important in management decisions, since they can greatly influence animal performance on high quality pastures.

5.4.2 Prediction of intake

Intake cannot be readily predicted from a knowledge of digestibility, nutrient content or even the concentration of cell wall material, except in general terms where forage species can be classified into broad quality categories, such as high, medium, low and poor (Table 5.4). Numerous other factors, in addition to management procedures, are in volved, complicating the prediction of intake for individual species. The data in Table 5.5 are nonetheless presented to provide some guide as to the expected intake of digestible organic material by sheep on a range of pastures.

Table 5.4   Criteria for the categorisation of different quality roughages.

Category	Digestibility (DOM[a], %)	Fibre (CWC[b], % of OM)	Lignification (Lignin, % of CWC)	Intake (g DM/kg.mass$^{0.75}$/day)	
				Cattle[c]	Sheep[d]
High	>70	<45	<5	>90	>75
Medium	55–70	45–65	5–10	70–90	60–75
Low	45–55	65–80	10–15	50–70	40–60
Poor	<45	>80	>15	<50	<40

[a]   DOM = Digestible organic matter
[b] ˙ CWC = Cell wall constituents determined as the insoluble fraction in neutral detergent (Van Soest 1982)
[c]   Cattle – Ellis *et al.* (1988)
[d]   Sheep – Meissner *et al.* (1989) and other data

Table 5.5   Correlation co-efficients between cell wall constituent and dry matter intake of sheep as influenced by region and the presence of subtropical crops, and predictions of intake at different digestibilities, based on the Van Soest analyses (Rohweder *et al.* 1978).

Hay type	Correlation co-efficient	Legume hay		Grass hay	
		In vivo DMD %	DMI g/kg. M$^{0.75}$	In vivo DMD %	DMI g/kg. M$^{0.75}$
Lucerne From northern USA From north and south	0.75 0.62	>70 66–70 58–65 <58 —	>80 75—80 68—74 <68 —	— >72 62–72 55–61 <55	— >69 65–69 59–64 <59

Grasses		*Predictive formulae:*
Temperate	0.94	
+ Pongola	0.80	Legume hays:
+ Bahia	0.76	DMD = 65.5 + 0.975 ADF% – 0.0277 ADF%^{2}
+ Bermuda	0.59	DMI = 39.0 + 2.68 NDF% – 0.041 NDF%^{2}

Variation in intake accounted for by variation in cell wall constituents of 35–89% (mean = 57%)

Grass hays:
    DMD = 34.8 + 2.56 ADF% – 0.0491 ADF%^{2}
    DMI = 54.8 + 1.22 NDF% – 0.0176 NDF%^{2}

Where:
    DMD = Dry matter digestibility
    DMI   = Dry matter intake
    NDF = Neutral detergent fibre
    ADF = Acid detergent fibre

As a guide, 30 kg lambs are expected to consume approximately 25 g/kg.$M^{0.75}$/day, i.e. 25 g of feed per kg of metabolic weight per day (or 58.5 g/day) for maintenance, 35 g/kg.$M^{0.75}$/day (81.9 g/day) when gaining 100 g/day and 55 g/kg.$M^{0.75}$/day (128.7 g/day) when gaining more than 200 g/day. For cattle, the intake of digestible organic matter is expected to be 5 g to 10 g/kg.$M^{0.75}$/day greater than for sheep on forage of equivalent quality, while requirements of lactating animals will be between 20% and 30% higher than those of non-lactating animals.

5.4.3 Voluntary (*ad lib*) intake of forage

The *ad lib* intake of an animal when offered an excess of a single feed or forage defines voluntary intake (Van Soest 1985). This is one of the measures used to estimate forage quality (Deinum 1984). Voluntary intake is not the same as palatability although palatability has some influence on it. Intake is mainly controlled by involuntary physiological reflexes within the animal, rather than its 'liking' for the feed or forage (Raymond 1969).

Intake in ruminants depends on the capacity of their digestive tracts, especially the rumen. They will eat until a certain degree of 'fill' is achieved (Raymond 1969; Van Soest 1985). As indicated previously, the level of 'fill' is influenced by digestion and the movement of food residues through the digestive tract – the more rapidly a feed is digested and passed through the animal, the greater the potential for a high intake.

Histological studies (Pond *et al.* 1984) of forage plants can shed light on differences in voluntary intake between forage types. Legumes have less cell wall than grasses and their fibre retention time in the rumen is shorter, contributing to the often observed greater intake of legumes than grasses. The sclerenchyma tissue above and below the vascular bundles of many grasses facilitates a strong attachment of the cuticle and epidermis to the interior tissues, making grasses more resistant to physical digestion mechanisms. Vincent (1983), for example, concluded that the observed lower intake of stem than of leaf tissue was as much a function of the different anatomy of fibres in these tissues, as of the difference in indigestible cell wall content.

The general relation between voluntary intake and digestibility is explained in terms of the 'bulk' of the feed, and its rate of digestion and rate of passage out of the rumen (Osbourn *et al.* 1966). While forage intake and digestibility may be closely correlated within a forage species, there is no generalised relation for all species (Deinum 1984). Ruminants are able to eat a greater amount of highly digestible forages because they occupy less volume, they are in the rumen for a shorter time, and there is less indigestible residue which has to be passed down to the hindgut; but digestibility is often depressed at high levels of dry matter consumption because the forage is in the rumen for only a limited period of time (McDonald *et al.* 1973). In forages of high digestibility, intake is limited by metabolic factors (blood concentrations of glucose, organic acids, etc.) rather than by rumen 'fill' (Raymond 1969).

5.4.4 Models of dietary selection

The observation that animals select a diet which is more nutritious than would be expected if it were selected at random (Weir & Torell 1959), has led to the suggestion that animals possess nutritional wisdom or euphagia. This assumes that animals have an innate ability to detect the presence of nutrients or toxins in plants (Provenza & Balph

1990). Euphagia also assumes that animals can recognise nutrient deficiencies in their metabolic systems and can detect forages which contain these nutrients. For example, cattle and sheep have been observed to rectify sodium deficiencies very precisely when offered sodium containing supplements (Denton & Sabine 1963; Bell 1984). Animals do not, however, appear to be able to compensate for at least some mineral imbalances, such as of P (Gordon *et al.* 1954). They also frequently select toxic plants such as *Senecio* spp. and *Morea* spp. in preference to non-toxic herbage. All classes of animals will readily consume *L. leucocephala* which contains the toxin mimosine (Maclaurin *et al.* 1981). Animals can, nevertheless, generally satisfy their nutritional requirements.

A second model, hedyphagia, assumes that animals select a nutritious diet by selecting foods which are immediately pleasing to the senses of taste, smell and touch, and reject those which are not (Arnold & Dudzinski 1978). This assumes that plant compounds which are nutritious also taste good, while those that are toxic have a bad taste (Rhoades 1979). Animals would therefore select a nutritious diet purely by association (Provenza & Balph 1990). Although animals generally accept sweet and reject bitter tastes, a number of exceptions are known to occur. Plants with a high concentration of alkaloids and tannins are, for example, often selected (Cooper *et al.* 1988), while others with characteristics that are immediately pleasing, at least to the human senses, are rejected (Provenza *et al.* 1990).

The major shortcoming of the hedyphagia model is that it does not consider the post-ingestive consequences to the animal of consuming a particular food item. These have the potential to either reinforce or undermine the sensations experienced by the animal when the food is eaten (Provenza & Balph 1990). An animal may therefore come to avoid a sweet tasting food item if its consumption is followed by nausea, while it may learn to accept bitter food items if their consumption is followed by some positive feedback.

Animals may also learn to discriminate between nutritious and toxic foods by a process termed dietary learning. Young animals may learn what foods to eat by watching older animals of the same species. Evidence for this is provided by lambs which, having been in the presence of their wheat-eating mothers when young, ate 10 times as much wheat when again exposed to wheat at an age of 34 months, than lambs which had not been given this 'learning experience' (Green *et al.* 1984).

A fourth model of dietary selection is based on the allometric relation between body size, on the one hand, and morphology and physiology on the other (Chapter 4). Large animals, because they require more food than small animals, have less time available, per unit of nutrient required, for feeding selectively (Bell 1971). But because their metabolic requirements per unit of body mass are lower than in small animals, they can subsist on lower quality food items than can small animals (Kleiber 1961). Added to this, they typically have wider muzzles than small animals and therefore cannot as effectively select out small, highly dispersed food items of high quality as can small narrow muzzled animals (Bell 1971). These different aspects are linked by the way in which they will affect food intake.

Optimal foraging theory assumes that animals attempt to optimise their intake of nutrients and/or minimise their intake of plant toxins. Foods are thus assessed according to their utility to the animal, based on the cost of harvesting (and processing) and the benefit of digesting that particular food, i.e. on the net profit to the animal of ingesting that food (Stephens & Krebs 1986). As an example, an animal may use any of the following criteria in deciding whether or not to accept or reject a particular food: its nutritive content; the rate at which it can be harvested (and digested); the animal's vulnerability to predation when harvesting such a food; and the energetic cost of harvesting the food.

None of the models presented here necessarily provides a complete and satisfactory explanation of dietary selection. Taken individually, the optimal foraging theory, while it has never been entirely successful in explaining dietary selection (Westoby 1978; Owen-Smith & Novelli 1982), nevertheless represents a major advance over other theories such as euphagia and hedyphagia, as it takes into account the many different constraints in which the animal is forced to operate. Perhaps, however, all the models must be seen as complementary to each other since they address different levels of the interaction between animals and the plants they feed on. For example, while body morphology and physiology provide a broad indication to the animal of the plants it should feed on (e.g. grass versus browse), this may be modified by euphagia (e.g. eat green grass), and then by hedyphagia (e.g. eat green grass which tastes good). This in turn will be modified by optimal foraging theory (e.g. eat species which allow for a high rate of intake), and finally by dietary learning (e.g. eat grass because the rest of the animals eat it).

5.5 DIGESTIBILITY

5.5.1 Relation to feed value

Digestibility is estimated as the difference between the amount of feed ingested and the amount excreted. If the energy contents of the forage ingested and of the excreta are known, the digestible energy content of the feed can be estimated. In effect, this amount represents only the apparent digestible energy content, since not all the energy recorded in the excreta is derived from the forage, nor is all energy exchanged, i.e. sweat, methane, breath, hair loss, etc., accounted for during feeding trials in metabolic crates. Note that both digestibility and digestible energy are usually expressed as a fraction or percentage of the original amount ingested.

The relation between digestibility and feed value is apparent from the definitions presented in section 5.1. It is usually positively related to the concentration of nutrients in the forage and to its intake (Table 5.2). This is so because the larger the quantity of nutrients in the feed, the more easily they can be digested. However, there are other factors which influence this association. Intake is, for example, more closely related to the rate of digestion in the rumen of ruminants (see section 5.4) than to the total digestibility of roughages. Because of this, two grasses having the same total digestibility may have different feeding values because of different rates of digestion. Indeed, in many situations total digestibility is not well correlated with intake (Meissner *et al.* 1989), nor with the growth rate of the animal (Clark & Bartyh 1970). Because of these complications, digestibility values should rather be seen as providing an index which can be used to separate roughages into classes (e.g. high, medium, low and poor, as in Table 5.4), than as an absolute measure of the feeding value of a forage.

5.5.2 Methods used to estimate digestibility

Various methods have been developed over the years to estimate digestibility. Conventional digestibility trials are undertaken using animals fed under controlled conditions. Such trials provide *in vivo* (in the body) estimates of digestibility. Unfortunately the procedure is too costly and time consuming for large-scale routine analysis of forage samples, but it does provide data against which other procedures can be calibrated. One such method involves the suspension of a dry sample of forage, enclosed in a bag made of indigestible material such as nylon, within the rumen of a rumen fistulated animal. The bag is removed after a specified period in the rumen and the loss of dry matter from the bag determined. This allows for the estimation of both the rate of fermentation and amount of fermentation which has taken place. The method is, however, difficult to standardise because of variations in the size of the pores of the materials used to make the bags and because results are partially dependent on the diet fed to the test animals.

The so-called *in vitro* (in glass) method of estimating digestibility is the most commonly used method worldwide for the routine analysis of large numbers of samples. This technique attempts to simulate *in vivo* digestion in the laboratory. The most commonly used *in vitro* procedure is that of Tilley & Terry (1963), which simulates digestion in both the rumen and in the lower gut of ruminants. It has provided correlations with *in vivo* digestibility of between 0.79 and 0.97 (Barnes 1973). The equation developed in South Africa between *in vitro* and *in vivo* digestibility appears to be reasonably reliable for non-tannin and non-toxin containing forages with organic matter digestibilities between 45% and 70%. This equation, from Engels *et al.* (1981), is as follows:

$$y = 16.4205 + 0.7892x \text{ (SE } x = \pm 1.31; r = 0.962)$$

where y = *in vivo* digestibility of the organic matter

x = *in vitro* digestibility of the organic matter of the solid fraction of oesophageal samples dried at 50°C. (This seems to hold also for hand-harvested pasture samples and hay.)

Digestibility may also be reliably estimated using a cellulase-based procedure (Zacharias 1986). Most agricultural regions in South Africa have laboratories which are capable of undertaking *in vitro* analyses of forages. In the absence of *in vitro* digestibility data, one can fall back on TDN (total digestible nutrients) values which are available for the forages produced over most of the country. These values are often well correlated with *in vitro* and *in vivo* digestibility.

A further technique that has been used with reasonable success in estimating digestibility of forages is based on cell wall or, alternatively, fibre content. Digestibility is negatively related to cell wall (fibre) content, especially lignin content (Table 5.4). In the Van Soest analysis (Van Soest 1982), the indigestible residue of the cell wall component, which is comprised partially of lignin, is effectively isolated. This fraction is referred to as acid detergent fibre (ADF) which, according to Rohweder *et al.* (1978), provides the best of the chemical estimates of digestibility.

5.5.3   Factors which influence digestibility

Digestibility is directly influenced by a number of factors and indirectly also by grazing management. Typically, the digestibility of tropical and subtropical grasses is lower than that of legumes and of grasses of temperate areas. There are a number of reasons for this. Tropical and subtropical grasses have a higher cell wall content than do grasses of temperate areas (Morris 1984) and the digestibility of their cell wall material (fibre) is lower than that of temperate grasses. This lower digestibility is due to specific structural characteristics such as greater lignification and a lower ratio of mesophyll to parenchyma and bundle sheath material (Akin 1979). Tropical and subtropical species also have a lower leaf to stem (culm) ratio than temperate species (t'Mannetje 1984). The relevance of this is that stem material is typically less digestible than leaf material, and its digestibility declines more rapidly with age than does that of leaf material (Fig. 5.2).

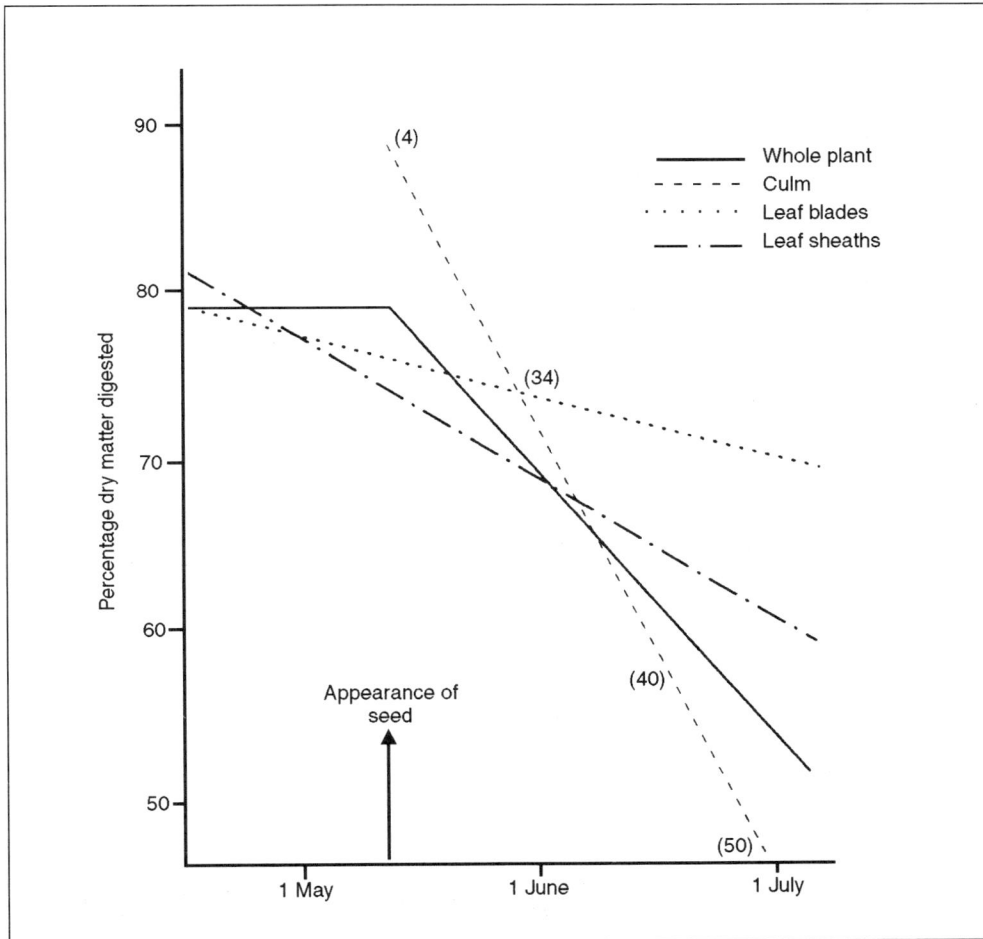

Figure 5.2 The *in vitro* digestibility of the dry material of the whole plant, leaf sheath, leaf blade and culm of the spring growth of S.37 Cocksfoot. Numbers in brackets indicate the percentage culm in the whole plant (data from Terry & Tilley 1964).

Plant age is an important factor affecting digestibility, although its influence varies across species, between primary growth and regrowth (Lascano 1979; Hacker & Minson 1981) and with season. What the data of Fig. 5.3 suggest is that the relation between the cell content and cell wall components differs for different species and for recovery growth harvested at different times during the year, even though these have the same digestibility. If this is related to the differences in the availability of nutrients in cell walls and their cell contents (Table 5.1), then we can expect large differences in the nutrient value of feeds which have the same digestibility.

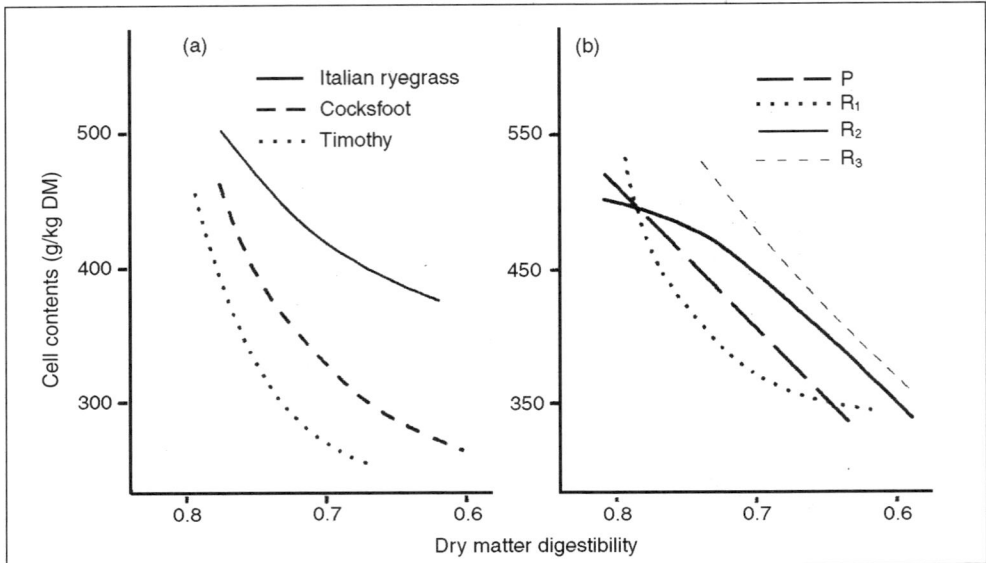

Figure 5.3   Change in cell content in grasses as the crop matures and the digestibility declines: (a) primary growth of three grass species; (b) primary growth (P), and regrowth following cutting of the primary growth on 12 April (R_1); 26 April (R_2) and 24 May (R_3) (Osbourn 1982).

Tables 5.2 and 5.6 provide South African data on the influence of season and stage of forage maturity on the feeding value of forages.

High temperatures result in low digestibility, which may partly explain why tropical and subtropical species are less digestible than temperate species. However, those relations are affected by both season (Tables 5.2 and 5.6) and the relative maturity of the plant. Fertilizer levels do not influence digestibility other than through their influence on growth rate and stage of maturity of the material (Raymond 1969).

5.5.4   Predictive ability of total digestible nutrient (TDN) and metabolisable energy (ME) values

As animal production is normally dependent on the ability of a feed to supply energy to the animal, feeding standards throughout the world are usually based on energy values, as for example in the ARC (1980) and NRC (1984) feeding standards. The allocation of gross energy (GE) to digestible energy (DE), metabolisable energy (ME) and net energy (NE) for maintenance and production for ruminants is shown in Fig. 5.4.

Net energy will provide the most suitable index of the energy value of a feed since it discounts all energy losses involved in food conversion within the animal. However, its determination is too sophisticated for routine use, and so metabolisable energy values are commonly used as an alternative. Unfortunately, the loss of energy as methane (which forms a component of ME) is difficult to measure, although in most roughages it can be assumed to account for about 8% of gross energy or 12% of digestible energy.

Table 5.6 Digestible organic matter intake (DOMI) of different pasture species by sheep (adapted from Meissner *et al.* 1989 and other unpublished data).

Description	DOMI (g/kg.mass$^{0.75}$/day)
Grasses	
Kikuyu	
Spring	35
Standing hay	30
5 weeks regrowth	35
10 weeks regrowth	25
Smuts' fingergrass	
Summer	35
Standing hay	25
Pangola grass	30
Guinea grass	35
Bermuda	30
Cocksfoot	40–45
Eragrostis curvula	35
Rye	
Annual	50–65
Perennial	45–55
Triticale	50
Legumes	
Lucerne	40–65
Clover	65
Grass/legume combinations	
Eragrostis/lucerne	
Late spring	45
Late summer	35

The feeding value of South African roughages have in the past been based largely on TDN rather than on ME values but it is possible to convert from TDN to ME values using the following equations:

1. $\text{DOM (\% of DM)} = \dfrac{\text{TDN (\% of DM)}}{1.05}$

$$\text{(Kearl 1982)}$$

    where DOM = digestible organic material

           DM = dry material of forage

2. DE (Mcal*/kg DM) = 0.04409 TDN (% of DM)

$$\text{(Kearl 1982)}$$

    where DE = digestible energy

3. ME (Mcal/kg DM) = 1.01 DE (Mcal/kg DM) – 0.45

$$\text{(Moe \& Tyrrell 1976)}$$

   * 1 Mcal = 4.184 MJ

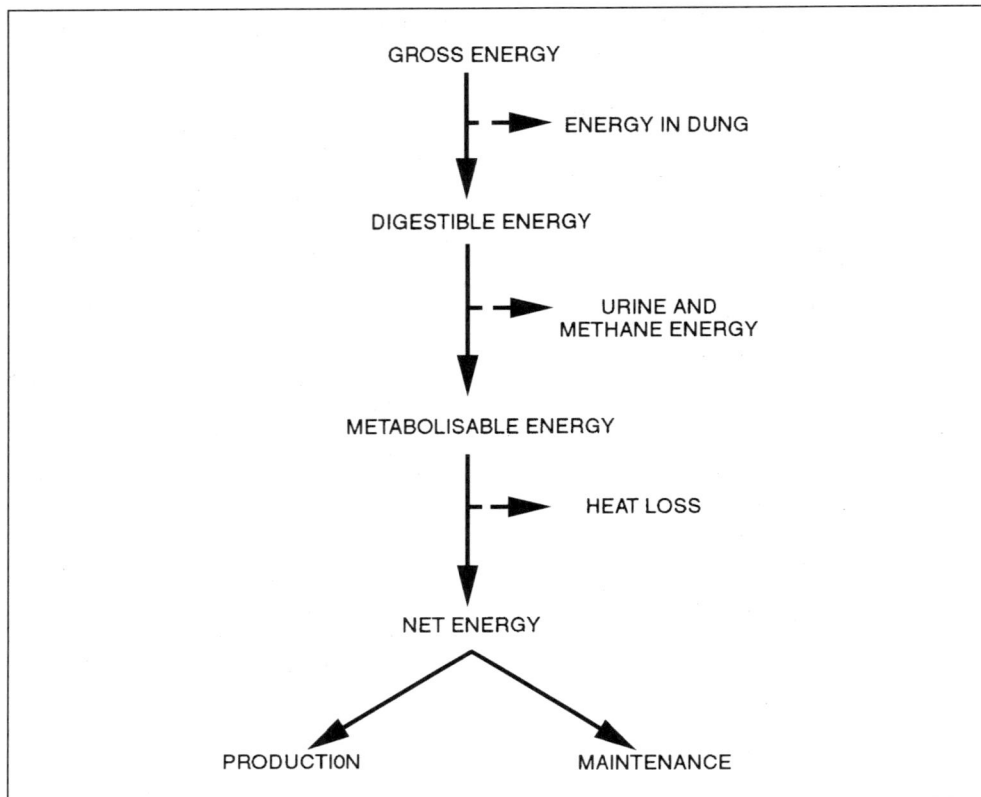

Figure 5.4 The allocation of the plant's (gross) energy during digestion and utilisation by ruminants.

The equations of Kearl (1982) are preferred to others in the literature (e.g. Anon. 1983) as they are based on several African studies involving grazing rather than pen fed animals.

5.6 EXAMPLES OF THE QUALITY OF PASTURE IN SOUTH AFRICA

5.6.1 Quality of grasses and legumes

Quality data with respect to some grasses and legumes in South Africa are shown in Tables 5.2 and 5.6. The data show large differences between species and within the same species during different seasons and at different stages of maturity, but it needs to be appreciated that, in practice, the quality of the forage provided by different species at any time will depend partially on the condition of the pasture at that time. Recent regrowth on a pasture type which on average produces relatively low quality material may, for example, be qualitatively superior to the mature forage from pastures capable of providing very high quality material. Seen in isolation, these data mean very little unless they can be related to the maintenance and production requirements of animals. Requirements of beef animals, dairy cows and small stock are therefore shown in Tables 5.7, 5.8 and 5.9.

Table 5.7 Nutrient requirements of beef animals at the minimum required digestible organic matter values (adapted from Kearl 1982).

Weight (kg)	ADG (kg)	DOM (%)	DM intake (kg)	Crude protein (g)	ME (MJ)	Calcium (g)	Phosphorus (g)
Oxen and heifers							
150	0.0	45	3.2	233	21.5	6	5.5
	0.5	55	3.9	493	34.0	15	11.0
	1.0	65	4.5	615	46.5	26	17.0
250	0.0	45	4.6	301	31.5	8	7.0
	0.5	55	5.7	595	49.5	16	14.0
	1.0	65	6.6	742	62.5	26	19.0
350	0.0	45	5.7	391	40.5	12	11.0
	0.5	55	7.3	684	63.5	18	16.0
	1.0	65	8.5	836	86.5	25	19.0
Finishing oxen							
400	1.2	70	9.4	967	105.0	33	25.0
	1.4	75	9.7	1009	116.0	34	26.0
450	1.2	70	10.3	998	114.0	31	28.0
	1.4	75	10.5	1038	125.0	33	29.0
Cows – last 3 months of pregnancy							
500	0.4	50	9.6	546	76.0	16	15.0
550	0.4	50	10.3	579	81.5	17	16.0
First 12 weeks of lactation							
500	—	55	10.5	821	85.5	28	27.0
550	—	55	11.3	857	91.5	29	28.0

Using these data in conjunction with those of Tables 5.2 and 5.6, it is possible to assess the potential of pasture and other roughages to support livestock production. Intake is invariably unknown, and so the expected DM intake must of necessity be estimated. The data of Tables 5.2, 5.4 and 5.7 can be used for this purpose. For example, summer kikuyu has a crude protein content of 150 g/kg and a ME value of 8.7 MJ/kg (Table 5.2) which equates to a digestible OM content of 55%. In order to gain 500 g/day, a 250 kg steer requires 5.7 kg DM, 595 g crude protein and 49.5 MJ ME (Table 5.7). Assuming an intake of 5.7 kg DM, kikuyu will supply 855 g crude protein (150 × 5.7) and 49.5 MJ (8.7 × 5.7).

The question is whether an intake of 5.7 kg DM will be achieved on kikuyu. According to the data in Table 5.4, the expected intake should be 70 g/kg.$M^{0.75}$/day for forage having a digestible organic matter content of 55%. This equates to 4.4 kg DM which would be insufficient for a gain of 500 g/day, unless energy were supplemented to make up the shortfall. The crude protein supply should be sufficient for this level of performance. However, using the oesophageal fistula values for DOM content of the kikuyu (approximately 65% for most of the season in Table 5.2) in these calculations, produces the following: DM intake would be 85 g/kg.$M^{0.75}$/day, which would be

Table 5.8 Feed requirements of dairy cows (adapted from NRC 1978).

Nutrients (concentration in feed DM)	Cow weight (kg) ≤400 500 600	Lactation rations				Other rations	
		Production level (ℓ/day)				Cow, dry late pregnancy	Heifers
		≤8 ≤11 ≤14	8–13 11–17 14–21	13–18 17–23 21–29	≥18 ≥23 ≥29		
DOM %		60	64	68	71	60	60
ME, MJ/kg		9.8	10.6	11.3	12.1	9.8	9.8
Crude protein %		13	14	15	16	11	12
Crude fibre %		17	17	17	17	17	15
ADF %		21	21	21	21	21	19
NDF[1] %		38	38	38	38	—	—
Calcium %		0.43	0.48	0.54	0.60	0.37	0.40
Phosphorus %		0.31	0.34	0.38	0.40	0.26	0.26
Magnesium %		0.20	0.20	0.20	0.20	0.16	0.16
Daily requirements							
ME[2], MJ	400	93	106	131	144	66	
	500	115	130	162	178	76	
	600	139	157	195	217	87	
Crude protein[2], kg	400	1.09	1.31	1.74	1.96	0.71	
	500	1.38	1.64	2.17	2.45	0.82	
	600	1.71	2.00	2.66	3.01	0.92	
Calcium[2], g	400	38	45	58	65	27	
	500	48	56	72	81	32	
	600	59	68	88	99	36	
Phosphorus[2], g	400	27	32	41	45	19	
	500	34	39	51	56	23	
	600	42	48	62	69	26	

[1] Mertens (1983)

[2] Daily requirements determined for 4% fat corrected milk. For 3.5% or 4.5% butter fat, subtract 5% or add 5%, respectively, to the above values.

equivalent to 5.3 kg kikuyu DM/day. This would provide 53.5 MJ ME (65% DOM is equivalent to 10.1 MJ ME/kg according to Table 5.2). This would be sufficient for a daily weight gain of 500 g.

The total feed requirements for dairy cows may be determined from the concentration of DM in the roughage as a starting point, as for example when hay is combined with concentrate feed to provide the complete feed requirement (Table 5.8). On pastures, of course, the concentrate feed is fed separately from the roughage. But how much milk can be produced from a specific pasture, and how much concentrate feed must be provided to achieve a specified level of production? Suppose we have a ryegrass pasture with an ME value of 10.1 MJ/kg (DOM = 65%), a crude protein content of 230 g/kg and Ca and P concentrations of 4.0 g/kg and 3.0 g/kg respectively (Table 5.2). Intake from ryegrass, if fed alone, would be expected to be approximately 85 g/kg.$M^{0.75}$/day (Table 5.4); increased by 25% for a lactating animal (section 5.4.2), which would mean an intake of 12.9 kg DM for a 600 kg cow. This intake would provide 130 MJ ME, 2.97 kg crude protein, 52 g calcium and 39 g P. This would be sufficient for the production of 14ℓ of milk with a butter fat content of

Table 5.9 Nutrient requirements of small stock at the minimum required digestible organic matter values (adapted from Kearl 1982).

Weight (kg)	ADG (g)	DOM (%)	DM intake (kg)	Crude protein (g)	ME (MJ)	Calcium (g)	Phosphorus (g)
Lambs							
20	0	50	0.49	43	3.90	2.8	2.0
	50	55	0.56	59	4.90	3.0	2.2
	150	60	0.81	108	7.75	3.3	2.4
	250	70	0.94	143	10.50	3.6	2.5
30	0	50	0.62	59	4.95	4.0	2.6
	50	55	0.69	81	6.05	4.3	2.9
	150	60	1.10	138	10.50	4.9	3.1
	250	70	1.12	174	12.50	5.3	3.3
40	0	50	0.79	75	6.25	5.2	3.0
	50	55	0.96	100	8.35	5.6	3.1
	150	60	1.41	167	13.40	6.0	3.3
	250	70	1.39	206	15.50	6.3	3.5
Dry ewes							
40	25	50	0.96	82	7.60	3.3	2.8
50	25	50	1.03	96	8.15	3.5	3.0
60	25	50	1.28	110	10.20	3.7	3.2
Last 6 weeks of pregnancy							
40	150	60	1.52	146	14.50	3.8	3.6
50	150	60	1.73	166	16.50	4.1	3.9
60	150	60	1.86	277	17.70	4.4	4.1
First 8 weeks of lactation							
40	− 20	60	1.63	168	15.50	10.4	7.4
50	− 20	60	1.84	193	17.50	10.9	7.8
60	− 20	60	2.04	219	19.40	11.5	8.2

3.5% (Table 5.8). However, note the excessive amount of crude protein in the diet.

A production target of say 25ℓ can be achieved only if the pasture is supplemented with concentrates. It then becomes difficult to determine the total intake from both pasture and concentrates, although this can be estimated from the following equation:

DM intake (% of cow weight) = 2 + (0.0506 × ℓ milk/day)

For a cow producing 25ℓ, DM intake should be approximately 3.265% of body mass, or 19.6 kg.

Where concentrates are fed, intake of pasture will normally be depressed, particularly on high quality pasture (Osbourn 1982). Intake of DM from pasture will normally decline to between 1.5% and 2% of cow weight, with the lower value (1.5%) applying to high quality ryegrass pasture. This means that pasture intake will decline to about 9 kg DM, which would provide 91 MJ ME. Dairy meal normally has a TDN of approximately 75%. According to the comparisons provided in section 5.5.4, this would equate to 12.1 MJ ME/kg DM. The ME requirement to produce 25ℓ of milk is 195 MJ (Table 5.8). Therefore, approximately 8.5 kg of dairy meal (195 − 91/12.1) needs to be provided.

The total intake from pasture and concentrates will therefore be within 2 kg of the intake capacity of the cow.

The disadvantage of this combination of 8.5 kg dairy meal and 9 kg ryegrass is that there would be insufficient roughage in the diet. According to Table 5.8, cell wall material should make up 38% of the diet. Approximately 40% of ryegrass forage is made up of cell wall material, while dairy meal contains very little of this material. This will mean that the diet will have only about half the required amount of roughage and it would therefore be advisable to add 2 kg of poor quality hay to this diet.

5.6.2   Quality of browse material

Tree leaves generally contain more crude protein than grass but have a lower digestibility because of the greater lignification of the cell wall tissues of browse plants (Owen-Smith pers. comm.). Tree legumes have not been widely researched or utilised as a component of pasture systems although Zacharias *et al.* (1991) have shown that *L. leucocephala* provides a cost-effective supplement to cattle grazing foggaged kikuyu.

6 Stocking rate

C.D. Morris, M.B. Hardy & P.E. Bartholomew

The rate at which a cultivated pasture is stocked is perhaps the single most important management factor affecting animal performance and the profitability of a livestock system (O'Reagain & Turner 1992). Of particular relevance to the manager is that it is one of the variables that is under his direct control and the stocking rate he chooses largely determines the degree of interaction between the grazing animal and the pasture which, in turn, determines production per animal and animal production per hectare.

6.1 DEFINITION AND EXPRESSION OF STOCKING RATE

Stocking rate can be defined as the number of animals, of a particular class, which are allocated to a unit area of land for a specified period of time, usually the growing period of the pasture in question (Bartholomew 1991). It can be expressed in a number of different ways (Turner & Tainton 1990) but most commonly in terms of either animal numbers per unit land area (ha) or as land area available for each animal. The former (animals/ha) is usually used when referring to cultivated pastures and the latter (ha/animal) when referring to veld. This to avoid having to use fractions of animals. Converting animals/ha to ha/animal, and vice versa, is achieved by simply calculating the reciprocal. So, for example, 5 animals/ha = 0.2 ha/animal and 0.25 ha/animal = 4 animals/ha. It should be noted that the two ways of expressing stocking rate differ markedly in their relation to the number of animals stocked (Danckwerts 1989). Stocking rate on a particular portion of land, expressed as animals per hectare, increases linearly with an increase in the number of animals stocked (Fig. 6.1). In contrast, stocking rate changes non-linearly with increasing stock numbers when expressed as hectares per animal (Fig. 6.1). These differences may give rise to some confusion and it is generally recommended that stocking rate be expressed as animal numbers per hectare rather than as hectares per animal.

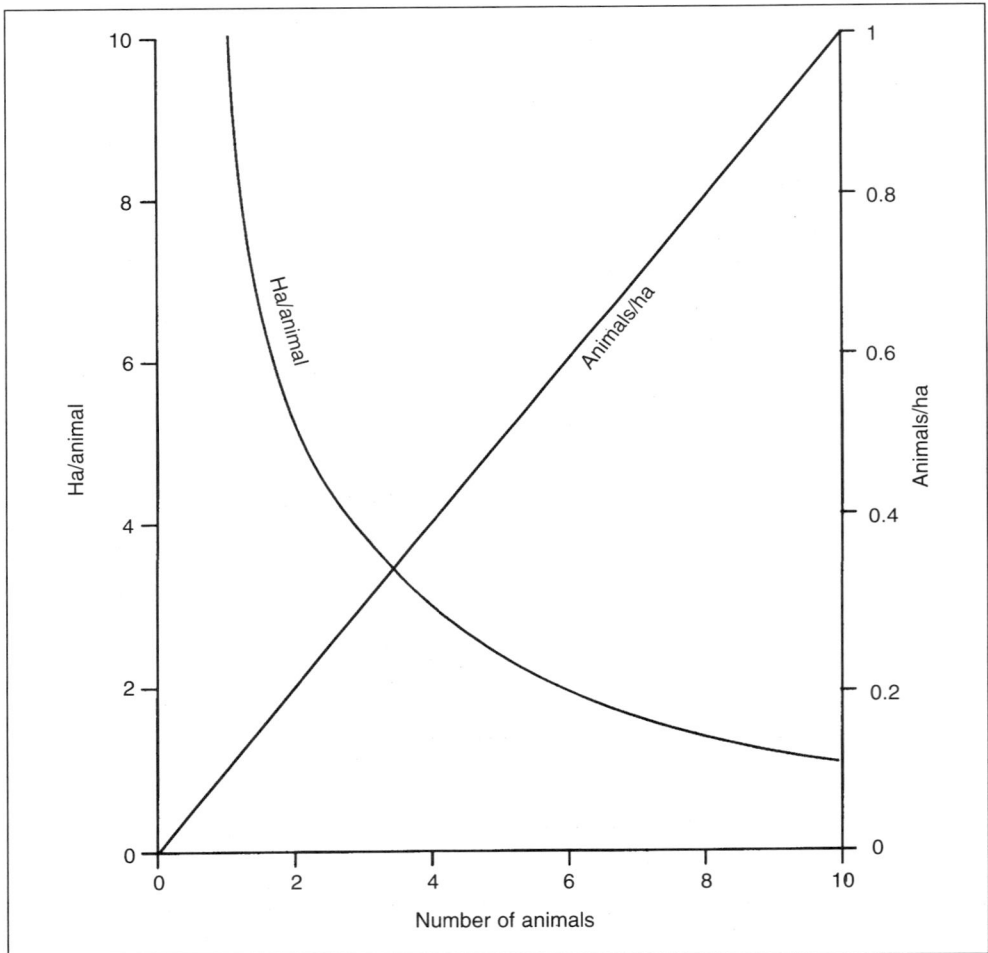

Figure 6.1   The relation of the units ha/animal and animals/ha with increasing numbers of animals (0–10) on a 10 ha area of land (after Danckwerts 1989).

6.2 STOCKING RATE–ANIMAL PERFORMANCE MODEL

Various models have been proposed to describe the relation between stocking rate and the performance of grazing animals measured in terms of saleable output: meat, milk, wool or hair (see Sandland & Jones 1975). One of the most widely used of these is that proposed by Jones & Sandland (1974) (Fig. 6.2).

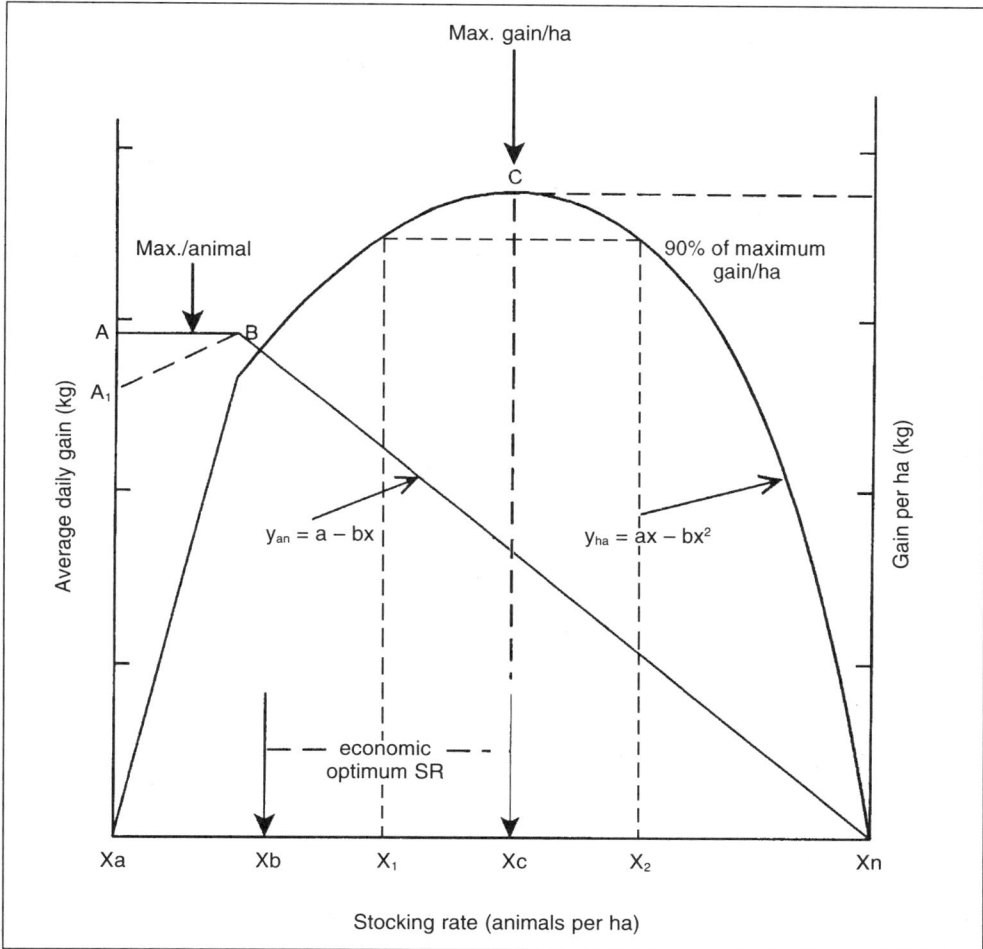

Figure 6.2   The theoretical relation between the stocking rate and average daily gain ($y_{an} = a - bx$) and between stocking rate and livemass gain per hectare ($y_{ha} = ax - bx^2$) (Bartholomew 1991).

(a)   The individual animal performance model

The relation between stocking rate and individual animal performance – expressed as average daily gain per animal (ADG), milk production, or some other measure ·of animal performance, is well described by two straight lines; represented by AB and BXn in Fig. 6.2. AB is parallel to the stocking rate axis. Here there is no improvement in performance with a reduction in stocking rate from Xb to Xa. At such low grazing pressures intake is not restricted so the amount of feed available per animal does not affect individual animal performance. Production here may be limited by the genetically-defined growth potential of animals or by pasture quality, and probably most often the latter. Animal performance may in fact decline at stocking rates lower than Xb due to the accumulation of forage of poor quality (Fig. 6.2: the dotted line A_1B) (Stobbs 1970).

At stocking rates greater than Xb, commonly referred to as the critical stocking rate (Hart 1978), animal performance declines as stocking rate increases. This decline results from an increase in grazing pressure, i.e. an increase in the ratio of forage demand to forage supply. Here quantity limitations, rather than quality limitations, reduce animal gains as competition for the available forage increases with an increasing stocking rate. Eventually a stocking rate is reached (Xn) where mass gains are zero and animals are at maintenance. Maximum individual animal performance on a particular pasture, for a particular class of animal, occurs at a stocking rate where quality and quantity limitations are balanced (Turner & Tainton 1990).

The portion of the ADG model between Xb and Xn is described by the linear model:

$$y_{an} = a - bx \quad \text{.. Equation 1}$$

where y_{an} = average daily gain per animal (ADG);
           x = stocking rate in animals/ha;
           a and b are constants (Fig. 6.2).

Since the model $y_{an} = a - bx$ is empirical, no biological determinants exist for a and b and their value must be determined experimentally for each pasture and class of grazing animal (Hart 1978). Data derived from any two stocking rates between Xb and Xn can be used to describe this relation. These points should ideally be as far apart as possible so as to accurately estimate the slope and intercept of the line. However, due to possible deviations from linearity at the extremes, the predictions of animal performance derived from the linear model are most accurate when made from points near the centre of the line.

Deviations from linearity in the ADG model have been reported (Peterson *et al.* 1965; Conolly 1976), particularly at high stocking rates where animal performance may drop off more rapidly with increasing stocking rate than at lighter stocking rates. This may result from increased energy requirements for foraging where forage is scarce and widely dispersed across the landscape (Heitschmidt & Taylor 1991). However, the linear model is generally a good approximation of the relation between stocking rate and individual animal performance.

(b) The gain per unit area model
The characteristics of the gain per unit area model are very different from those of the ADG model (Fig. 6.2). This model shows a sharp linear increase in animal production per hectare with each unit increase in stocking rate from Xa to Xb, a slower rate of increase at stocking rates beyond Xb until a maximum gain per hectare is reached at Xc, and then an accelerating decline in gain per hectare at stocking rates beyond Xc. Finally, point Xn is reached where production per unit area is zero.

The animal gain per unit area model at stocking rates between Xb and Xn is a quadratic of the form:

$$y_{ha} = \text{days} (ax - bx^2) \quad \text{.............................. Equation 2}$$

where y_{ha} = gain per ha;
           a, b and x are as before;
           days = number of days over which y_{an} (Equation 1) was determined.

This model is simply derived by multiplying the individual animal performance model (Equation 1) by the stocking rate expressed in animals per hectare. It should be noted that the same production per hectare can be achieved at different stocking rates. For example, from Fig. 6.2 it can be seen that 90% of maximum production per hectare can be achieved at stocking rates of X_1 and X_2 but that the animals at these two stocking rates will have vastly different daily gains, represented by the intercept of the gain per animal line (BXn) and the perpendicular from the stocking rate axis.

(c) Maximum animal gain per unit area

The stocking rate providing for maximum animal gain per hectare (SR_{max}) occurs where the slope (*viz.* the derivative) of the gain per hectare curve is zero (i.e. at Xc). SR_{max} can be derived from Equation 1:

$$SR_{max} = \frac{a}{2b}$$

The ADG at this stocking rate is determined by substituting SR_{max} for x in Equation 1. Gain per hectare at SR_{max} is then determined by multiplying the ADG at SR_{max} by SR_{max} and by the number of days for which Equation 1 was derived.

(d) An example of the use of the stocking rate–animal performance model

Steers with an initial mean mass of 200 kg were run at three stocking rates on a stargrass pasture for 180 days (Edwards & Mappledoram 1979). A linear model ($y_{an} = a - bx$; Jones & Sandland 1974) was derived, by regression, from the data which are presented in Table 6.1. This allowed the derivation of values for a and b.

Table 6.1 Data used in calculations in (d) (1–6). Mean values for 180 days (after Edwards & Mappledoram 1979).

Stocking rate (Steers/ha)	Initial mean mass (kg)	Final mean mass (kg)	ADG (kg/day)
6	200	315	0.639
10	200	270	0.389
12.86	200	244	0.244

ADG = 0.9810 – 0.579 x ... Equation 1 (R^2 = 0.997)

1) Stocking rate providing for maximum gain per hectare (SR_{max})

$$SR_{max} = \frac{a}{2b}$$
$$= \frac{0.9810}{2 \times (0.0579)} = 8.5 \text{ steers/ha}$$

2) ADG at SR_{max}

Substitute SR_{max} in Equation 1
$$= 0.9810 - (0.0579 \times 8.5)$$
$$= 0.489 \text{ kg/animal/day}$$

3) Gain/ha/day at SR_{max}
$$= SR_{max} \times ADG \text{ at } SR_{max}$$
$$= 8.5 \times 0.489$$
$$= 4.16 \text{ kg/ha/day}$$

4) Gain/ha/180 days at SR_{max}

gain/ha/day at $SR_{max} \times 180$
$$= 4.16 \times 180$$
$$= 749 \text{ kg/ha}$$

5) Gain per animal at a particular stocking rate

This is determined by substituting the desired stocking rate for x in Equation 1.

6) Stocking rate required to provide a specific ADG

This is calculated by substituting the desired ADG for y_{an} and solving for x in Equation 1.

6.3 APPLICATION OF THE STOCKING RATE–ANIMAL PERFORMANCE MODEL

A number of points should be noted when using an empirical model, such as the Jones & Sandland model, for predicting animal performance and gains per hectare at various stocking rates. First, the models are empirical and their predictive ability is dependent on the quality of the data used to derive the models. Second, because of the empirical nature of the models, they are specific to the particular pasture or vegetation type and to the particular type and class of animal used in deriving the data and cannot be extrapolated to predict animal performance under different conditions. The models will also vary as plant growth fluctuates within a season and between years. Variation in herbage production resulting from rainfall fluctuations will alter the intercept of the ADG line rather than its slope (Bartholomew 1985; Heitschmidt & Taylor 1991) and will shift the position of SR_{max}. Therefore, the data used to develop a stocking rate–animal performance model should represent as wide a range of conditions as possible, and be representative of a period sufficiently long to incorporate climatic variation if it is to describe the potential range of variation in the ADG–stocking rate relation.

Separate models can be developed for various portions of the grazing season to account for possible variations in animal performance resulting from fluctuations in forage quality (Edwards & Mappledoram 1979). Such seasonal models can be useful for planning grazing strategies to achieve target animal masses at target dates. To derive such models, animal production should be determined at regular intervals, or at least at the start and end of each period in question.

The objectives of the livestock system and the type and class of animals used in the system will determine which portion of the stocking rate–animal performance model is most useful for prediction. For example, if the objective is to achieve maximum performance of high value animals (with high overhead costs), such as in a stud or dairy enterprise, then a stocking rate where individual animal performance is high is generally of interest (Fig. 6.2; point Xb). For livestock systems which aim to maximise production per hectare by running as many animals as possible, the portion of the model from Xb to Xc is likely to be of interest. Although maximum animal production per hectare (SR_{max}) will be achieved at Xc, this is usually not the most economic stocking rate to operate at because of the variable costs of animals (inoculation, dosing, dipping, labour, drinking water, etc.). With established fixed costs, maximum economic returns will be obtained at a stocking rate intermediate between that which gives maximum production per animal and that which gives maximum production per hectare (Fig. 6.2).

It is normally not economically viable to stock at rates above Xc unless animals are to be finished in some other system (feedlot or other pasture) or unless there is a large price differential between the start and end of the grazing period. In certain livestock systems, e.g. communal grazing systems, the aim is often to maintain a large population of livestock, irrespective of the condition of the individual animals (Mentis 1984; Hardy

& Mentis 1986). In such systems, non-consumptive products of livestock (i.e. draft, fuel, ceremonial purposes, etc.) may be more important than saleable products (i.e. meat, wool, milk, hair, etc.) and the system is stocked at, or close to, Xn (Fig. 6.2). Here animals are at or near maintenance.

Extreme caution is needed in extrapolating beyond the limits from which the model was derived to predict animal performance or animal production per hectare. Care should also be exercised in extrapolating from stocking rate–animal gain regression equations derived from a specific grazing system to a different system. At low stocking rates (in the vicinity of Xb, Fig. 6.2), there is adequate herbage available per animal and intake is unlikely to limit animal performance. The grazing system is therefore likely to have little effect on animal performance at such low stocking rates. However, at moderate stocking rates (in the vicinity of Xc and higher, Fig. 6.2), animals will compete for herbage and intake will largely determine animal performance. Also, defoliation height and frequency are unlikely to be the same under different grazing systems. At high stocking rates plants are likely, for example, to be severely and frequently grazed under continuous grazing and severely but much less frequently grazed under rotational systems. Both herbage production and intake, and therefore animal performance, will differ under the two systems. At these stocking rates, therefore, it would be unwise to extrapolate from one system to another.

6.4 LONG-TERM EFFECTS OF STOCKING RATE ON PASTURE COMPOSITION AND PRODUCTIVITY

Stocking rate has an immediate effect on the quantity of forage available to the grazing animal, thereby affecting intake and animal performance. There are, however, also important long-term effects of stocking rate on the pasture which in turn influence the productivity of livestock and the economic viability of grazing systems. Repeated severe grazing induced by using high stocking rates will reduce the vigour of forage plants. McKenzie (1997) reported, for example, that frequent grazing, particularly in summer, induced serious weed invasion in perennial ryegrass pastures.

Such overgrazing reduces the ability of the pasture to produce, and to continue producing, herbage while prolonged overgrazing can lead to a change in botanical composition (Harris & Brougham 1968), as can frequent and close cutting (Harris & Thomas 1972). The latter reported serious invasion of temperate pastures under continuous close grazing, but little infestation under moderate and lenient grazing. There is therefore an inferred link between stocking rate and pasture composition, i.e. that high stocking rates will result in frequent and severe grazing and that such treatment, especially in an inappropriate season, will favour certain species over others, resulting in altered, and often undesirable, pasture composition. Generally, these changes constitute a reduction in the palatable and productive plant species and their replacement by unpalatable and less productive grasses and forbs. This in turn leads to reduced animal performance and a marked reduction in profitability per hectare.

7 Nutrient supplementation of the grazing animal

H.H. Meissner

CONTENTS

The range of pasture types and the associated range in species composition, forage production and nutritive value of the forage produced is probably as great in southern Africa as in any other region in the world. This variability is further compounded by the unreliable rainfall. The ruminant is therefore confronted with forage which fluctuates in both quality and quantity. This necessitates effective supplementation if animals are to perform adequately.

There are two primary reasons for supplementing minerals and energy to ruminants: firstly, to correct deficiencies in the diet, and secondly, to stimulate intake. These two objectives are seldom mutually exclusive. Also, supplementation does not always produce positive results. Whereas the correction of deficiencies in the diet often leads to improved digestion and intake of forage, negative associative effects can lead to a reduced digestion of fibre and reduced intake. Incorrect supplementation can lead to nutrient imbalances which may neutralise any potential advantages of supplementing livestock.

Basic to successful supplementation is the provision of the appropriate amount and ratio of nutrients as economically as possible to each type of animal at the appropriate time. Decisions in this regard depend in the first instance on the nutrient requirements and goals of any particular system, but also on the circumstances on any property, recent feeding history, the production system being adopted and the financial position of the producer. In grazing ruminants the precise requirements of animals is difficult to determine because of difficulties in estimating the quality and quantity of material consumed (De Waal 1990). This leads to discrepancies in recommendations given by different advisers and often to poor animal response to supplementation.

7.1 PROTEIN AND ENERGY

The availability of energy and protein declines as the forage ages and the cell walls lignify. This reduces both digestibility and intake. Low digestibility and low protein contents restrict the amount of forage that a ruminant will consume. As an example of this, data from Smith (1962) showed that oxen daily consumed 1.2% of their body weight when the digestibility of the organic material (DOM) was 50%, but only 0.8% of body weight when the DOM declined to 38% later in the season.

Deficiencies of both protein and energy can reduce fertility and cause weight loss. However, the primary deficiency among the subtropical pasture species during winter is protein, and not energy. Energy supplementation on protein deficient forage can in fact negatively affect animal performance (Van Niekerk 1978). In contrast, energy rather than protein is usually deficient during the summer because of the low dry matter content of green, succulent forage. Good growth and conception rates in cattle have resulted, for example, from the provision of 1.5 kg to 1.75 kg of maize per head per day during the summer period (Bishop & Kotze 1965; Pieterse & Preller 1965). However, the need for energy supplementation will depend on circumstances and it is important to appreciate that responses to supplements cannot always be justified economically (Lishman *et al.* 1984).

7.1.1 Synchronising supplementation of protein and energy with animal requirements

Conception and calving should be planned to fall in the rainy season when forages contain high levels of protein and energy. The requirements of the cow are then at their peak (refer to Table 5.7 in Chapter 5). Weaning can then be timed for the winter when its requirement is at its lowest. This principle is illustrated in Fig. 7.1 in which the seasonal energy and protein requirements of beef animals are compared with the provision of these nutrients by sweetveld (veld which remains nutritious and palatable year-round) and sourveld (veld which becomes unpalatable and loses its nutritive value in winter). The sweetveld situation in Fig. 7.1 would equate with the patterns of nutrient provision by palatable pastures (such as Smuts' fingergrass) in arid and semi-arid areas, and the sourveld with less palatable pasture types (such as *Eragrostis curvula*) in high rainfall areas.

7.1.2 Protein sources

Potential protein sources are of three types: non-protein nitrogen (NPN), plant sources and industrial by-products. Decisions on which of these to use will depend on their local availability, their price, the production system in which they are to be used and the requirements of the animals. For over-wintering there is little reason to use protein

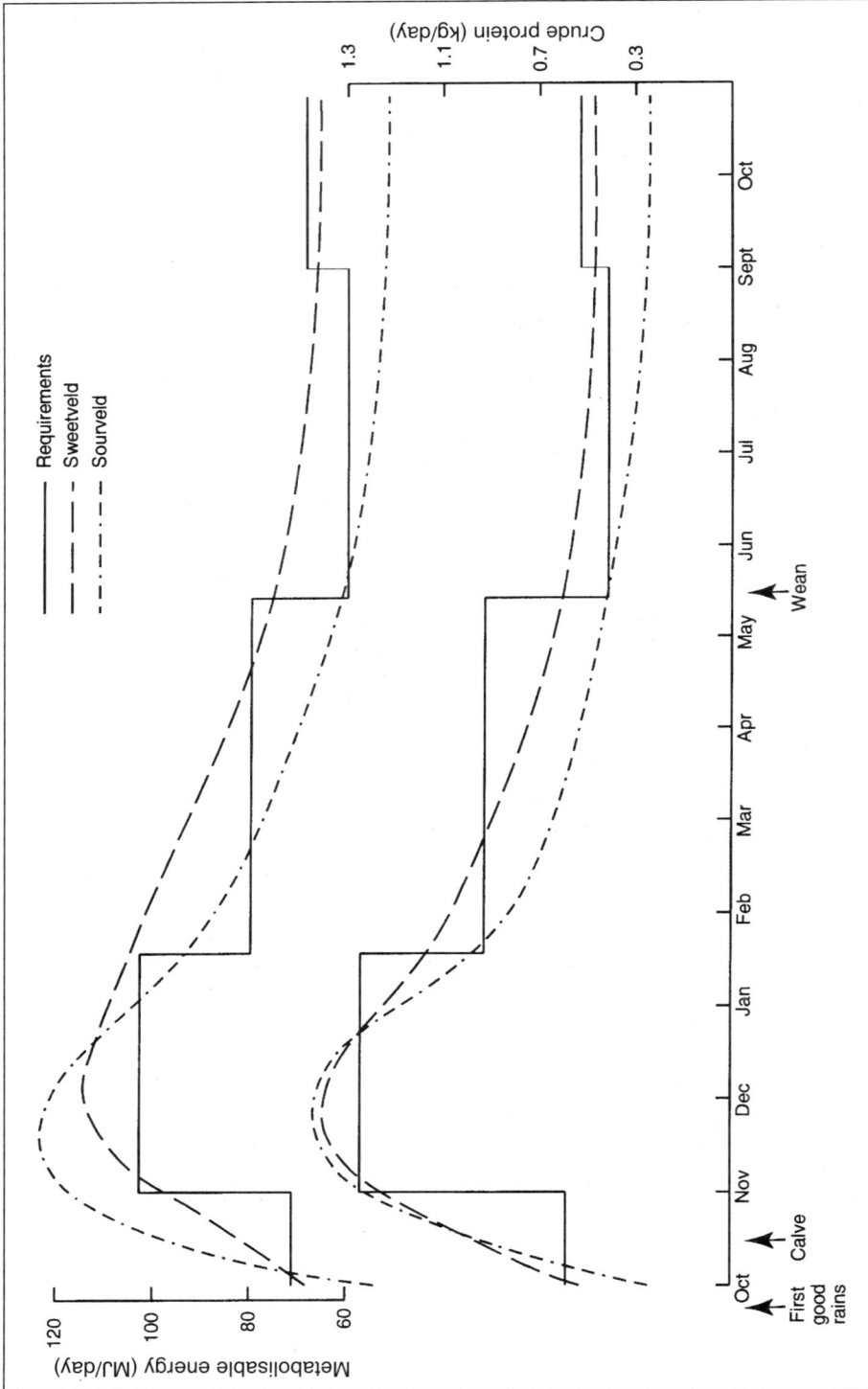

Figure 7.1 Comparison of the energy and protein requirements of beef cows and their provision by veld or summer dryland pasture for a spring calving season.

sources other than NPN, but for growing animals, heifers and young females approaching calving or lambing, it may be advisable to use protein-rich plant or industrial sources. Such protein sources would also supplement energy and some other nutrients which may be important for both the animal and the microbes and they also contain escape (or bypass) protein which would supplement microbial amino acids in the small intestine, to the benefit of the animal.

7.1.2.1 NPN sources

Potential non-protein nitrogen sources are shown in Table 7.1. The rumen micro-organisms are able, through hydrolysis, to convert the N contained in these materials to ammonia which can in turn be converted to microbial protein. For NPN to be effective, animals must have access to an adequate supply of digestible organic matter (from roughage), readily digestible carbohydrate (from such sources as molasses or maize) in a ratio of 8 to 10 units of carbohydrate to one unit of NPN (Groenewald & Boyazoglu 1980), an adequate mineral supply (for example, one unit of sulphur to each 10 to 15 units of N (Shirley 1986)), as well as to phosphate and sometimes trace minerals. NPN provides an inexpensive source of N and can often be used as the only supplementary N source.

Table 7.1 Potential NPN sources
(Briggs 1983 – table shortened*).

Urea	1 Acetyl-3-methyl-urea
Biuret	Acetyl urea
Triuret	1 Acetyl-2-thio-urea
Ammonium acetate	Ammonified molasses
Ammonium phosphate	Isobutidine di-urea
Aminoguanidine-bicarbonate	

* Only the sources which have been tested are included.
Sterilised chicken manure is an additional source of protein that can be classified as NPN.

7.1.2.1.1 Urea

Urea is a common source of NPN fed to grazing animals. Feed quality urea contains approximately 460 g N per kg and is able to provide the equivalent of 2 875 g protein per kg (460 × 6.25) (refer to Chapter 5).

A considerable amount of information is available on animal response to urea supplementation (Bembridge 1963; Kreft 1963, 1966; Pieterse 1966; Nel & Van Niekerk 1970). The general recommendation is that approximately one-third of the protein requirements for maintenance can be supplied as urea so that urea can constitute about 1% of intake. The total safe urea intake of cattle is about 100 g per day and of sheep about 15 g per day. These amounts are in fact specified in Act 36 of 1947.

The micro-organisms in the rumens of animals need to become adapted to urea before they can make effective use of the N it contains. Such adaptation normally takes two to three weeks (Golding 1985).

If ruminants consume too much urea, the rate of its hydrolysis may exceed the micro-organisms' ability to convert the ammonia to microbial protein. The excess ammonia may then be absorbed into the bloodstream resulting in ammonia, or so-called urea, poisoning. There are two alternative procedures of guarding against such poisoning. Either a slowly hydrolysing source of NPN must be used, or the rumen pH must be lowered. The first of these options has led to the development of biuret as a replacement for urea (Briggs 1983), but it has not been widely used in South Africa because it is expensive and there have been problems associated with microbial adaptation. The second option, that of lowering the rumen pH, can be facilitated by providing readily available carbohydrate such as starch (maize meal) or sugar (molasses) with the urea. In addition to their influence on pH, these products also serve as a source of energy for the micro-organisms and so improve their efficiency.

It is widely accepted that NPN (urea) is not as useful a source of supplementary N for animals on high quality roughage, i.e. roughage containing more than about 60 g to 70 g per kg crude protein, as it is for animals on low quality roughage (Oltjen *et al.* 1974; Shirley 1986).

7.1.2.1.2   Chicken litter

Broiler litter and residues of layer houses are sold under the general term 'chicken litter' or 'chicken manure'. The composition of the two differs considerably, as specified by Act 36 of 1947 (Table 7.2).

Table 7.2   Specifications for broiler chicken and layer litter according to Act 36 of 1947 (g/kg unless otherwise indicated).

	Broilers	*Layers*	*Limit*
Moisture	120	120	Maximum
Crude protein (CP)	24	22	Minimum
Crude protein from uric acid	60	60	Maximum
Fat	1.5	1.5	Minimum
Fibre	15	15	Maximum
Ash	15	25	Maximum
Feathers	1	1	Maximum
Calcium	3.5	8	Maximum
Phosphorus	1.5	2	Minimum
Sodium	0.5	0.5	Maximum
Silica	0.5	0.5	Maximum
Copper (mg/kg)	50	50	Maximum

Pathogen free: maximum of 20 000 micro-organisms per gram

Chicken manure contains 500 g to 600 g NPN per kg in its crude protein fraction (Bhattacharya & Fontenot 1966; Blair 1975), approximately half of which is uric acid (Bhattacharya & Fontenot 1966; Swingle *et al.* 1977). This should provide for the production of adequate amounts of microbial protein (Oltjen *et al.* 1968). Oltjen & Dinius (1976) report that steers receiving chicken manure as a supplement to roughage performed better than animals receiving either urea or biuret, but the relative response to these different products remains a contentious issue. In any event, chicken litter is widely used in South Africa largely because it is generally less expensive than other equivalent products.

Chicken litter is used very successfully as a winter supplement at rates of 1.5 kg to 2 kg per day for cattle and 200 g to 400 g per day for sheep. During droughts, as much as 4.5 kg to 6 kg is fed to cattle (Van Ryssen 1990).

There are, however, dangers associated with feeding chicken litter to animals. The use of unsterilised material, for example, is extremely dangerous and needs to be strongly discouraged. Health risks include fungi, bacteria such as Clostridia (e.g. *Clostridia botulinum*, which causes botulism, but which can be inoculated against), Salmonella, antibiotic residues, coccidiosis and the possibility of high levels of copper, arsenic and hormone residues (Van Ryssen 1990). Non-health problems include variations in the composition of the material, its high and variable moisture content and the very high concentrations and/or imbalance of minerals.

7.1.2.1.3 Ammonium sulphate

Ammonium sulphate is an attractive supplement because it is less toxic (Huston & Eng 1974) and less expensive than urea. In addition to supplying 210 g N, it provides 240 g sulphur per kg. However, the efficiency of its utilisation by the rumen microbes is still questionable.

7.1.2.2 Plant protein sources

Expensive high quality protein supplements can sometimes be justified, as for creep-feeding calves and lambs, when replacement heifers or young ewes are being grown out, for young lactating ewes and during droughts. Lucerne is widely used on these occasions, providing both protein and energy.

7.1.2.2.1 Oil seeds

Oil seeds are sometimes available in a whole, unprocessed form, as for example cottonseed. This contains approximately 200 g crude protein per kg and also contains high levels of fat, so that it is also a rich source of energy. Whole cottonseed, however, contains gossypol, and so is unsuitable for feeding to young ruminants whose rumen is not yet fully functional (Wolf 1983).

The most general use of oil seeds is as processed oil cake produced from such crops as sunflower, groundnuts, cotton and soya beans. Biologically all the oil cakes provide a high quality supplement, and the choice between them is normally based on availability and cost. Depending on such factors as feeding level, the carrier used and the way it is processed, a variable amount of the true protein can escape fermentation in the rumen (the so-called escape protein) and will be available for digestion and absorption in the duodenum (Owens & Bergen 1983). Heat treatment will increase the amount of escape protein. The amino acids produced replace those of microbial origin and such treatment may increase, and even improve, the quality of the amino acids available for absorption.

Oil cakes contain approximately 400 g crude protein per kg. Cottonseed cake and heat treated soya meal contain the highest levels of escape protein.

7.1.2.2.2 Legumes

Legume pastures can make an extremely effective winter supplement, but few legumes grow well during the winter over much of South Africa. Lucerne hay is very useful as a winter supplement and as a drought feed in arid regions often, unfortunately, at completely unrealistic prices. Good quality lucerne hay contains 140 g to 150 g crude protein per kg, a reasonable amount of energy and can also serve as a useful source of minerals.

7.1.2.2.3 By-products

Useful by-products include fish meal, carcass meal and blood meal. They contain high levels of escape protein and should be used only when high levels of production, rather than just maintenance, are required. An analysis by Dennison & Phillips (1983) showed that these protein sources are better able to provide amino acids to the duodenum than are plant protein sources. Indeed, in some circumstances so much of the protein escapes degradation in the rumen that a readily fermentable N source, such as urea, also needs to be supplied to satisfy the requirements of the rumen micro-organisms (Krause & Klopfenstein 1978).

Because of their high cost and their often poor availability in South Africa, they are seldom fed as supplements. In the past fish meal was used with good, although variable, success to stimulate wool growth of sheep.

Other well known by-products are brewers grain derived from barley and sorghum brewing. In a dry form it contains approximately 250 g crude protein per kg, much of it escape protein (Cronje 1983; Erasmus *et al.* 1988), but the drying costs are too high for it to be a competitive supplement. It can, however, be used in the undried form, but transport costs will be high and there is a risk of the material deteriorating.

7.1.3   Energy sources

7.1.3.1   Principles of energy supplementation

Grain and by-products such as bran, germ meal and molasses, all of which contain moderate to high levels of available carbohydrates, are widely used as winter supplements and for drought feeding. As supplements on young summer grazing they can be used to maintain liveweight or to provide for moderate liveweight gain in young ewes or heifers, for young lactating animals and for finishing.

Because of compensatory growth of animals on the high quality spring forage, it is normally advisable to supplement only sufficient energy to promote the digestion and intake of the low quality grazing during the winter and during drought periods. If high levels of supplementation are provided, animal response is often disappointing in spring, when the advantages of compensatory growth will be lost.

The following principles apply to the supplementary feeding of energy:

(a) supplementation with readily available carbohydrates can reduce roughage intake (substitution) or reduce the digestibility of fibre (Doyle 1987). These two effects are not mutually exclusive;

(b) any supplementation of energy is likely to reduce intake of animals on a low quality roughage (Hennessy *et al.* 1983) if not accompanied by a protein source (Coombe & Tribe 1962; Kondos & Mutch 1975), because N is likely to be more deficient than energy. Increasing levels of energy will increase the level of substitution (Crabtree & Williams 1971a, 1971b; Mulholland *et al.* 1976) but the reaction is not linear because certain energy sources escape fermentation in the rumen, with starch being digested in the small intestine (Waldo 1973; Sutton 1980). The microorganisms in the rumen are therefore not affected;

(c) roughage quality also influences the level of substitution. Fibre digestion is depressed and the substitution effect increased as roughage quality increases (Gulbransen 1974; Meissner *et al.* 1991). Above levels of 550 g to 600 g per kg cell wall content (representing low quality forage), supplementation has little influence on fibre digestion, but its effect increases as the cell wall content declines below these levels. As on low quality roughage, substitution will increase when the energy levels increase, but the energy source and mode of processing will influence this. It is anticipated that by-products such as bran and germ meal, which contain little starch, will not have as great an effect as will grains, and that coarsely milled or whole grain will have less effect than finely ground grain; and

(d)   to gain maximum benefit from energy supplements, those nutrients in short supply, such as N and minerals, should be included in the concentrate mixture in amounts appropriate to the levels of performance which are anticipated. To limit substitution, the amount of energy which is supplemented should not exceed 15% to 20% of daily intake.

7.1.3.2   Sources

The most widely used grains in South Africa are maize, grain sorghum, barley and, to a lesser extent, wheat and oats. Of these, maize has the highest energy value (14.7 MJ ME (metabolisable energy) per kg DM), followed by wheat (13.8) and oats (12.3). Maize has the lowest crude protein concentration (100 g per kg DM), with grain sorghum (125 g per kg DM), oats (130 g per kg DM) and wheat (140 g per kg DM) containing increasing amounts. Large variations do, however, occur in the above values, depending on the cultivar, area and year in which the crop is grown, soil type and fertilizer levels used. Processing (e.g. milling or rolling) also affects feed value. However, sheep and cattle respond differently to grain processed in different ways. Grain should preferably not be milled too fine and under certain circumstances can be fed whole. All the grains contain reasonably high levels of P but are poor in calcium, which should therefore be supplemented.

7.1.3.2.1   By-products of the milling industry

Maize bran comprises only the outer layers of the maize grain and therefore contains no starch, is high in fibre and reasonably high in protein. It is used primarily as a filler for other concentrate grain feeds, such as hominy chop or maize germ meal. Hominy chop is a mixture of maize grain and maize germ meal. Although it has more fibre and less starch than maize, it has a higher protein and fat content (about 115 g and 50 g per kg DM respectively). The high fat content gives it an energy value equal to or better than that of maize.

Maize germ meal is usually a mixture of hominy chop and maize germ and so its composition will vary according to the ratio of the two in the mixture. It should contain approximately 120 g crude protein and 150 g to 200 g crude fat per kg DM. This would make it a high energy feed.

Maize gluten is available in two products which can also be used to supplement protein. One contains 200 g crude protein per kg, and the other 600 g. It is not normally freely available and is expensive and therefore not readily used.

Wheat bran is similar to maize bran but is higher in protein, and particularly also in P. Pollard is produced from a mixture of fine particles of wheat bran, wheat meal and other residues of the milling process and is a better product than wheat bran. Middlings comprise a mixture of wheat bran and pollard. The energy value of these by-products ranges from about 75% to 100% of that of maize. Because of their lower starch content, higher crude fibre, crude protein, crude fat and mineral (particularly P) concentrations, they are usually more suitable than maize for use as supplements.

7.1.3.2.2 Molasses

Molasses is the dark liquid of high viscosity which remains after the sugar has been extracted from sugarcane. It contains approximately 250 g to 300 g water and approximately 500 g sugar per kg and has an energy value on a dry matter basis of 12 MJ ME. It is low in crude protein but is a good source of minerals, particularly calcium, magnesium, potassium and sulphur. It is very palatable and is a useful binding agent.

7.1.3.2.3 Brewers grain

This has a high energy value of approximately 12.5 MJ ME per kg dry matter but is either difficult to handle (when wet) or very expensive (when dry).

7.2 MINERALS

7.2.1 Requirements and toxicity

To date 21 minerals have been identified as being essential in the diets of animals. Thirteen of these can cause poisoning. A few minerals belong to the 'heavy metal and contaminant' group which are a danger to human and animal health. Table 7.3 lists the minerals.

Table 7.3 Essential and toxic minerals for ruminants.

Essential macro-nutrients	Essential trace elements	Toxic if certain limits are exceeded
Calcium (Ca)	Iodine (I)	Sodium (Na)
Phosphorus (P)	Iron (Fe)	Iodine (I)
Magnesium (Mg)	Copper (Cu)	Copper (Cu)
Potassium (K)	Cobalt (Co)	Cobalt (Co)
Sodium (Na)	Manganese (Mn)	Zinc (Zn)
Chlorine (Cl)	Zinc (Zn)	Selenium (Se)
Sulphur (S)	Selenium (Se)	Molybdenum (Mo)
	Molybdenum (Mo)	Fluorine (F)
	Fluorine (F)	Lead (Pb)
	Chromium (Cr)	Cadmium (Cd)
	Silicon (Si)	Mercury (Hg)
	Vanadium (V)	Arsenic (As)
	Nickel (Ni)	Aluminium (Al)
	Tin (Sn)	Others
	Lithium (Li)	
	Cadmium (Cd)	
	Rubidium (Rb)	
	Strontium (Sr)	
	Boron (B)	
	Others	

Sources: Loosli 1978; Williams & McDowell 1985

Of the essential minerals, calcium (Ca), phosphorus (P), magnesium (Mg), sodium (Na), iodine (I), iron (Fe), copper (Cu), cobalt (Co), zinc (Zn), manganese (Mn) and selenium (Se) are the most likely to be deficient in the diets of ruminants. A calcium deficiency has not yet been identified on veld but does occur in dairy cows and young, growing ruminants fed in intensive systems (Loosli 1978). It is also unlikely that an iron deficiency will occur under grazing conditions (McDowell 1985b), while most rough-ages contain sufficient quantities of potassium (K).

Particularly important to pasture farmers in many parts of South Africa is the very low Ca:P ratio of kikuyu, an extremely widely used pasture species, due to its low Ca content. For much of the season this ratio may remain under 1.0 and it may drop as low as 0.65 in mid-summer (Miles pers. comm.). This compares with the recommendation that it should generally be greater than 1.5.

Of the trace elements, chromium (Cr), silicon (Si), vanadium (V), tin (Sn) and nickel (Ni) are necessary for small animal species (Loosli 1978) and probably also for ruminants. Cadmium (Cd), rubidium (Rb) and strontium (Sr) are probably also involved with rumen fermentation (Van Niekerk 1987). Nonetheless, from the point of view of supplementation of grazing animals, it is only the trace elements iodine, copper, cobalt, zinc, manganese and selenium that are important, although there are interactions with other trace elements which need to be taken into account. Requirements, tolerance and toxic levels of these and other minerals are shown in Table 7.4.

7.2.2   Interactions

All nutrients (minerals, vitamins, amino acids and the protein and energy fractions) are, to some extent, interrelated. There is an optimum level of each in relation to all others. Some inter-relations are more critical than others and form the basis of much intensive research today, often with little success.

One of the first relations between minerals to be identified was that between calcium and P. It is now generally accepted that the appropriate ratio (1.5:1 to 2:1) (see Table 7.4) between calcium and P is more important than the actual amounts of these two minerals, so long as P is not deficient. Complicating this relation is the one with magnesium, as well as that between iron and P, copper and iron, molybdenum and copper, sulphur and both copper and molybdenum, and so on. Ammerman (1965) produced an illustration showing 16 interrelated minerals, while Miller (1979), some 14 years later, showed 22 interrelated minerals (both quoted by Ammerman *et al.* 1989).

Complete texts can be devoted to the interactions between different minerals. For the purposes of supplementation we can perhaps do no more than supplement those minerals that have been shown to be deficient in a particular area.

7.2.3   Deficiencies and excesses in southern Africa

It is widely known that forage produced from veld in South Africa is often deficient in P and one would expect the same to apply to subtropical pasture species grown on unfertilized or poorly fertilized soils having low levels of available P. *Eragrostis curvula*, for example, will grow well on soils which have little available P. Here it will produce forage which is extremely deficient in this element. However, where soils have

Table 7.4   Requirements, tolerance levels and toxic levels of macro and trace elements for cattle and sheep (values in g or mg per kg feed).

Mineral	Situation	Requirement	Tolerance	Toxic/damaging
Macro minerals (g per kg)				
Ca	Cattle	1.8–10	—	>20
	Sheep	2.1–5.2	—	—
P	Cattle	1.8–6.0	—	>20
	Sheep	1.6–3.7	—	>6
Ca:P	—	1.5–2:1	>1:1, <7:1	—
Na	On grazing	0.4–1.8	5–7 g/ℓ water	—
Cl	Dairy cows	1.0–1.8	2.7	—
Mg	Growth	1.0	} 4.0	20–40
	Lactating	1.8–2.0		
	Heavily fertilized pasture	3.0–3.8	—	—
K	Normal	5.0–8.0	} 40	—
	Stress	7.0–15		
S	—	2.0	3.0–4.0	—
N:S	Cattle	12–15:1	—	—
	Wool sheep	10–12:1	—	—
Trace elements (mg per kg)				
F	Cattle	—	30–50	>50
	Ewes	—	60	>70
	Slaughter lambs	—	150	>170
Cu	Sheep : normal	5–7	—	>20
	Sheep : lime soils	10	—	—
	Cattle	8–10	—	>70–100 <0.1–0.2 Mo in soil
Mo	—	<2	—	>5–6
Cu:Mo	—	2:1	—	—
Co	—	0.10	5	>60
I	—	0.05–0.8	—	>50
Se	On pasture	0.05–0.3	—	—
	Lambs and calves	0.1–0.2	4–5	5–40
Fe	Cattle	} 30–100	1 000	—
	Sheep		500	—
Mn	—	20–40	200 chronic 1 000 acute	2 000–4 000
Zn	Normal	20–50	500 (cattle)	} 1 000–1 700
	Per 1.0 increase in Ca	+16	300 (sheep)	

Sources: Loosli 1978; McDowell 1985b

been provided with adequate amounts of P fertilizers, the subtropical pasture grass appears capable of producing forage having adequate dietary levels of the mineral. Data from the Kokstad Research Station in KwaZulu-Natal have shown that lambing ewes, which spend three months a year on arable fields which have been well fertilized with phosphate, need not be supplemented with this mineral. Cattle fed foggage produced from well fertilized pastures in winter similarly showed no response to phosphate supplementation (Murray pers. comm.).

Deficiencies or excesses of a number of trace elements are found in certain regions. So, for example, copper, cobalt, zinc, magnesium and manganese have been identified as being deficient in various parts of the country (Van Niekerk 1987). More specifically, the following problems have been identified:

copper and cobalt: deficiencies of these two minerals are commonly found in the west and south-western Cape coastal regions (Van der Merwe & Perold 1967), with copper deficiencies also reported from some parts of the Karoo. An excess of copper has been alleged in the Beaufort and Frazerburg regions but this has not yet been confirmed.

iodine: there are unconfirmed reports of iodine deficiencies in the KwaZulu-Natal Midlands and in the northern areas of Namibia.

selenium: deficiencies of selenium have been reported from the Cape coast, certain parts of the Free State and KwaZulu-Natal. In the Beaufort West district, concentrations of selenium in forages is reported to be excessive.

zinc: possible zinc deficiencies have been reported from northern KwaZulu-Natal and from the Northern Cape. A deficiency of this mineral is associated with foot-rot, which suggests that zinc deficiencies could be much more widespread than has been reported. The success which has been achieved in combatting this disease with zinc injections supports this view (Vosloo pers. comm.).

manganese: deficiencies may arise locally, specifically in areas treated with alkaline fertilizers.

magnesium: deficiencies have been reported from planted pastures, where it may be associated with high levels of potassium and N fertilizer and with the high lipid content of spring grass (Suttle 1987).

It would be unwise to use the above information (except perhaps for the copper and cobalt deficiencies in the Western Cape) as reliable evidence of mineral deficiencies in South Africa. There is still an urgent need for further work on this aspect.

phosphorus: because P deficiency is so widespread in South Africa, it is appropriate to examine this mineral in some depth. Research on P dates back to Theiler (1912, 1920) who examined the cause of botulism and stiff-sickness. A number of reports followed, showing P to be widely deficient in South Africa (e.g. Du Toit *et al.* 1940). In spite of the success which was achieved with supplementation of P (see review by Van Niekerk 1978), researchers in Australia in the 1970's and in South Africa in the 1980's seriously questioned the need for P supplementation, particularly to sheep which show no response to P supplementation on much lower P intakes than cattle (Van Niekerk 1987). Sheep have, however, been shown to suffer a P deficiency, but to a lesser extent than cattle (Read *et al.* 1986a, 1986b). In both cattle and sheep, the response to P supplementation varies considerably from place to place.

There remain doubts as to the need to supplement P during the winter if the producer's objective is mainly one of maintaining the weight of his animals (Van Niekerk & Jacobs 1985; Jacobs pers. comm.). Current recommendations in many areas are that winter supplementation is unnecessary. In summer, however, when animals are in a positive energy balance and when their requirements are high (because of high levels of performance at this time), P supplementation is often essential (Van Niekerk 1978) and will often improve responses to supplementation of both protein and energy.

7.2.4   Sources and availability

Producers need to be familiar with the effectiveness of the various salts and mixtures available on the market if supplementation is to be effective. While the chemical composition of any product provides information on the concentration of minerals in the material, it gives no indication of the availability of these minerals to the animal. No mineral is ever totally available since some is always lost during normal digestion and metabolic processes. The actual extent to which any mineral is utilised can, however, be gauged by measuring its true digestion. This value is related to the extent to which the mineral is retained by the animal, i.e. its biological availability (Table 7.5), but these values have been poorly quantified. The best we can do is categorise their availability into broad classes of good, moderate, poor and unavailable.

The most important mineral sources used for supplementation are shown in Table 7.5. Also shown is the availability of the minerals they contain, where this is known. These data show that the true digestibility of the minerals, with few exceptions, is not much greater than 50%, and is often much less. It is nonetheless greater than that of most minerals in plant material but is sometimes lower than that of the minerals contained in animal products such as fish meal.

The reason for the low availability of minerals in inorganic salts is that they form insoluble associations with natural chelates such as phytic acid, amino acids, proteins and other organic compounds in the digestive tract. Where there is a reaction with the proteins of the walls of the digestive tract, mucosal blockage occurs. This severely depresses uptake from the digestive tract.

A further reason is that the divalent ions such as copper, manganese, cadmium and iron compete for absorption sites on the intestinal mucosa, so reducing uptake (Underwood 1981).

Because of the relatively low availability of minerals, the terminology used to describe the value of mineral supplements (moderately or poorly available) may cause confusion because such terms describe the relative rather than the absolute value of a product. They provide an index whereby the salt which provides the highest true digestibility (e.g. dicalcium phosphate in the case of P supplements) is rated as 100, and other P supplements are rated against this. Values above 70% to 80% are rated as good, above 50% as moderate, and less than this as poor.

Also apparent from Table 7.5 is the lack of a good relation between the actual solubility of an inorganic salt in water and its availability to the animal. Availability is, rather, more closely related to the salt's solubility in the liquid of the digestive tract (Fritz 1983), which may be either acid or alkaline, depending on where in the digestive tract the mineral is absorbed. Sulphate forms of trace elements are normally readily available to the animal and they are also soluble in water.

In conclusion, the minerals contained in true chemical compounds used for supplementing animals are more available than those contained in natural ores.

7.3   SUPPLEMENTATION PRACTICES

7.3.1   Salt licks

Salt licks are the most widely used form in which minerals are supplemented in South Africa. They are both convenient and relatively inexpensive and at the same time

Table 7.5　Mineral sources in use and the availability of the minerals.

Mineral	Source	Chemical formula	Mineral in source (g/kg)	Solubility in water	Actual digestibility (g/100g)	Availability
Calcium	Calcium carbonate	CaCO₃	Pure – 400	Insoluble	46	Moderate
	Limestone powder	CaCO₃	Commercial – 360	Insoluble	41	Moderate
	Bonemeal	Ca₃(PO₄)₂	220	Effectively insoluble	61	Good
	Bone ash	Ca₃(PO₄)₂	Commercial – 360	Effectively insoluble	—	—
	Monocalcium phosphate	Ca(H₄PO₄)₂.7H₂O	Commercial – 180	80% Insoluble	59	Good
	Dicalcium phosphate	CaHPO₄.2H₂O	220	Effectively insoluble	54	Good
	Defluorided rock phosphate	—	290	—	48	Moderate
Phosphorus	Bonemeal	Ca₃(PO₄)₂	95	Effectively insoluble	46	Good
	Bone ash	Ca₃(PO₄)₂	Commercial – 150	Effectively insoluble	—	—
	Monocalcium phosphate	Ca(H₄PO₄)₂.7H₂O	210	80% Insoluble	—	Good
	Dicalcium phosphate	CaHPO₄.2H₂O	Commercial – 120	—	—	—
	Rock phosphate	—	140, 160	Effectively insoluble	50	Moderate
	Defluorided rock phosphate	—	130	—	47	Moderate
	Monosodium phosphate	NaH₂PO₄.H₂O	220	Soluble	—	Good
	Soft phosphate	—	90	—	14	Poor
	Phosphoric acid	H₃PO₄	310	Soluble	—	Good
	Diammonium phosphate	(NH₂)₂HPO₄	230	Soluble	—	—
Sodium	Sodium chloride	NaCl	390	Soluble	—	Good
	Sodium sulphate	Na₂SO₄.10H₂O	140	Soluble	—	Moderate
Potassium	Potassium chloride	KCl	520	Soluble	—	Good
	Potassium sulphate	K₂SO₄	410	Soluble	—	Good
Magnesium	Magnesium oxide	MgO	540	Soluble	40	Good
	Magnesium sulphate	MgSO₄.3H₂O	140	Soluble	37	Good
	Magnesium carbonate	Mg(CO₃)₂	210	Soluble	39	Good
Sulphur	Sodium sulphate	Na₂SO₄.10H₂O	100	Soluble	82	Moderate – good
	Flowers of sulphur	S	960	Insoluble	40	Poor
Copper	Copper sulphate	CuSO₄.5H₂O	255	Soluble	—	Good
	Copper oxide	CuO	800	Insoluble	—	Poor
	Copper carbonate	CuCO₃.Cu(OH₂.H₂O)	530	Insoluble	—	Moderate
Cobalt	Cobalt sulphate	CoSO₄.7H₂O	210	Soluble	—	Apparently good
	Cobalt chloride	CoCl₂.6H₂O	250	Soluble	—	Apparently good
Iodine	Potassium iodide	KI	760	Soluble	—	Good
Iron	Iron oxide	Fe₂O₃	700	Insoluble	—	Unavailable
	Iron sulphate	FeSO₄.7H₂O	200	Soluble	—	Good
Manganese	Manganese sulphate	MnSO₄.H₂O	325	Soluble	—	Good
	Manganese oxide	MnO	630	Insoluble	—	Good
	Manganese chloride	MnCl₂	440	Soluble	—	—
Selenium	Sodium selenite	Na₂SeO₃.5H₂O	300	Low solubility	—	Good
	Sodium selenate	Na₂SeO₄	420	Soluble	—	Good
Zinc	Zinc sulphate	ZnSO₄.7H₂O	230	Soluble	—	Good
	Zinc sulphate	ZnSO₄.H₂O	360	Soluble	—	—
	Zinc chloride	ZnCl₂	480	Soluble	—	Moderate
	Zinc oxide	ZnO	800	Insoluble	—	Good

Sources : Shirley (1978); Thompson (1978); Groenewald & Boyazoglu (1980); McDowell (1985b) supply the animal's need for sodium. Their use is based on the assumption that salt can be used to regulate intake. It has proved reasonably effective for supplementing NPN and macro-elements (especially P) and has also been used successfully to supplement small quantities of trace elements such as selenium (Paulson *et al.* 1968; Jenkins *et al.* 1974). However, note should be taken of certain problems.

7.3.1.1 *Factors which affect lick intake*

The reader is also referred to reviews by Underwood (1977), McDowell (1985a) and Doyle (1987) for information on this topic.

(a) Urea poisoning: urea which is moistened can cause poisoning, particularly if animals drink water in which it has dissolved, when they may take in too much urea. Also, wet conditions can reduce the palatability of licks.

(b) Soil fertility and type of roughage: lick intake usually declines as the fertility of the soil from which the roughage is produced increases.

(c) Amount and quality of forage: lick intake is usually high on pasture which is overgrazed or when its quality is low, as in winter.

(d) Composition of licks: the palatability of the formulation used will influence lick intake. Intake is usually increased by the inclusion of molasses or maize meal in the formulation, when oil cake is used rather than NPN as a protein source and when bonemeal is used instead of dicalcium phosphate or feed lime.

(e) Individual requirements and preferences: these will vary depending on growth rate, milk production and maintenance requirements and will result in different levels of intake. Individuals vary, however, in their preference for licks to the extent that it can result in as much as a 100% difference in intake between them. This problem appears to be most serious with salt-based licks.

(f) Salt content of drinking water: intake of brackish water will tend to reduce the intake of salt-based licks. Animals have a natural craving for salt but when this craving is satisfied by the salt in the drinking water, lick intake will decline and palatable constituents will need to be added to stimulate intake.

(g) Palatability of the mineral mix: animals do not have a preference for minerals other than salt. Through satisfying their craving for salt, the intake of minerals can be reasonably accurately controlled using formulations containing 300 g to 400 g salt per kg of lick. Even higher concentrations of salt can be used as, for example, in a 1:1 ratio of salt and dicalcium phosphate or bonemeal. Bonemeal, if well processed, and monosodium phosphate are the most palatable sources of phosphate and studies have shown that dairy cows prefer dicalcium phosphate to fluoride-enriched rock phosphate. Magnesium oxide is unpalatable and needs to be mixed with such products as cottonseed cake or molasses.

(h) Previous access to licks: the previous diet or access to licks will influence lick intake in the short term. Unless animals have regular access to salt licks, for example, they may take in excessive amounts when given free access. This can result in poisoning, caused not only by urea but also by certain trace elements. In the case of urea, it is advisable to condition the animals by initially providing them with a pure salt lick to satisfy their salt craving before adding urea to the lick, or by initially including only low concentrations of urea and minerals in the lick.

(i) Physical consistency of lick: lick intake is approximately 10% lower when it is presented in block form than when it has a loose consistency. Intake from blocks is, however, dependent on the form of the block.

Licks should be replaced regularly and lick containers should be designed to protect the lick from rain. Some products such as maize meal and molasses may ferment, but 200 g to 400 g salt per kg lick acts as a preservative and prevents the development of mustiness.

Licks will be visited by the animals more regularly if they are placed near watering points, in shade, in dry places or near good grazing. The containers must allow for easy access by young animals and there should be a sufficient number of containers to allow for unimpaired access. A general recommendation is one container for each 30 head of cattle.

7.3.1.2 *Dangers inherent in excessive intake of salt*

Salt can be toxic when taken in excessive quantities (see Na in Table 7.4), but in practice this should not occur when normal salt licks are fed, provided water is readily available. High levels of salt disrupt some rumen processes and can reduce weight gain and production, probably because of the development of an imbalance between sodium and potassium in the body fluids (De Waal *et al.* 1989a, 1989b). Potassium should therefore be added to licks where salt intake is likely to be high.

7.3.1.3 *Guidelines for the formulation of salt licks*

All commercial licks in South Africa must comply with the specifications set out in Act 36 of 1947 and there is therefore little risk attached to their wise use. Many farmers, however, formulate their own licks. Here the following guidelines should be used:
(a) The final mixture should contain 60 g to 80 g of P per kg, increasing to 80 g to 100 g P where the P concentration in the forage is less than 2 g per kg.
(b) The calcium:phosphorus ratio should not exceed 2:1.
(c) A reasonable proportion (50% to 100%) of the trace element needs of the animal should be added where specific deficiencies occur.
(d) The minerals contained in the constituents used should be readily available to the animal and constituents which contain contaminants (such as those from natural ores), should be avoided.
(e) Intake will be improved by adding palatable constituents like molasses or maize meal.
(f) The constituents used should have a relatively similar particle size to promote a uniform distribution throughout the mixture.
(g) The formulation must be appropriate to the particular region, the required level of animal production (maintenance, growth or lactation), the climate (particularly temperature and humidity) and be as cost effective as possible.
The use of these guidelines in formulating mineral licks is shown in Table 7.6.

7.3.1.3.1 *Summer or growing season supplementation*

When animals are gaining weight only P needs to be supplemented, sometimes with energy. Phosphorus can be provided in a salt-phosphate lick – the animals are attracted to the salt which, in turn, limits intake. This lick is commonly produced from a 1:1 mixture of salt and dicalcium phosphate. Molasses can be added at a rate of 50 g to 100 g per kg to improve its palatability and to bind the mixture. To promote re-conception and lactation, energy should also sometimes be added.

Table 7.6   Composition of a mineral lick for cattle and sheep (adapted from Groenewald & Boyazoglu 1980).

Constituent	Amount g/kg	Mineral per kg lick	Daily requirement	
			Sheep (min)	Cattle (min)
Bonemeal	630			
or		60 g P	1.5 g P	10 g P
Dicalcium phosphate	380			
Molasses	150			
Copper sulphate	1.0	250 mg Cu	5 mg Cu	25 mg Cu
Cobalt sulphate	0.25	50 mg Co	0.3 mg Co	2.5 mg Co
Manganese sulphate	2.0	600 mg Mn	10 mg Mn	50 mg Mn
Zinc sulphate	3.0	750 mg Zn	15 mg Zn	75 mg Zn
Plus salt to	1 000g		20 g total	100 g total

7.3.1.3.2   Maintenance or winter (dry) season supplementation

When protein is the most limiting nutrient as it often is during the winter, a number of different supplements are commonly used but most are based on NPN. A commonly used mixture is one of 100 g to 250 g urea, 250 g to 400 g maize meal, 250 g dicalcium phosphate and 250 g salt. Because of the high concentration of urea in this mixture, its intake has to be carefully controlled. Chicken litter has recently become a popular protein supplement because it is relatively inexpensive, but it does introduce potential health problems (see section 7.1.2.1.2). A mixture of 800 g chicken litter and 200 g salt, with or without molasses, gives good results.

High quality protein sources such as oil cake or dried brewers grain can also be used, but their cost can seldom justify their use. The need for P supplementation to animals on maintenance is being seriously questioned (refer to section 7.2.3).

7.3.1.3.3   Production licks

For growing animals and for those which are being finished for the market on relatively poor quality material and for lactating cows and ewes, it may often be advisable to provide a protein-energy lick. Such production licks are often not based entirely on NPN but also make use of some escape protein.

An example of a protein-based lick which can be used is as follows:

    50 g to 100 g urea
    100 g to 150 g dicalcium phosphate
    150 g to 200 g cottonseed cake
    50 g molasses
    300 g to 400 g maize meal
    10 g of a trace element mixture

Cattle should take in between 500 g to 1 000 g and sheep between 100 g to 200 g of lick per day. Other appropriate products can be used to replace the cottonseed cake and maize meal, such as dried brewers grain or hominy chop. Lucerne meal may also be added to improve the protein content and the palatability of the mixture.

7.3.2   Liquid supplements

Liquid supplements have been developed as a means of using liquid molasses without having to incur drying costs. In practice, urea, P and minerals are often added to the molasses to provide both protein and minerals, depending on need.

Liquid supplements are normally fed *ad lib*, with intake being regulated with a lick-wheel. The intake of cattle normally ranges between 1.0 kg and 1.5 kg per day and of sheep between 50 g to 100 g per day. Regulation is also achieved by designing an appropriate formulation, by placing the licks at appropriate positions in the camp, such as near water, and by providing a sufficient number of containers.

Intake is affected by the prevailing weather, the palatability of the formulation, the availability of roughage and other factors and has proved difficult to regulate. It is known, for example, to even vary widely among members of the same flock. There are, however, a number of advantages to providing supplements in a liquid rather than in a solid form:

(a)   many of the constituents are less expensive, largely because there are no drying costs;

(b)   costs of storing, transporting and mixing liquid supplements are usually lower;

(c)   nutrients are usually more evenly distributed through the material. Liquid supplements will remain stable and evenly mixed and animals should therefore receive a balanced supplement;

(d)   wastage should be lower;

(e)   sources of phosphate used in liquid supplements are usually more, or at worst equally available to the animal, while urea is usually well utilised by animals;

(f)   because of the stability of liquid supplements, large batches can be mixed and stored for some time.

7.3.3   Free-choice supplementation

Free-choice intake of individual minerals or mineral mixtures is based on the false assumption that animals know what minerals they need, and in what amounts, and in effect possess nutritional wisdom. One reason for this is that animals suffering from mineral deficiencies, of P in particular, show pica and, it is believed, will seek out sources of the mineral to alleviate the problem. However, such animals will eat almost anything in an effort to alleviate the disorder and are unable to select only those materials which will correct any specific deficiency. There is a considerable volume of evidence which shows that free-choice mineral supplementation is unrealistic because animals are not able to balance their own mineral needs when given a choice (Arnold 1964; Coppock *et al.* 1972; Muller *et al.* 1977).

7.3.4   Other methods which can be adopted to supplement minerals

7.3.4.1   Doses and injections

These have the advantage of absolute control of the amounts of minerals provided to each animal. Injections have the additional advantage of the mineral being readily available when injected directly into the tissue of the animal. However they should generally be used only for those trace elements which accumulate in body tissue, and not for those which need to be provided at regular intervals.

Dosing has to date been used to supplement selenium and cobalt, while sub-cutaneous and intra-muscular injections are used to supply vitamin B_{12} (instead of co-balt), copper and selenium (Underwood 1977) and, more recently, to also provide manganese and zinc (Vosloo pers. comm.). The main disadvantages of these methods are that they are time consuming and labour demanding, and therefore costly.

7.3.4.2 Heavy capsules

Ruminants can be dosed with slow-release capsules containing specific trace elements. These capsules are retained in the rumen, where they slowly release trace elements which become available to the animal over an extended period of time. They have to date been developed for the trace elements copper, selenium and zinc and have proved to be particularly useful in areas which have clinical deficiencies in any of these trace elements.

7.3.4.3 Glass capsules

These are produced from a variety of phosphate-glass compounds (Knott *et al.* 1985) which allow for different rates of release of copper, cobalt and selenium. When dosed, the capsules settle in the reticulum of the animal. Alternatively, the capsules may be implanted either behind the ear or intra-muscularly in the neck of the animal (Allen *et al.* 1981). Good results have been obtained with dosing (Doyle 1987) but implants have not proved to be particularly successful.

7.3.4.4 Copper oxide particles

This product, when dosed to animals, settles in the abomasum of ruminants (Costigan & Ellis 1980). Copper is slowly released from the copper oxide particles, but not always at a rate sufficient to satisfy the animal's needs in all situations.

7.3.5 Vitamin supplements

Powders of the vitamins A and D are enclosed in gelatin capsules and, on release, are biologically available to the animal. Alternatively, powders can be added to feed mixes or to liquid supplements without any danger of loss of activity during periods of normal storage. Losses can, however, occur during feed processing, should the vitamins be exposed to high temperatures, and in milled feed, because of a greater exposure to oxygen. For example, Schields *et al.* (1982) measured a 30% to 40% loss of activity in vitamin A during the pelleting process. Other factors which can reduce the activity of vitamin A are long periods of exposure to high temperatures and humidity, exposure to ultraviolet light and interactions with trace elements, but with the addition of anti-oxidants and their enclosure in emulsified powders or gelatin capsules, losses should be negligible.

Tocopherol acetate is available in gelatin capsules or as granules or powders for the supplementation of vitamin E. Granules or powders can be mixed into feed rations, or a liquid dispersion of the vitamin can be added to liquid supplements. Vitamin E is usually included in a supplement together with vitamins A and D, but the addition of such a vitamin mixture to animal feeds should be seen as giving support to, rather than replacing, vitamin injections.

Injecting vitamin A, D and E is the most reliable method of ensuring that animals are adequately supplied with these vitamins. One or more intra-muscular injections may be needed each year. The vitamins are rapidly taken up in the bloodstream and deposited in the liver (Baurenfiend & De Ritter 1983), whereas oral presentation results in a measure of destruction of the vitamins by micro-organisms. Nutrients in the rumen may also react negatively with vitamins presented in this way, particularly when nutrient levels are high.

In practice, vitamin A, D and E supplements need be added only to production licks aimed at promoting growth or lactation.

Examples of liquid vitamin supplements which can be added to feeds are shown in Table 7.7. The formulations include previously emulsified mixtures of the three vitamins, with a minimum of 100 000 units of vitamin A per kg of lick. It also includes a trace element mixture which, at an intake of 500 g per day, would provide 50% of the minimum requirements for cobalt, calcium, zinc, manganese, iron and iodine (Baurenfiend & De Ritter 1983).

Table 7.7   Liquid supplements fortified with Vitamins A, D and E and trace elements.

	Formulation 1	*Formulation 2*
Molasses	554	519
Water	178	209
Ammonium polyphosphate	90	—
Phosphoric acid	—	64
Urea	178	209
Plus Vitamin A, D and E and a trace element mixture		

Although there would seem to be good reasons for supplementing producing animals with vitamins A, D and E, there is little evidence suggesting that animals respond to such supplements.

7.3.6   Ionophores

The incorporation of ionophores into summer licks and production licks fed to animals being finished on pasture brings about improved growth and improved utilisation of feed and an improvement in economic terms of up to 20% (Van Niekerk pers. comm.). These improvements can be attributed to the control exerted by the ionophores on coccidiosis in lambs and calves and to its favourable influence on the pattern of fermentation in the rumen.

The most commonly used ionophores are monensin and lasalocid. Because of the absence of reliable information, recommended dosages range widely from 0.5 mg to 2.0 mg per kg live weight. In practice, 50 mg provided to 25 kg lambs and 200 mg to weaned calves per day has given good results.

8 Radical veld improvement

N.M. Tainton, R.H. Drewes, N.F.G. Rethman & C.H. Donaldson

CONTENTS

8.1 GENERAL INTRODUCTION[1]

The natural grazing lands (veld types) of South Africa have been described in detail by Acocks (1988) and their ecology by Tainton (1999) and Bosch (1999). They vary considerably but all have one thing in common – they constitute the major feed resource for the domestic livestock population. Of course, this was not always so. Before technologically-advanced man arrived on this subcontinent and before the development of the agricultural industry, these natural grazing lands supported a population of diverse forms of wild game. These natural veld/game systems were self-regulatory and, while fluctuations in structure and vigour would have occurred, they were probably relatively stable and permanent, at least over a long time scale. Then man entered the scene, settled on the land and started to exploit and manage the system.

8.1.1 Exploitation, restoration and intensification

Man's exploitation of the veld/game system constituted nothing other than game harvesting but soon, and usually simultaneously, he introduced domestic livestock (cattle, sheep and goats) and so the process of domestication, or game replacement, began (see Fig.

---

[1] Acknowledgement is made to Booysen (1981) for many of the basic principles related to radical veld improvement discussed in this chapter.

8.1). The veld/game system became the veld/game/livestock system and in most cases the veld/livestock system soon followed. The introduction of livestock into the system inevitably implied control by man. Livestock numbers and stock movement were subject to decisions by the operator and the execution of these decisions constituted aspects of veld management.

The complexity of the little understood veld/livestock system together with subjectivity, and often ignorance, ensured that man's management decisions varied greatly. They ranged from bad to good. Good veld or grazing management maintains the veld in good structure and sound vigour and therefore maintains, or even increases, its productive capacity, while bad management inevitably leads to the deterioration of the structure and vigour of the veld and consequently of its productive capacity. The end result of this process is denudation and erosion. Fortunately, it is within man's power to halt this process of deterioration and to reverse it. All his activities in this vein are collectively referred to as 'restoration' (see Fig. 8.1), which has as its eventual objective the return of the vegetation to its original state.

It may be convenient in certain circumstances to recognise two facets of the restorative process – reclamation and renovation. These two cannot always be separated in practice or objective. They are both restorative in purpose, but reclamation implies the winning back of land from a waste or denuded condition to a vegetated condition irrespective of the nature of that vegetation. On the other hand, renovation implies the re-establishment of vegetation of good structure and vigour, as near as possible to the original condition. While good grazing management may improve the vegetation and bring it back to its original condition, the techniques of restoration usually include soil disturbance, seeding and fertilization. However, veld deterioration and restoration are not the subject of this chapter.

Man is not only concerned with the restoration of veld to a good condition. As the demand for food increases and land becomes limiting, he is forced to adopt procedures designed to increase the productivity of the land above that of the natural vegetation, where this is feasible. This process of increasing agricultural production per unit area of land is referred to as 'intensification'. The three major procedures involved in this process are fertilization, reinforcement and replacement. While fertilization may be practised alone, reinforcement and replacement are usually not attempted without the addition of fertilizer. Veld fertilization and veld reinforcement are collectively referred to as 'radical veld improvement' (RVI). Sometimes partial veld replacement is also included in this definition. However, complete veld replacement is hardly a form of RVI but is better seen as pasture establishment. Such replacement represents the final stage of pasture resource intensification. It will be considered in Chapter 10.

8.1.2 Productivity and intensification

Veld intensification practices and procedures have been defined as those designed to increase the productivity of the land above that of the natural veld. Herein lies the first problem. How is the productive level of pastoral land best measured and expressed? Clearly this matter must first be understood before it is possible to consider the intensification process.

The productivity of a grazing system (veld/livestock or pasture/livestock) is determined by the relations among:

EXPLOITATION

Natural veld/Game system

↓

GAME HARVESTING

↓

Exploited veld/Game system

↓

GAME REPLACEMENT
(DOMESTICATION)

↓

Veld/Livestock system

BAD GRAZING
MANAGEMENT

GOOD GRAZING
MANAGEMENT

| Denuded land (no production) | Deteriorated veld (decreased production) | | Good veld (increased production) |

RECLAMATION   RENOVATION   FERTILIZATION

Fertilized veld
(increased production)

Vegetated
land
(some
production)

Improved veld
(increased
production)

REPLACEMENT   REINFORCEMENT

Sodseeded
pasture
(increased
production)

Reinforced
veld
(increased
production)

RENOVATION

PASTURE ESTABLISHMENT

Deteriorated
cultivated pasture
(decreased
production)

BAD PASTURE
MANAGEMENT

Good cultivated
pasture
(increased
production)

RENOVATION

GOOD PASTURE
MANAGEMENT

RESTORATION   INTENSIFICATION

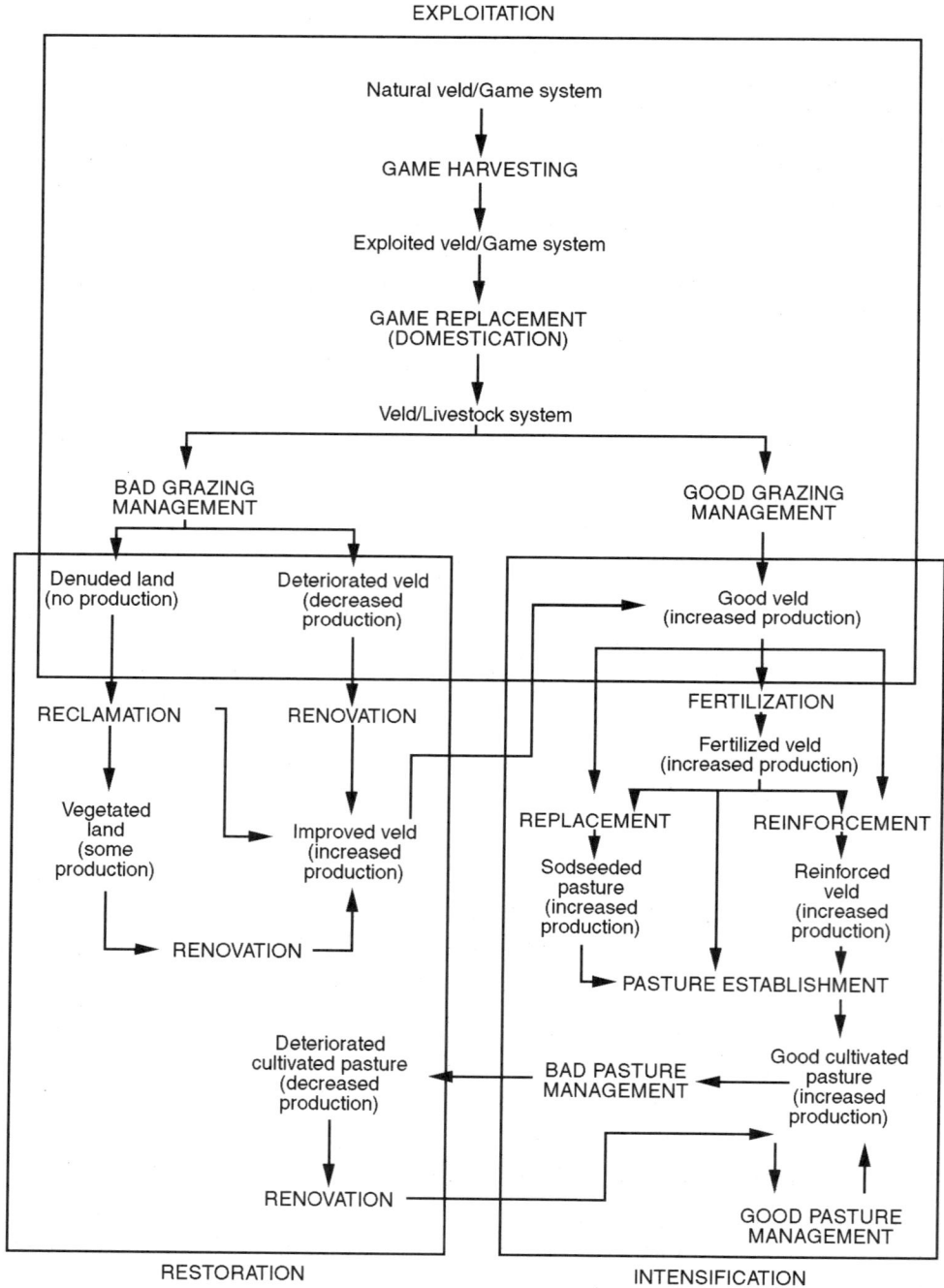

Figure 8.1   The processes involved in grassland restoration, exploitation and intensification.

(a)   area of land;
(b)   the amount of grazeable and nutritive herbage produced;
(c)   the number of animals;
(d)   the quantity of final product; and
(e)   marginal profit.

The relation between successive pairs in the series is the concern of a particular function within the system. Pasture production is concerned with the most favourable relation between area of land and the amount of grazeable and nutritious herbage produced. Grazing management attempts to develop the optimum relation between the amount of grazeable and nutritive herbage and the number of animals (without resource degradation). The endeavour of animal production is concerned with providing for the optimum relation between number of animals and the quantity of the final product. Finally, product marketing is the activity concerned with the relation between the quantity of the final product and marginal profit (Booysen 1972).

Because the objective of all free enterprise agricultural systems is profit to the operator and because land eventually (though not necessarily initially) becomes the limitation to production, in the final analysis the most useful measure of productivity of entire grazing systems is profit per unit area of land. In certain circumstances other factors (e.g. capital, labour supply) may temporarily limit production and constitute a better basis for expressing profitability while that situation persists. Also, in certain circumstances one may be primarily concerned with only part of the system and a different relation may better characterise the level of success of that part. For example, different pasture production situations can conveniently be compared by a consideration of the quantity of grazeable and nutritious herbage produced per unit area of land.

The process of veld intensification is, in the first instance, concerned with increasing pasture production. Therefore, in the context of this discussion, veld intensification practices and procedures are those designed to increase the productivity of the land in terms of the quantity of grazeable and nutritious herbage produced per unit area of land.

8.1.3   Limiting factors and intensification

In order to describe the practices and procedures which will increase the productivity of the veld, it is necessary to first identify those factors which limit the production of the veld system. All the factors of any veld system are conveniently classed as either edaphic, climatic or biotic. Of the edaphic (soil) factors which may limit the growth and productivity of veld plants and which can be modified on a practical scale by man, the most common is soil fertility. There may be other factors but these are less frequently encountered. In South Africa, the climatic factor most likely to limit production is rainfall, although temperature may play an important role in some areas, particularly in winter. Of the biotic components, it is the genetic constitution of the plants themselves which is most likely to limit production. So the major limitation to increased productivity in our veld systems is likely to be one or more of soil fertility, rainfall, and plant yield potential. Therefore, if intensification practices are to be effective and economic, they should be directed at alleviating one or all of these limiting factors, starting with that factor which is most limiting in any particular situation. However, since the edaphic, climatic and biotic factors have interacted over millions of years in the development of the systems as we know them today, a balance amongst them exists in natural systems so that more than one factor frequently needs simultaneous alteration if the system is to maintain a high level of production in the long term.

8.2   VELD INTENSIFICATION BY RADICAL VELD IMPROVEMENT

8.2.1   Humid regions

For convenience, the classification of Edwards & Booysen (1972) will be used to define this zone. These authors recognised the important roles played not only by the total rainfall, but also by its seasonal distribution and the evaporative capacity of the atmosphere as determinants of the moisture regime. They suggested that the productive capacity of an area can be increased above that of the natural vegetation in the following areas:

(a)   in the winter rainfall region, where rainfall is in excess of 300 mm per annum;
(b)   in the all-year rainfall zone where rainfall is in excess of 400 mm and the evaporative rate less than 1 778 mm per annum, or where rainfall is in excess of 500 mm where the evaporative rate is greater than 1 778 mm per annum; and
(c)   in the summer rainfall zone where the rainfall is in excess of 500 mm and the evaporative rate less than 2 032 mm, or in excess of 600 mm where the evaporative rate is greater than 2 032 mm per annum.

These areas are shown in Fig. 8.2 (Edwards & Booysen 1972). Here, radical veld improvement includes two primary procedures: veld fertilization and veld reinforcement.

8.2.1.1   *Veld fertilization*

According to the earlier argument, applying fertilizer to veld is likely to be economically feasible only where neither moisture nor the growth potential of the plants will limit the growth response of the plants.

The fact that positive yield responses to fertilizers have for the most part been recorded in areas identified by Edwards & Booysen (1972) as suited to intensification, is clear evidence that the yield potential of the plants does not limit production in these areas. The lack of response in drier areas than these could, of course, be due to either insufficient moisture or to the low yield potential of the plants, with the former being the most likely.

Since the area in which the vegetation is likely to respond to fertilizer in the summer rainfall zone is considerably greater in extent than that of the winter rainfall and all-year rainfall zones, and because most of the research on veld fertilization has been undertaken in the summer rainfall areas, the remaining discussion on veld fertilization will be confined to the summer rainfall areas which for the most part are covered by grassveld of various kinds.

Topography and soil depth do not generally limit responses of veld to fertilizer, but may influence the methods used to apply fertilizer. However, sites with gentle slopes and deep soils can be used for annual field crops and sites of intermediate slope and soil depth are suited to high-producing, intensive, cultivated pastures. As agricultural production intensifies, these more favourable sites are likely to be used for these potentially high-producing systems. So, while at our present stage of intensification it may be

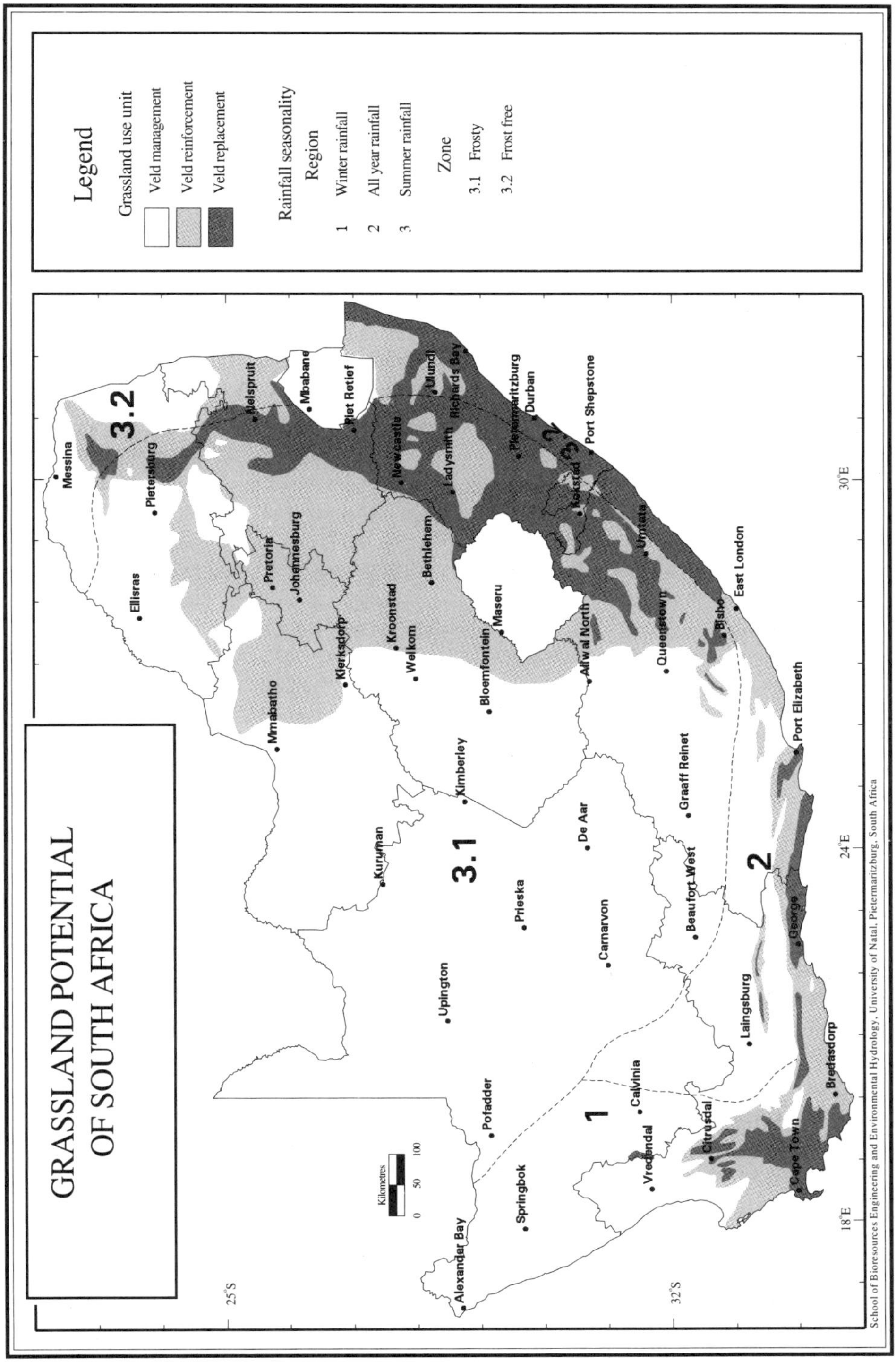

Figure 8.2 Grassland potential and potential for reinforcement and replacement in South Africa.

School of Bioresources Engineering and Environmental Hydrology, University of Natal, Pietermaritzburg, South Africa economical to apply fertilizer to such sites, it must be seen in the long term as a method of intensification for sites with steep slopes and/or shallow soils where cultivation is not feasible (see section 8.2.1.2.4).

The objective of applying fertilizer to grassveld is to increase the productivity of the veld. To achieve this, the goal must be to increase one or more of (a) the yield of herbage, (b) the nutritive value of the herbage, and (c) the acceptability of the herbage to the animal. If one or more of these three objectives is achieved without detriment to the others then the practice becomes potentially feasible. However, if the procedure is either uneconomical, or leads to botanical changes which are undesirable, then it is not tenable. All five of these factors therefore need to be examined.

8.2.1.1.1 Herbage yields

Experiments examining the effect of various fertilizers on grassveld yields have been conducted in a variety of veld types (Booysen 1954; Grunow *et al.* 1970) and in most such trials nitrogenous and phosphatic fertilizers have been found to increase yield. Responses to phosphate have usually been observed only when it is applied together with N. Responses to lime and potassium fertilizers have been recorded only rarely, although they have not been well studied. Lime responses may, however, be expected after long periods of addition of acidifying materials such as ammonium sulphate, while a deficiency of potassium may be induced by long periods of yield promotion with nitrogenous and phosphatic fertilizers.

The responses of grassveld to nitrogenous and phosphatic fertilizers in a variety of experiments have been very similar, but with a tendency for greater responses in the regions of high rainfall and leached soils than in the lower rainfall areas with more fertile soils. A typical example of this pattern is provided by the results of a continuing veld fertilizer experiment in the Tall Grassveld of KwaZulu-Natal situated at Ukulinga Research Station of the University of Natal, which experiences a mean annual rainfall of 710 mm. The herbage yields of the 1952/53 (third) and 1953/54 (fourth) seasons of fertilization reveal positive responses to N alone up to approximately 100 kg N/ha, when the yield increase is of the order of 40% (Fig. 8.3). Both ammonium sulphate and ammonium nitrate were used as sources of N. When N was applied together with annual dressings of 336 kg/ha of single superphosphate (8.3% P), positive yield responses were recorded up to approximately 175 kg N/ha and at this level yield increased by approximately 100%. Maximum responses to nitrogenous and phosphatic fertilizers appear to commonly occur at levels of N in the region of 100 kg to 200 kg N/ha but the response reported varies considerably and is often as much as 250–300% of the yield of an unfertilized veld (Visser 1966; Grunow *et al.* 1970).

The importance of rainfall in determining herbage yields of veld is illustrated by the results of the Ukulinga experiment in the 1952/53 and 1953/54 seasons (Table 8.1). The veld yielded twice as much herbage in 1953/54, a year of good rainfall (756 mm), than in 1952/53, a year of low rainfall (608 mm). However, the pattern of response to both N and P remained essentially the same in both seasons (Fig. 8.3).

8.2.1.1.2 Nutritive value of herbage

There have been relatively few studies of the influence of applying fertilizer to grassveld on herbage quality and the studies that have been done have for the most part been directed at the crude protein content. This is not unreasonable since protein is

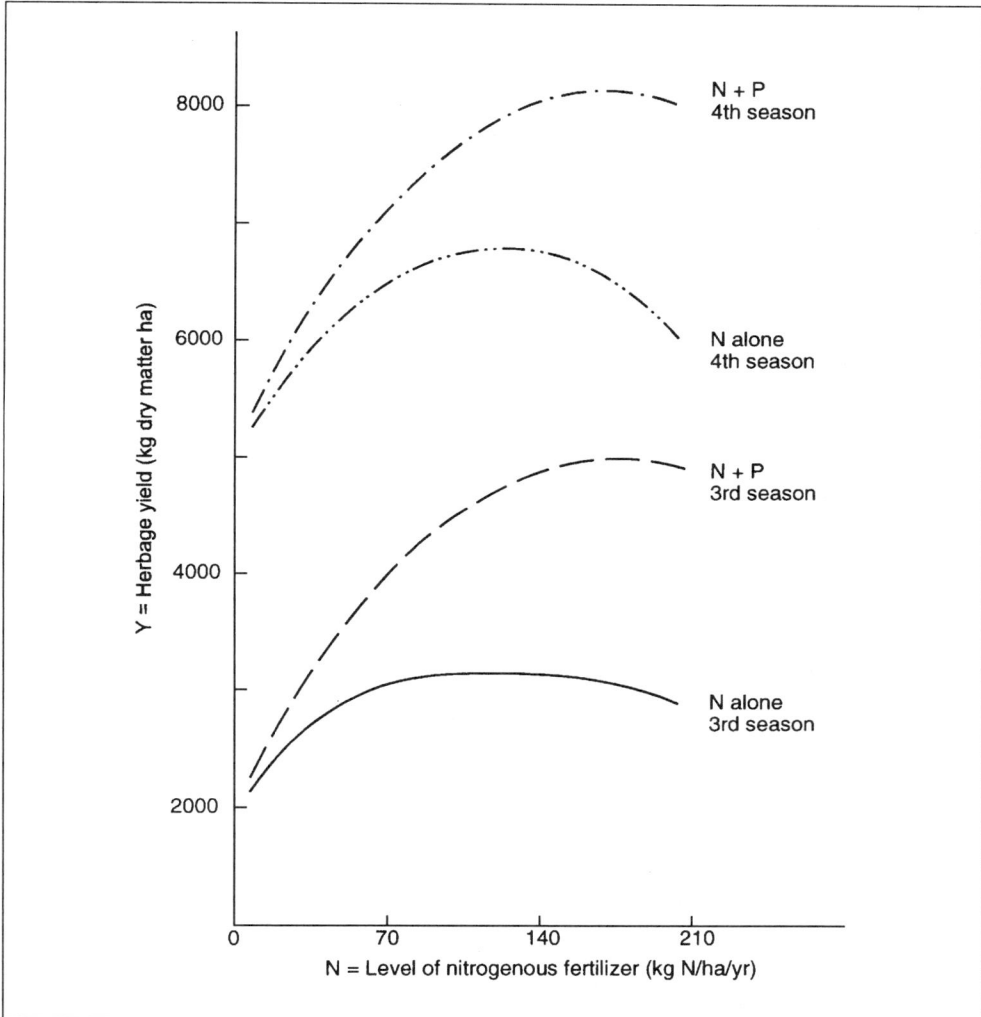

Figure 8.3 Yield response of veld to nitrogenous fertilizer in the presence and absence of applied phosphate in seasons of contrasting rainfall.
In the third season, which experienced low rainfall:
N alone: $Y = 2\,234.3 + 18.7\,N - 0.073\,N^2$
N + P: $Y = 2\,187.8 + 18.7\,N - 0.107\,N^2$
In the fourth season, which experienced high rainfall:
N alone: $Y = 5\,257.0 + 28.7\,N - 0.119\,N^2$
N + P: $Y = 5\,271.9 + 39.7\,N - 0.114\,N^2$.

most often the nutrient which most limits animal production. Phosphate is often also deficient in grassveld but this can readily be supplemented (see Chapter 7). In poor quality veld the digestibility of the herbage may be so low that the daily intake of energy is below requirement. For this reason, and because fertilizer additions should increase the digestibility of the feed, it is unfortunate that this relation has not been investigated more fully.

Table 8.1   Influence of rainfall on yield of grassveld and the response to nitrogenous fertilizer in the presence of phosphatic fertilizer (28 kg P/ha).

Season of experiment	Rainfall July – June (mm)	Control yield (kg DM/ha)	Level of N producing maximum yield (kg N/ha)	Response to nitrogen (kg DM/ha)	Response to nitrogen (kg DM/kg N)
1952/53	608	2 188	164	2 873	17.5
1953/54	756	5 272	174	3 456	19.9

The early work on the influence of fertilizer on the crude protein content of grassveld herbage was not very encouraging (Booysen 1954). However, Hall *et al.* (1940) had previously shown substantial increases in the crude protein content of fertilized veld. At Frankenwald Research Station in the Gauteng Highveld, they observed an increase in protein content from 6.78% to 9.46% (approximately 40%) when veld was fertilized with 673 kg/ha of ammonium sulphate, 448 kg/ha of super and rock phosphate and 90 kg/ha of muriate of potash. Since then other experiments have shown increases of between 25% and 50%. The mean crude protein content for the first and second hay cuts of the 1952/53 and 1953/54 seasons in the Ukulinga trial mentioned previously increased by 0.01% for every 1 kg of N applied per hectare per annum up to 210 kg N/ha.

8.2.1.1.3   Acceptability of herbage to livestock

The available evidence suggests that nitrogenous fertilizer applied to grasses increases the crude protein and moisture content and decreases the crude fibre content of the herbage. These changes should increase the acceptability of the herbage to livestock. The preference shown by livestock and game for fertilized areas of pasture and grassland has frequently been observed and would lend support to such an assumption. If, however, the fertilizers change the species composition of the sward, a different picture may emerge (see section 8.2.1.1.5)

8.2.1.1.4   Economic considerations

A question of vital concern to the agriculturalist is whether or not the benefits derived from applying fertilizer to veld more than offset the cost of the operation. Whether this is so or not will depend on the particular circumstances on any farm and so the economic viability of fertilizing veld needs to be determined for each individual situation.

8.2.1.1.5   Botanical effects

Despite any short-term economic benefits which may be derived from applying N and P to veld, the feasibility of such a practice depends on whether or not it induces long-term undesirable changes in the vegetation. For example, if fertilization causes (a) the species composition of the vegetation to change so that the herbage produced is a poorer animal feed, or (b) the cover to change so that the rate of soil loss and water runoff is substantially increased, then the advisability of veld fertilization must be seriously questioned.

Changes in plant composition as a consequence of the fertilization of grassveld have been frequently recorded. The consistent pattern is for species of the genera of the tribe

Andropogoneae, for example *Themeda, Hyparrhenia, Diheteropogon, Cymbopogon, Trachypogon, Elionurus, Andropogon* and *Heteropogon*, which dominate veld which is considered to be in good to moderate condition, to be replaced, in the humid grasslands, by members of the tribe *Eragrosteae*, and more particularly by *Eragrostis curvula* and *Eragrostis plana*. In some situations *Cynodon dactylon* commonly occurs with the *Eragrostis* species and in others forbs are a significant constituent of the pioneer cover. In the drier areas it seems that members of the *Paniceae* (such as *Panicum* and *Digitaria*) are important pioneer invaders after fertilization (Visser 1966; Grunow *et al.* 1970).

In the early years of fertilization, the *Andropogonoid* grasses are greatly stimulated. The large plants become larger and the small plants are shaded out. Thus the stand becomes more tufted and, while the total basal cover may change little, it is composed of fewer larger plants with large bare areas between them. With continued fertilization, some of the bigger plants also die out and the total basal cover then declines. At this time, annual and perennial forbs, together with some *Eragrostoid* grasses, invade the bare areas. The *Eragrostoid* species increase in abundance with time at the expense of the forbs, although forbs may remain a significant component of the grassland. As the individual tufts of the invaders increase in size, the total basal cover usually increases again and may, in fact, eventually increase to above the level of the original grassland.

Two explanations are usually advanced for the change in species composition caused by nitrogenous fertilizer. Firstly, there is evidence that *Andropogonoid* grasses are more sensitive to soil N than *Eragrostoid* grasses. While both are stimulated by moderate concentrations of soil N, the growth of *Andropogonoid* grasses is retarded by high levels of soil N which still stimulate the *Eragrostoids* (Roux 1954). Secondly, the stimulation of shoot growth of *Andropogonoids* by fertilizer causes more soil water to be utilised by the plant and therefore a more rapid drying out of the soil. This is likely to favour the xerophytic *Eragrostoids* relative to the mesophytic *Andropogonoids* (Visser 1966). These two explanations are not mutually exclusive and it is not unlikely that both processes are operative.

The time it takes to induce the above compositional changes varies with treatment. The rate of change is largely a function of the amount of N applied (Visser 1966). It can occur after only one season when high levels of N are applied but it can take much longer where applications are small. Other factors do, however, also seem to be involved. Treatments which most promote growth and production of the *Andropogonoid* grasses will cause the most rapid change in composition. The application of phosphate and lime either alone, or in combination, had no effect on productivity and did not result in any botanical changes after 27 years of treatment in the Ukulinga experiment. The treatment involving a high level of N (210 kg N/ha/annum) together with phosphate (28 kg P/ha/annum) became very tufted and with large bare areas after only about five seasons of fertilization. Soon afterwards the *Eragrostoids* started to invade and constituted a significant and noticeable component of the vegetation by the 10th year and were dominant by the 15th year. By the 20th year the change to *Eragrostoids* was complete.

Are these changes in basal cover and species composition such that they seriously detract from the feasibility of the exercise? While it is true that during the period of change the basal cover declined considerably, this may not be serious as a reasonably good cover subsequently developed. However, this modified grass mantle is usually extremely tufted and not as effective in preventing runoff and in conserving the soil as is the original cover. On level sites this may not be a serious disadvantage but in steep landscapes the cover induced by fertilization may not be sufficiently dense to prevent soil erosion. It is particularly in these steep, non-arable areas that veld fertilization would most likely play a role in the intensification process. A possible solution to this is to sow seed of some suitable sod-forming grass into the fertilized area in order to improve the productivity of the veld while still providing a dense cover to the soil, as discussed in section 8.2.1.2.

The matter of fertilizer induced change in species composition of the veld may also present other problems. While it has been shown that fertilizers increase the yield, protein content and acceptability of *Andropogonoid*-grasses, how does the induced cover compare in terms of these parameters? The Ukulinga experiment affords a comparison of the response curves of the fertilized *Andropogonoid*-dominated veld of the 1952/53 (3rd) and 1953/54 (4th) seasons and the fertilized *Eragrostoid*-dominated cover of the 1972/73 (23rd) and 1974/75 (24th) seasons (Fig. 8.4). Clearly the induced cover does not have the productive capacity of the original veld. This may seem strange as it is well known that *E. curvula* (the dominant grass in the induced cover) is frequently sown as a cultivated pasture where it produces yields of 10 000 kg dry matter per hectare in the Ukulinga environment. However, these pastures are sown to ecotypes selected for high production, while the invaders into fertilized veld are a mixture of low-producing ecotypes. Also, other low-producing species of *Eragrostis* often invade with *E. curvula*.

The crude protein content of the induced cover increases in response to N in much the same way as that of the original grass cover. Bartholomew & Cross (1973) reported a linear response in crude protein content of *E. curvula* to N from approximately 8.5% in the absence of fertilizer N to 13.5% with 500 kg N/ha. The function of the crude protein response in the *Andropogonoid*-dominated veld at Ukulinga is $Q = 6.82 + 0.01\ N$ (where Q is the percentage crude protein), while that of *E. curvula* reported by Bartholomew & Cross (1973) is $Q = 8.49 + 0.01\ N$. So while the crude protein content of unfertilized veld is somewhat lower than that of unfertilized *E. curvula*, in both situations it increases by 0.01 for every kilogram of N applied.

There is every reason to believe that the acceptability of the herbage produced by the cover induced by N fertilizers is related to the level of N applied, as is general for other pastures. This cover is nonetheless likely to be less acceptable than the original cover which it replaces, even at high N levels. For example, well fertilized *E. curvula* is likely to be less acceptable to animals than unfertilized *Themeda triandra*.

Thus, while the change in composition induced by the fertilizer application does not alter the positive response in the crude protein content to N, the relatively low yield potential of the induced cover, assuming potentially more productive species are not oversown into the area, is cause for concern and will obviously have long-term consequences on the economics of fertilizing veld.

In conclusion it can be said that the application of N and phosphate to veld in the high rainfall areas results in increased yields of dry matter, increased crude protein content, and greater acceptability of the herbage, at least in the short term. However, even where the practice proves to be profitable, it can be recommended only where its impact on basal cover is not such that both runoff and soil loss will be accelerated to unacceptable levels.

8.2.1.2 *Veld reinforcement*

Veld reinforcement is aimed at releasing the system from genetic limitations to yield. This is achieved by introducing seed of appropriate pasture species which are genet-

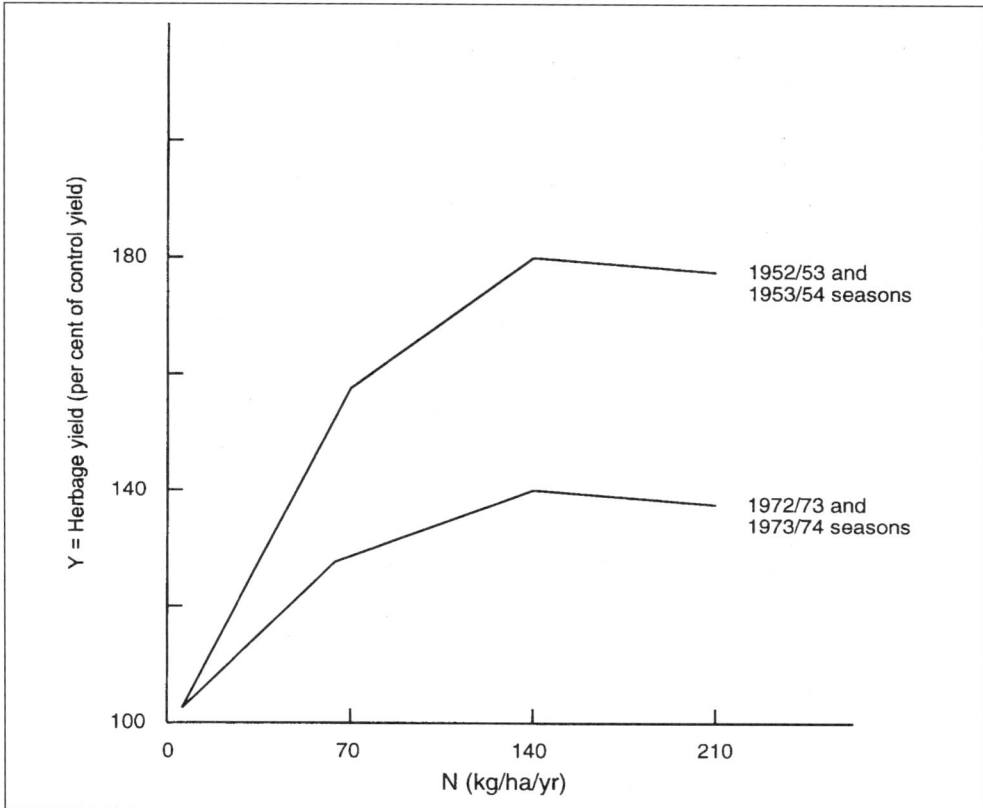

Figure 8.4 Yields of KwaZulu-Natal Tall Grassveld in response to the application of nitrogenous fertilizer. Yields are for the 1952/53 (3rd) and 1953/54 (4th) seasons and the 1972/73 (23rd) and 1973/74 (24th) seasons. 28 kg P/ha was applied to all treatments.

ically able to outyield the species of the natural veld of the area. This will occur only where both soil fertility and the moisture regime will permit such higher yields. Under natural conditions, highly fertile soils and high rainfall do not normally occur together. Thus, for the most part, veld reinforcement in humid regions is likely to be successful only where fertilizer is also applied.

Veld reinforcement, like veld fertilization, is more easily applied to level sites than to steeper slopes. As there are many such level sites available at present, farmers are more likely to reinforce these sites than those with steeper slopes. However, for reasons already outlined, veld reinforcement is likely in the long term to find its place only in less favourable sites.

A number of classifications of the degree of intensification which should be permitted on land of different slopes have been published (Edwards 1970, 1978; Anon. 1977; Edwards & Scotney 1978), an example of which is shown in Table 8.2. The most comprehensive of these are those of Edwards (1978) and Edwards & Scotney (1978). The former also takes into account the degree of rockiness and soil drainage characteristics in defining nine landscape classes. This classification is illustrated in Fig. 8.5.

Bottomlands are separated into three classes on the basis of their moisture status and drainage, namely vleis, other poorly drained bottomland (not permanently wet and more common in drier regions) and drained alluvial soils. The uplands are subdivided on the basis of slope and rockiness.

Table 8.2   Effect of slope on land use capability (Anon. 1977).

Slope class	Slope (%)			Land use suitability
	Humid	Sub humid	Sub arid, arid	
A	0 – 3	0 – 3	0 – 2	Annual cropping
B	4 – 8	4 – 7	3 – 5	Annual cropping with occasional ley or special tillage
C	9 – 15	8 – 12	6 – 8	Rotation of ley and crops
D	16 – 25	13 – 20	9 – 15	Permanent cover crop e.g. pasture
E	26 – 35	21 – 30	16 – 20	Natural veld grazing or afforestation with special treatment
F	> 36	>31	> 21	Natural veld grazing or afforestation with special treatment or total protection from agricultural use e.g. wildlife

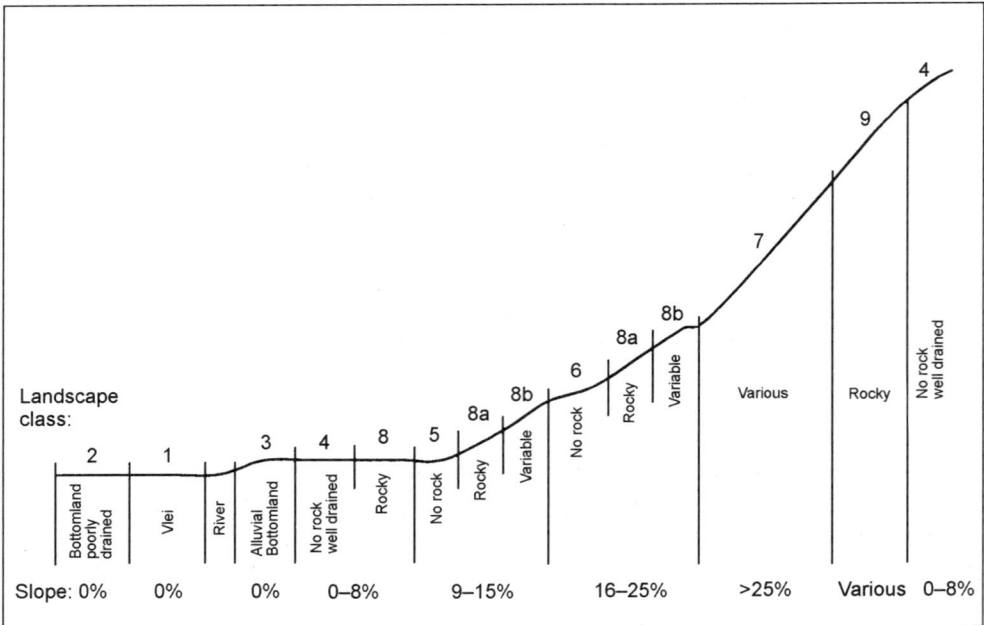

Figure 8.5   Proposed pasture landscape classes (Edwards 1978). 1) vlei; 2) poorly drained bottomlands; 3) alluvial bottomland; 4) flat (0–8% slope) without undue rockiness, well drained, not alluvial; 5) medium slope (9–15%) without undue rockiness; 6) steep slope (16–25%) without undue rockiness; 7) very steep slope (>25%); 8) rocky sites or very variable slopes over small distances on all classes except 7) and 9); 9) rock.

The classification of Edwards & Scotney (1978) is described in some detail in Chapter 10 section 10.1.2 where it identifies areas suited to conventional pasture establishment. The future potential for reinforced veld would seem to lie mainly in areas with slopes in excess of 12–15%. The influence of landscape class on the reinforcement method used is discussed later (refer to section 8.2.1.2.4).

There are three basic questions which need to be considered in discussing the mechanics of veld reinforcement. How should the seed be introduced into the veld? How should the fertilizer be applied? How can the competition from veld plants be most effectively reduced? Each of these questions will be considered in turn before a number of veld reinforcement systems are discussed. Matters such as the choice of species for introduction and the choice of type and amount of fertilizer to apply are covered in Chapters 9 & 11.

8.2.1.2.1 Seeding procedures

While there is no standard procedure of reinforcing veld, in all methods the seed of superior species is introduced either by 'broadcasting' or 'drilling'. The former procedure places the seed on the soil and the latter places the seed in the soil. When seed is broadcast into existing vegetation the process is known as 'overseeding' or 'oversowing'. When it is drilled into existing vegetation it is referred to as 'overdrilling'. If the existing vegetation is veld or pasture the process is known as 'sodseeding'. Thus the two basic procedures for veld reinforcement are overseeding and sodseeding.

(a) Veld can be overseeded either manually or mechanically. A variety of agricultural implements, including the fertilizer distributor, can be used for mechanical overseeding. These procedures will be discussed more fully in Chapter 10.

All overseeding procedures aim to place the seed on the soil. Some of the seed will almost inevitably be trapped in the herbage or on the crowns of plants where it is unfavourably placed for germination and establishment. Thus as much as possible of the veld herbage should be removed prior to seeding. However, even the seed which is placed on the soil surface is less than ideally situated for germination as it is subject to desiccation and temperature extremes. Because of this, and because some of it may be consumed by insects, termites, rodents and birds, the proportion of wasted seed is high and so seeding rates must be high.

Various techniques have been employed to improve the environment for oversown seed and thus increase the germination percentage when reinforcing veld. Two basic procedures are involved – seed pelleting and soil disturbance.

Pelleting involves coating the seed with, or embedding it in, some material such as soil or fertilizer. The procedure can have one or more of a number of objectives and the materials and procedures used will depend on the main objectives of the exercise. Pelleting may be directed primarily at:

1) improving the moisture regime surrounding the seed;
2) providing a source of plant nutrients in close contact with the rootlet of the germinating seed;
3) protecting the seed against predation; or
4) ensuring a supply of viable rhizobia to the germinating rootlet of legume seed.

Often all four objectives may be achieved simultaneously. When overseeding grass into veld, pelleting would be aimed at the first three objectives only. If legume seed is used all four objectives become important. The basic material for pelleting is usually either lime, superphosphate or rock phosphate. An essential ingredient is an adhesive such as carboxy methyl cellulose (CMC) or gum arabic. The procedure is normally to mix the appropriate quantity of adhesive with the seed before mixing in the pelleting material. The seeds should be evenly coated (Anon. undated (a)). With legume seed the inoculum is added to the adhesive before the pelleting material is introduced. With grass seeds and legume seeds which are not to be inoculated, fungicides may be added to the adhesive.

A second procedure for improving the environment of the oversown seed involves scarifying the soil (using a variety of disc, tine or rotary implements), while leaving the vegetation largely intact or by mob grazing with small-hooved animals. When either pelleted or unpelleted seed is then applied to land treated in one of these ways, much of it becomes lodged in the crevices formed and so the chances of germination and successful establishment are greatly increased.

(b)   Sodseeding involves drilling seed into existing vegetation. Here the soil needs to be prepared to receive the seed. Both this and the placement of seed in the soil require some or other implement. These two processes can be accomplished separately by discing or ploughing and then drilling, but such a double operation is usually too costly to be practical. Consequently, machines (sodseeders) have been developed to both prepare the soil and place the seed in one operation.

The most common means of soil preparation by sodseeders involves either a tine fitted with sweeps to loosen the soil and remove the vegetation, or rotary cutters to simultaneously destroy the vegetation and cultivate the topsoil. In either case the cultivating attachment is followed by a shoe to draw a furrow, then a fertilizer tube to band-place the fertilizer in the furrow, a seed tube to place the seed and, finally, a press wheel to cover the seed and compact the soil (Krog *et al.* 1969).

(c)   Veld may also be vegetatively spot planted with material of sod-forming species such as kikuyu to improve its productivity, as described in Chapter 10 section 10.1.5.2.

Veld is usually reinforced in spaced rows with strips of existing vegetation being left to provide erosion control. If the objective is complete veld replacement, the rows must be contiguous. Veld reinforcement may therefore be viewed as the first step in a programme of veld replacement. The initial operation is that of sodseeding in spaced rows. When the introduced plants are well established, the inter-row vegetation is sodseeded in one or more successive stages.

Because the sodseeder places the seed in the soil in close proximity to fertilizer, it is not usual to pellet grass seed intended for sodseeding. Legume seed should be inoculated with rhizobia prior to seeding.

8.2.1.2.2   *Fertilizer application*

Fertilizers need to be applied where veld is reinforced in areas with a high rainfall and leached soils. The procedures used to apply the fertilizer will, however, differ between overseeding and sodseeding operations.

With overseeding, the fertilizer is usually applied with a fertilizer distributor. However, the timing of the first application of fertilizer in relation to the seeding operation varies. Seed and fertilizer can be broadcast together. If the seed is pelleted and fertilizer is added to the pelleting medium, then the first broadcast application of fertilizer can be delayed until the introduced plants are well established. This is likely to result in more efficient use of the fertilizer nutrients but does necessitate an extra field operation. Where grasses are sown, the seeding and fertilizing operations must, in any event, be separated since grass pastures would require N fertilizer, which if mixed with the seed is likely to reduce germination. Another procedure is to apply nitrogenous fertilizer to the veld for one or more seasons before seeding. The purpose of this is to first capitalise on the increased yields of the fertilized veld and then, when the veld cover starts opening up, to reinforce it with the seed of the species to be introduced. This procedure may, however, promote the early invasion of pioneer grasses such as *E. plana* and *E. curvula*, which may then provide strong competition to the seedlings of the later introduced species (Mappledoram 1989).

When sodseeding procedures are used, the fertilizer and seed are normally applied in the same operation. Because the fertilizer is applied in bands, this procedure leads to extremely efficient use of the fertilizer and so less is required than in the broadcast operation. Where grass is introduced, regular N dressings will be required.

Whether veld is reinforced by overseeding or sodseeding, the resultant pasture is relatively low producing and so the reinforcement process must be accomplished as inexpensively as possible. Thus the number of mechanical operations and the amount of N fertilizer used must be kept to a minimum. Because of this, the use of a legume either alone or together with a grass in the reinforcement mixture has considerable economic advantages.

8.2.1.2.3   Reduction of competition

A factor of great importance in determining the success or otherwise of veld reinforcement is the amount of competition offered to the introduced plants by the existing veld plants. In the higher rainfall grassveld areas the cover is usually dense and individual plants grow strongly so that competition is strong. For reinforcement to succeed here the competitive ability of the existing veld plants needs to be reduced, at least until the introduced plants are well established. A variety of procedures can be used to accomplish this, either alone or in combination – fertilizing, burning, trampling, soil scarification, the application of chemical herbicides and sodseeding.

As already indicated (refer to section 8.2.1.1.1), applied N stimulates plant growth. In time, however, the climax grasses become weaker and the cover opens up (refer to section 8.2.1.1.5). Under natural conditions, pioneer species then invade. This sequence of events can be used to advantage in reinforcing veld. When the competitive ability of the climax grasses weakens and before the weedy pioneer grasses invade, the veld can be overseeded with species with a higher yield potential and better quality than the natural pioneer grasses of the area. High levels of N may reduce the vigour of the climax grasses within one or two seasons, in which case seed may be oversown in the season following the first fertilizer application. Table 8.3 shows the short-term results of a fertilizing and overseeding experiment conducted by Edwards (1970).

Growth of veld is typically reduced by about 50% in the early summer following a winter or spring fire (Everson 1999). This early season reduction in growth rate and competitive ability should assist in the establishment of introduced plants seeded in spring. The most harmful effect of fire on grassveld is, however, when the burn is applied in the autumn.

Table 8.3 The influence of nitrogenous fertilizer on the yield in the season following reseeding previously fertilized veld with *Trifolium repens and Trifolium pratense* (clover) and *Paspalum dilatatum* (paspalum) (after Edwards 1970).

	Fertilizer treatment		
	Control	158 kg N/ha	316 kg N/ha
Clover yield			
kg DM/ha	125	572	1 410
% of total yield	2.9	10.5	19.4
Paspalum **yield**			
kg DM/ha	33	79	180
% of total yield	0.8	1.4	2.5
Veld yield			
kg DM/ha	4 215	4 808	5 666
% of total yield	96.3	88.1	78.1
Total yield			
kg DM/ha	4 373	5 459	7 256

The use of livestock trampling to 'hoof cultivate' the soil has already been mentioned. This requires an extremely high density grazing for a short period. Stock numbers on the farm will usually limit the area that can be treated each year. Mob grazing should be of such intensity and duration that the growth rate and competitive ability of the veld plants are reduced sufficiently to allow for successful establishment of the introduced plants without destroying the existing vegetation. This method has the advantage of also adding nutrients to the soil from the dung and urine deposited in the area. Stock trampling can also be used to bury previously applied seed. Edwards (1970) applied three different intensities of hoof cultivation to Highland Sourveld in late winter after a mid-winter burn. The veld was fertilized and oversown with *Paspalum dilatatum, Trifolium repens* and *Trifololium pratense* in November, after which it was mob grazed. The yields of the various components of the reinforced veld are illustrated in Table 8.4. The highest intensity of hoof cultivation resulted in a pasture which produced about equal proportions of herbage from veld plants and introduced plants in a total yield threefold greater than overseeded veld which had not been mob grazed.

Table 8.4 The influence of six days of hoof cultivation in late winter on the late summer yields of veld fertilized and oversown in spring with *Trifolium repens and Trifolium pratense* (clover) and *Paspalum dilatatum* (paspalum) (after Edwards 1970).

	Intensity of hoof cultivation (steers/ha)			
	Nil	175	350	525
Clover yield				
kg DM/ha	50.8	719.1	1 066.4	1 455.1
% of total yield	3.5	28.8	29.0	32.5
Paspalum **yield**				
kg DM/ha	23.3	100.1	175.8	829.2
% of total yield	1.6	4.0	4.8	18.5
Veld yield				
kg DM/ha	1 382.0	1 675.3	2 430.4	2 191.1
% of total yield	94.9	67.2	66.2	49.0
Total yield				
kg DM/ha	1 456.1	2 494.5	3 672.6	4 475.4

Implements which scarify the soil also cause damage to the existing vegetation. Thus such treatment may be applied not only to improve seed placement and, therefore, germination, but also to reduce competition from the veld plants. A variety of disc, tine or rotary implements may be used for this purpose. Edwards (1970) compared five different methods of soil disturbance and vegetation removal on the establishment of clover and paspalum in veld (Table 8.5). Complete rotovation, wide strip rotovation and narrow strip rotovation removed plants and tilled the soil on 100%, 72% and 33% of the land area, and resulted in relative dry matter production from introduced plants of 100%, 46% and 25%, respectively. Cultivating first with a disc and then with a heavy tine implement resulted in a relative dry matter yield of 59%, while the figure for discing alone was 35%.

Table 8.5 The influence of various degrees of soil disturbance applied to veld before overseeding in January, on the dry matter yield of oversown *Trifolium repens and Trifolium pratense* (clover) and *Paspalum dilatatum* (paspalum) during part of the following season (after Edwards 1970).

Soil treatment	Clover yield (kg DM/ha)	Paspalum yield (kg DM/ha)
Disced once with open discs	306	51
Disced and tilled with heavy tine	484	105
Rotovated in strips 15 cm wide × 30 cm apart	210	42
Rotovated in strips 76 cm wide × 30 cm apart	371	90
Complete rotovation	788	218

The application of systemic herbicides at below lethal concentrations will retard the growth of plants without killing them. Their potential usefulness in a veld reinforcement programme is therefore clear. In these situations the veld must be overseeded soon after treatment so that the herbicide must have at most a very short residual effect if the introduced seed is not to be affected. For maximum effect, the herbicide should be applied late in the growing season (January–February in the summer rainfall grassveld areas) to veld grazed in early summer so as to reduce canopy density, but it should not be applied too late for overseeding to be successful (Cross & Theron 1970). Maximum absorption and translocation of the chemical by the plant and thus greatest affectivity is obtained when herbicides are applied in overcast conditions in the late afternoon.

The sodseeder, in the process of preparing the soil to receive the seed, destroys a strip of the existing vegetation where the seed is to be placed. Thus, the row of introduced plants grows in the middle of a strip from which competing plants have been removed.

8.2.1.2.4 Reinforcement systems

The different techniques for applying seed and fertilizer and reducing competition can be used in various combinations to effect veld reinforcement. These systems vary from simple to sophisticated and from inexpensive to costly. Systems which have been tested with some success are listed below in ascending order of sophistication and expense, and are briefly described and discussed. The landscape classes referred to are those of Edwards (1978) as depicted in Fig. 8.5.

(a) Overseeding alone. This aims to introduce genetic material of high production potential into the veld (as do all other methods subsequently listed) but poor seed

placement, the unfavourable seedbed, low soil fertility, high rates of seed predation and competition from the existing veld plants limit the success achieved using
this method. It can be applied to all landscape classes.

(b) Seed fed to grazing animals. This procedure improves seed placement and nutrient
conditions for early germination but seed placement and seedbed conditions are
still not ideal. Competition from existing plants remains a problem. This system can
be applied to all landscape classes except class 9.

(c) Fertilize and overseed. Here the seed is still unfavourably placed on a poor seedbed. Competition from the existing plants and seed predation are also likely to reduce the chances of successful establishment. This system too can be applied to all
landscape classes except 9.

(d) Scarification, fertilization and overseeding. Here the seed is reasonably favourably
placed in a loosened soil. However, the relatively poor seedbed, competition from
the existing plants and seed predation are counteractive. The scarification operation cannot be applied to landscape classes 7, 8 and 9.

(e) Prior fertilization with later scarification and overseeding. In addition to providing
for reasonable seed placement on a fertile soil, fertilizing with N prior to seeding
reduces competition from the existing plants. The poor seedbed and seed predation are potential problems. Again, because of the scarification operation, this system can be applied only to landscape classes 1 to 6.

(f) Trample, fertilize and overseed. Here success is usually more assured than in the
methods already discussed, but only a limited area can normally be covered in any
one season and the seedbed is still not ideal. Only landscape classes 7 and 9 are
excluded from this system.

(g) Burn and/or herbicidal treatment, fertilize and overseed. Similarly, all the objectives of veld reinforcement are reasonably well achieved apart from the fact that
the hard seedbed is not ideal and the seed is exposed to predation. This system can
be applied to all landscape classes except class 9.

(h) Sodseeding. Here the seed is better placed than in the previous systems. Band
placement of the fertilizer makes for efficient and economical use of fertilizer,
while competition is completely removed from the immediate vicinity of the introduced plants. Only landscape classes 2 to 6 are suited to this system of reinforcement.

Irrespective of the methods used, reinforced veld presents serious management problems because of the mixture of plants which results. The veld plants are normally low-
producing and not always highly acceptable to livestock, while the introduced plants
are usually selected for high production and are usually reasonably to very palatable.
Grazing therefore tends to be very selective. Fairly intensive and sophisticated grazing
management procedures must be used if the introduced plants are to survive.

The importance of maintaining the reinforced sward must be stressed, particularly
since this has proved difficult in many areas. If for any reason the introduced plants fail
to survive, they are usually replaced by low-producing unpalatable pioneers. In this
event, the outcome is a poorer vegetation than the original veld. There are, however,
instances where initial reinforcement has been successful (Graven 1967; I'Ons 1967,
1969, 1973; Rethman & Beukes 1973; Clatworthy 1973; Theron *et al.* 1974), suggesting
that the procedure does show some promise. Murray (1975) successfully introduced
additional cocksfoot into humid high altitude grassland (rainfall of over 1 000 mm per

annum and an altitude above 1 800 m asl.) into which the grass had already invaded in areas where sheep had slept for a number of years and where fertility had accumulated. Here he further reinforced the veld by sowing additional cocksfoot seed in the autumn following a spring application of N and P fertilizer (designed to open up the sward). The result was a useful extension of the grazing season into the early spring, due to the early growth of the temperate species, and into the winter through the autumn foggaging of the cocksfoot (Murray pers. comm.).

8.2.2 Semi-arid regions   *R.H. Drewes, N.F.G. Rethman & C.H. Donaldson*

In the semi-arid regions of South Africa, radical veld improvement is seen as a means of improving the productivity and conservation status of degraded veld, rather than as a means of replacing good condition veld with introduced species more able to make effective use of the environment than the existing vegetation. Here, therefore, radical veld improvement is intimately tied to veld degradation.

Old-land sites or veld which has been very badly degraded will not normally recover by natural means even when good management, incorporating resting periods, is introduced. Here various procedures need to be implemented to assist veld recovery.

Two distinctly different situations, associated with either heavy clay and lighter sandy soils, arise over much of this area. This often gives rise to localised degradation, as for example on the potentially very productive heavy bottomland soils which are normally heavily used by animals. While such veld is relatively stable, it lacks resilience and once damaged it recovers only slowly. In contrast, the veld on sandy soils is relatively unstable and will degrade rapidly when poorly managed, probably because the grass tufts are poorly anchored in such soils. Heavy grazing therefore promotes 'sourness' by destroying the more heavily grazed sweeter grasses in the sward. The veld of sandy soils is, however, much more resilient than that of heavy clay soils and will recover rapidly if given the opportunity to do so.

Because of the considerable impact of soil type on the behaviour of the veld in these areas, it would seem necessary to separately camp off veld on sandy soils from that on the heavier ironstone soils. This should be done before any attempt is made to revegetate the veld.

Since overgrazing and the concentration of grazing on selected areas is, in most instances, responsible for the degradation of veld in this region, steps will first need to be taken to reduce grazing pressure and improve the level of grazing management if any success is to be achieved in regenerating this veld. Termite populations may also need to be controlled, since they have often contributed to veld degradation and will retard recovery. Even then, however, soil conditions on previously poorly managed areas and on old-land sites are seldom conducive to seedling establishment. There is, therefore, often a need to treat the soil surface before attempting regenerative procedures.

8.2.2.1 Suitable species for use in oversowing programmes

Species such as *Digitaria eriantha*, *Panicum maximum*, *Panicum coloratum*, *Anthephora pubescens*, *Cenchrus ciliaris* and *Chloris gayana*, all except the latter of which are indigenous to the semi-arid regions, have shown promise in overseeding operations. Smuts' fingergrass (*D. eriantha*) and blue buffalo grass (*C. ciliaris*), in particular, are able to establish readily in oversowing programmes. They both produce large quantities of seed and grow rapidly, and so offer substantial competition to other species growing in the area. Their ability to spread rapidly means that they need to be sown lightly over the degraded areas, sufficient only to allow mother populations of the plants to develop. Because their spread is dependent on seed production, it is necessary to rest oversown veld to allow the plants to seed freely from time to time.

Other indigenous grass species of the region also have the capacity to spread within degraded veld where appropriate management is applied. Included among these are *Brachiaria nigropedata, Eustachys paspaloides, T. triandra* and *Sporobolus fimbriatus.* Seed of these species is, however, not commercially available, but where it can be harvested locally it can be used to good effect in oversowing programmes. Annual grasses like *Eragrostis tef* and *Eragrostis lehmanniana* can also be usefully added to the seed mixture, either to serve as a nurse crop or, on badly eroded areas, to initially improve the habitat for the seedlings of the less hardy species. They rapidly provide a cover to the soil and may persist for a number of years if they are allowed to seed.

8.2.2.2   *Reduction of competition of existing vegetation*

Success in any overseeding programme depends, as in the humid areas, on a low level of competition from existing unwanted plants. Where such plants occupy degraded veld they will need to be removed using either chemical, mechanical or biological means, before the area is overseeded. Because of the extensive nature of livestock production in the semi-arid regions, the costs involved in any such operation will need to be carefully weighed against the potential benefits before any programme is embarked upon.

8.2.2.3   *Season in which to oversow*

Seed can be planted into dry soil in any season of the year although March and April seeding is not generally recommended because of the high probability of severe frost or extremely dry conditions in the ensuing winter months.

8.2.2.4   *Seeding techniques*

8.2.2.4.1   *Hand seeding*

The simple procedure of distributing a seed/soil mixture by hand under dead grass tufts, on old cow pats and under loose boulders, can produce good results.

8.2.2.4.2   *The walking-stick planter*

This method makes use of an instrument comprising a hollow pipe which has a specially designed pointed structure at one end which releases a small quantity of seed whenever the point makes contact with the soil. Used properly, it deposits the seed at a predetermined depth in the soil and can be effective in reseeding bare patches in the veld.

8.2.2.4.3   *Mulching*

Hay made from stands of appropriate grasses when in full seed can be fed out on bare areas in the veld. Here the seedlings are likely to benefit from the much improved seedbed provided by the residual mulch.

8.2.2.4.4   Mechanical disturbance

Where the soil surface is badly compacted, some form of soil disturbance will be needed if overseeding is to be successful. The provision of a mulch, as previously described, makes mechanical disturbance of the soil unnecessary, but this procedure is not suited to the treatment of extensive areas. Where mechanical procedures are required, the operations involved can often be combined with procedures designed to remove much of the existing vegetation. Here some of the procedures recommended for the humid grasslands (section 8.2.1.2.4) may be appropriate – such as methods (b), (d), (g – herbicidal treatment only) and (h). Also, a spring-toothed harrow fitted with small shears has proved effective in stone-free areas to break up the soil surface and sever the crowns of unwanted plants. There seems to be no advantage in working the soil to any appreciable depth with, for example, a sub-soiler. The summer rains merely reseal the furrows drawn by such an implement. The 'pit' or 'basin plough' (an implement comprising a sub-soiler with a rounded disc mounted over an oval wheel) has proved useful in that it scoops soil from the surface at regular intervals to create a series of shallow 'basins'. These collect water during rains and provide improved moisture conditions for seed germination and seedling establishment. This implement therefore both reduces runoff and promotes infiltration and water storage. It can also be used to regenerate vegetation on old roads and pathways, which otherwise often remain unvegetated for extended periods of time.

Where the soil has been worked with a toothed harrow, the area can normally be seeded with a normal arable planter. Alternatively, seed can merely be fed by hand into a delivery tube which deposits the seed in front of the rear wheels of a tractor.

Where severe erosion has taken place, it might often be advisable to institute measures to control the flow of rain water across the overseeded areas. Here stone packs or berms constructed along the contour may be useful in slowing down water movement and so increasing infiltration. Alternatively, Vetiver grass (*Vetiveria zizanioides*) can be planted along the contour to form a 'hedge' to slow down water movement.

8.2.2.5   Management of overseeded areas

8.2.2.5.1   Grazing

It is advisable to withdraw any overseeded area from grazing at least until the oversown species have produced their first seed crop. For best results, oversown areas should be grazed only during the winter period, at least until they have recovered to a reasonable condition. Where only a small proportion of a camp needs to be overseeded, a useful procedure is to brush-pack the area after seeding. The brush-pack will provide protection from grazing as well as some shade to the developing seedlings.

Where the principle of providing for long rest periods for seed production after each grazing has been applied or where winter grazing alone has been practised, the numbers of individuals of the better quality grasses have often increased dramatically and are now contributing quite substantially to the forage resources of the area.

8.2.2.5.2   Fertilizer application

The grasses typically used in overseeding mixtures respond well to high fertility so that, where necessary, their production levels can be increased by applying fertilizer. However, fertilizer addition does not always seem to be necessary for the persistence of these species, except where there is a severe deficiency of P, in particular, or where erosion has removed most of the topsoil. In areas with inherently relatively fertile soils, fertilizer application should normally be governed by the forage demand from the seeded area, and not by the need to fertilize to maintain the species in the sward. In other areas, such as in the Molopo, where the inherent fertility of the soil is low, fertilizer addition may be necessary for the persistence of at least some of the grasses likely to be used in overseeding operations. Fertilizer does not seem to be needed by *Anthephora*, whose persistence appears to decline where fertilizers are applied, or by *T. triandra*.

8.2.3   Arid regions   *C.H. Donaldson*

In the 200 mm to 350 mm rainfall zone of the Karoo region, veld reinforcement is normally targeted at either regenerating particularly degraded veld or at establishing plants which will provide a fodder reserve in times of drought, or both. According to Steynberg & De Kock (1987), for example, over 2.5 million ha of the approximately 28 million ha of veld in the Karoo region are suited to the cultivation of Old Man Saltbush (*Atriplex nummularia*), and that as many as 8.5 million sheep (about 685 000 more than at present carried in the Karoo region as a whole) could be carried on saltbush plantations alone during normal rainfall years. This species, together with American aloe (*Agave mexicana*) and Spineless cactus (*Opuntia aurantiaca*), can act as important drought fodder crops and can be used for this purpose in the reinforcement exercise. Of these, Old Man Saltbush is the most suited to direct grazing. It is, however, slow to germinate, establishes poorly when sown directly into the veld and establishment costs tend to be high.

None of the above drought fodder crops provide a complete feed to ruminants and so they are unlikely to play a major role in the general feeding strategy of these karoo areas. They may nonetheless play an important role in providing fodder during the inevitable droughts which plague these areas and they can play a potentially important role following rain in either spring or autumn by relieving grazing pressure from the veld at these times (refer also to Chapter 12 section 12.3).

A number of grasses (*A. pubescens, S. fimbriatus* and *E. curvula* var. *conferta*) have been successfully established on degraded veld in this area. However, information is still needed on their persistence under these conditions.

9 The selection of forage species

J.R. Klug & J. Arnott

CONTENTS

The most appropriate grass, legume or tree species for use as forage depends on a number of factors. There is no 'wonder' species that can meet all needs in a wide variety of habitats, and the choice of species or even cultivars within species will be guided by the environmental conditions of the area concerned, the intended use to which the forage is to be put, the time of year that it is to be utilised and the number of years for which it is to be grown. Add to this the various properties which are desirable in a good pasture species, not all of which are likely to be found in any one species. Such properties include:

(a) good germination of seed;
(b) rapid development and growth of seedlings;
(c) high yields in the first and subsequent seasons;
(d) the ability to produce highly palatable and nutritious forage;
(e) good production in a particular season of the year;
(f) suitability for grazing, hay, silage or foggage; and
(g) a strongly perennial habit (where long-term pastures are required).

There is no individual species that has all the above attributes and so it is necessary to carefully select species or cultivars that are adapted to the environment and that have as many of the properties required for any particular situation as possible.

9.1   CRITERIA USED IN CLASSIFYING FORAGE SPECIES

Different species have become adapted, over evolutionary time, to different environments and the main environmental factors to which such evolutionary development has taken place need to be considered in the choice of species. The most important of these are:

(a)   climate of the area
      1)   rainfall   – mean annual
                           – effective
                           – seasonal distribution
                           – variability
      2)   temperature – excessively high temperatures
                           – intensity and duration of frosts
      3)   light   – photoperiod
      4)   wind
(b)   topography of the site
      1)   slope
      2)   aspect
(c)   soil on the site
      1)   physical properties
      2)   chemical properties

If the environmental requirements of the different plants are known, then a knowledge of the environment of the area will immediately allow many species to be eliminated as possible candidates for that particular area.

The next step is to carefully consider the purpose for which the pasture is required. Here there are two factors which should guide one's choice – the longevity required of the pasture and the proposed method of utilisation. Concerning the first of these, pastures may be divided into three categories:

(a)   the permanent pasture;
(b)   the perennial ley for inclusion in a crop rotation; and
(c)   the annual pasture.

For the permanent pasture, attention must be focused on only those perennial plants suited to the environment which are capable of persistent vigorous growth over many years. For the ley pasture persistence for more than four years is unimportant but properties such as rapid establishment and high production in the first year, a dense root system and ease of eradication are important. Obviously only annual plants are needed for annual pastures as these are invariably quick-growing, and are generally easy to establish and to eradicate.

Thus, by considering the longevity required of the pasture, the choice of species can be narrowed and the final step in making a choice is dependent on the way in which the pasture will be used.

Permanent pastures have five common uses:

(a)   as grazing;
(b)   for the production of conserved feeds such as hay or silage, or as foggage; and
(c)   for soil protection.

The perennial ley usually has four possible methods of utilisation:

(a)   as grazing; or

(b) for the production of conserved feeds such as hay or silage or as foggage.

The extent of the root systems of different species is, however, of much interest in leys since the more extensive the root system, the greater will be its ability to improve soil structure (an important requirement of most ley pastures). Also, the susceptibility of the root systems of the pasture plants to nematode attack may be important, since ley pastures are often used to break the cycle where nematode susceptible crops are grown in the cropping phase.

In the case of annual pastures there are five possible methods of utilisation:
(a) as grazing;
(b) for the production of conserved feeds such as hay or silage, or as foggage; and
(c) for green manuring.

Whatever the longevity required of the pasture, if it is to be used for grazing the list of possible pasture plants can be narrowed still further by considering the properties which have proved advantageous in grazed swards. A good grazing grass should be:
(a) resistant to trampling;
(b) resistant to intense defoliation;
(c) palatable to stock when green;
(d) leafy;
(e) uniformly productive over a long growing season;
(f) resistant to weed competition; and
(g) nutritious.

The list of potential plants may be narrowed still further if grazing is required primarily at one particular season of the year.

Irrespective of the requirement for longevity, if the pasture is required for the production of hay, only those plants which have the following properties need be considered:
(a) a high proportion of leaf;
(b) fine hollow stems;
(c) high yields;
(d) an erect growth habit; and
(e) the forage produced should be palatable and nutritious when dry.

Again, irrespective of the duration of the pasture, if it is required for silage, desirable characteristics include:
(a) high yields at each cut;
(b) high moisture content;
(c) high carbohydrate content; and
(d) high enough nutritive value to warrant ensiling.

If soil protection is the primary objective, then the selected plant should be able to form a dense sod on the area to be protected. Strongly rhizomatous or stoloniferous grasses meet this requirement best. There are, however, different types of areas needing protection and each will require different properties of the selected plant.

The main types of areas requiring protection are:
(a) gully eroded areas – plants must be adapted to growing in poor soil and under extremes of soil moisture;
(b) contour banks and grass strips – plants must give good coverage and yet not be so aggressive as to invade adjacent cultivated areas;
(c) sand dunes – suitable plants must be adapted to sandy soils and be able to withstand high soil temperatures, extremes of moisture supply and coverage by sand;

(d) water meadows and spillways – plants must be able to withstand considerable ve-
     locities of water flowing down such meadows, periodic flooding, and submersion
     followed by dry periods;
(e) moist vlei areas – plants must be able to withstand repeated and prolonged flood-
     ing and poor drainage; and
(f) saline areas – suitable plants should be adapted to extreme salinity and to the par-
     ticular moisture conditions of the area concerned. This may vary widely from site
     to site.

The revegetation of open cast mining areas, and coal mines in particular, has become
particularly important in recent years. Here conditions may be extremely variable, both
within and between sites, and a variety of pasture species may be required for any particular site.

Where a plant is required for green manuring, annuals with the following properties
must be sought:
(a) the plant must germinate, grow and mature after one cash crop has been removed
     from the area and before the next is planted. The time will vary according to the
     crops used in the rotation;
(b) it must be capable of rapid growth;
(c) it must develop a large bulk of material in order to supply an abundance of organic
     matter to the soil (as this is the main aim of the crop); and
(d) it should preferably be a legume so as to increase the N content of the soil and to
     reduce the possible impact of a negative N period.

There are also large areas under high rainfall conditions in southern Africa where the
topography is such that the usual methods of pasture establishment are impractical and
here species must be selected not only for their good grazing characteristics, but also for
ease of establishment. The most important features here are:
(a) the seed must be capable of germinating and establishing in a rough seedbed.
     Large seeds are often better suited than small seeds to such conditions;
(b) seed must be suited to drilling and hence smooth-coated (glabrous) seeds are pre-
     ferable;
(c) plants must have a high competitive ability so as to establish in competition with
     other seedlings and be able to spread out from the planted rows; and
(d) they should preferably be N-fixers, i.e. legumes.

If this process of selection is followed on the basis of environment, required longevity,
method of use or purpose of the pasture, and time of utilisation, it should be possible to
eliminate all but a few species with the required characteristics. Those remaining
should then, if possible, be compared in test plots and the final selection based on their
relative performance.

9.2 CLASSIFICATION OF SPECIES

To select the appropriate species for a particular site and a particular purpose according
to the method outlined above, the various properties and requirements of the many
possible forage plants in existence need to be known. The most important of these will
be listed according to their suitability in different environments, their growth characteristics and the method of utilisation to which they are best suited. Seeding rates and es-
tablishment-season guidelines are also given.

In section 9.1 the main environmental factors were classified under climatic, topo-graphic, and soil (edaphic) factors. Of the climatic factors, the one for which data are most readily available is mean annual precipitation (MAP). However the efficiency with which the plant is able to make use of what rain does fall varies because it is influenced by a large number of additional environmental factors. Therefore, effective rainfall is far more important than total rainfall, but unfortunately no regional classification of rainfall effectiveness in South Africa has as yet been undertaken. Total rainfall data must there-fore be adjusted according to the factors which determine its effectiveness.

These factors include:
(a)　distribution throughout the year;
(b)　form in which the rain falls, i.e. rainfall intensity;
(c)　soil and air temperatures and thus evapotranspiration;
(d)　the aspect of the slope and thus evapotranspiration; and
(e)　the physical properties of the soil and hence runoff, infiltration and water-holding capacity.

It is usually accepted that high quality pastures can be produced only where the mean annual precipitation exceeds about 750 mm per annum. Good pastures can, however, also be grown in areas with a rainfall of between 600 mm and 750 mm, particularly where environmental factors are such that the rain which falls is highly effective. There-fore special consideration must be given to those environmental factors which affect rainfall effectiveness. Where MAP is less than 600 mm, pasture should be grown only on soils with a favourable moisture storage capacity and then planting should be in 1 m to 1.5 m rows. There are few forage crops that do well under these conditions and in most cases all other environmental factors must be ideal so as to maximise the effectiveness of rainfall. The above statements are generalisations, but these limits will be used in the tables which follow.

Other environmental factors, such as temperature, should also be taken into consid-eration. As a general rule, temperatures decline with an increase in both altitude and latitude, but they may be greatly influenced by local conditions such as aspect, slope and prevailing winds. The subdivision of the country into temperature zones is, there-fore, not particularly meaningful. The first difficulty is that of deciding which character-istic to use – daily, monthly or annual, maximum, minimum, mean or range. The ex-tremes of temperature rather than the averages are the more important. In Table 9.1 wet hot and wet cold areas have been treated separately.

If good local climatic data are not available, an extremely useful source of data is the *South African Atlas of Agrohydrology and -Climatology* (Schulze 1997).

Light is not usually considered an important factor in the selection of pasture species except where photoperiodism is found to affect seed production. Wind likewise is not a very important factor except where it influences the effectiveness of rainfall and the choice of species on sand dunes.

As far as topographic factors are concerned, there are good topographical maps of South Africa but it is not possible to classify forage plants according to either slope or aspect as these indirectly influence the growth of plants through their effect on moisture and temperature. Maps showing the physical properties of soils are also available for South Africa, but since the treatment applied to a soil may affect its physical properties, such as the amount of moisture it can hold, these are also not useful as a basis for spe-cies selection. The same can be said of a knowledge of the soils' chemical properties since these can be modified by fertilizer applications to give favourable conditions for plant growth even where soils are inherently infertile.

Therefore, despite its limitation, mean annual rainfall is currently the only factor of the environment that can generally be used in the selection of suitable species on a regional basis. Provided the other factors are borne in mind, and their influence on the effectiveness of the rainfall is appreciated, this classification can be extremely useful.

The species best suited to high summer rainfall (>1 000 mm per annum), cool inland and warm coastal areas are listed in Table 9.1, together with the longevity and growth form, the most appropriate method of utilisation, and the establishment season of each species. The high rainfall cool inland regions include the high altitude regions of the Drakensberg mountains in KwaZulu-Natal, the Gauteng Province and parts of Mpumalanga and parts of the Northern Province, while the warm coastal regions include the coastal belt of KwaZulu-Natal and the Eastern Cape coast.

The forage species best adapted to the 750 mm to 1 000 mm summer rainfall area are listed in Table 9.2. This region comprises most of the high-lying fire-climax grassland regions east and west of the Drakensberg range in Mpumalanga, KwaZulu-Natal and the Eastern Cape, and includes the subcoastal mistbelt of KwaZulu-Natal. Here summer temperatures are generally moderate but winter temperatures tend to be low. Cultivated pastures are likely to play an ever-increasing role in the livestock system in this region.

Species best adapted to the 600 mm to 750 mm summer rainfall region, which comprises much of the Free State and parts of the North West Province, Gauteng, the Northern Province and Mpumalanga, KwaZulu-Natal and the Eastern Cape, are listed in Table 9.3. This region lies at a lower altitude and experiences warmer temperatures than the former region, so that the cool-season grasses and legumes are less important. Here successful pasture species are largely those of tropical and subtropical origin.

In areas with a summer rainfall below 600 mm per annum some drought resistant summer-growing grasses can be used relatively successfully. In the 400 mm to 600 mm summer rainfall areas, which comprise part of the bushveld of the North West Province, the Northern Province, Mpumalanga, the western Free State and the Northern and Eastern Cape, species listed in Table 9.4 may be moderately productive, while in the still drier regions of the semi-arid Karoo, where rainfall is less than 400 mm, the species listed in Table 9.5 may be planted, particularly on denuded areas which require special treatment or in the development of a fodder reserve for periods of drought.

In the winter rainfall areas of the Western Cape, the dry summer period is an exceedingly difficult one for any perennial pasture plant. For this reason, annual cool-season plants are particularly important, and the perennial plants that are used must be capable of withstanding the dry summer period, except under irrigation. In general, summer feed must be fed as foggage from spring grown material, so that an ability to cure well is important. Also, many of the pasture plants used in this region have the ability to produce large quantities of edible seed in the spring, and it is this seed, rather than the herbage material, which serves as a useful source of food for livestock in the summer.

The species best adapted to areas which experience a winter rainfall of more than 500 mm per annum are listed in Table 9.6, and those suited to areas with a rainfall of less than 400 mm per annum in Table 9.7.

Table 9.1 Forage species suited to humid regions which experience a summer rainfall of more than 1 000 mm per annum.

| Botanic name | Common name | Growth characteristics | | Method of utilisation | | | | | | | Establishment Season |
		Life[1]	Form[2]	Hay	Silage	Foggage	Spring	Summer	Autumn	Winter	Months
Cool inland regions											
Grasses											
Acroceras macrum	Nile grass	P	S,R	2	2	1		1	1		Oct–Feb
Bromus wildenowii	Rescue grass	B	T	2		1	1		1	2	Feb–Mar
Dactylis glomerata	Cocksfoot	P	T	2	2	1	1		1		Feb–Mar
Echinochloa crusgalli	Japanese millet	A	T		2			2	3		Oct
Eragrostis curvula	Weeping lovegrass	P	T	1				2			Feb
Festuca arundinacea	Tall fescue	A	T	3	2	1	1	3	1	3	Feb–Mar
Festuca X Lolium	Festulolium	A	T	2	1		1		1	1	Feb–Mar
Lolium multiflorum	Italian ryegrass types	A	T	2	1		1	3	1	1	Aug, Feb–Mar
Lolium multiflorum	Westerwolds ryegrass types	B	T	2	1		1		1	1	Feb–Mar
Lolium perenne	Perennial ryegrass	B	T	2	1	1	1		1	2	Feb–Mar
Lolium boucheanum	Hybrid ryegrass	P	T		2		2	2	2	2	Feb–Mar
Paspalum dilatatum	Dallis	P	T		3	3	2	2	2		Feb
Paspalum notatum	Bahia	P	S	3	2	2	2	2	2		Feb
Pennisetum clandestinum	Kikuyu	A	S,R		2	2		1	2		Oct–Feb
Pennisetum glaucum	Babala or pearl millet	A	T					2	2		Oct
Sorghum hybrids	Forage sorghum	A	T		2			2	2		Nov
Root crops											
Beta vulgaris	Fodder beet	A							1	1	Jan–Mar
Brassica napus	Fodder rape	A							1	1	Jan–Mar
Brassica rapa	Fodder turnip	A							1	1	Jan–Mar
Raphanus sativus	Fodder radish	A							1	1	Jan–Mar
Legumes											
Coronilla varia	Crown vetch	P	S				1	1	1	1	Feb–Apr
Lotus corniculatus	Birdsfoot trefoil	P	S			1	1	1	1	1	Feb–Apr
Lotus pendunculatus	Big trefoil	P	S			1	1	1			Feb–Apr
Lupinus alba	White lupin	A	T		2						Feb–Mar
Lupinus luteus	Yellow lupin	A	T		2	1					Feb–Mar
Trifolium repens	White clover	P	S		2	1	1	1	1	3	Feb–Mar
Trifolium pratense	Red clover	B	T	3	2	1	1	1	1	3	Feb–Mar
Trifolium versiculosum	Arrowleaf clover	A	T	3	2	1	1	1		3	Feb–Mar
Cereals											
Avena sativa	Oats	A	T	2	2				1	1	Jan–Apr
Secale cereale	Rye	A	T	2	2				1	1	Jan–Apr
Tritico secale	Triticale	A	T	2	2				1	1	Jan–Apr

Table 9.1   *Continued*

Botanic name	Common name	Growth characteristics Life[1]	Growth characteristics Form[2]	Conserved Hay	Conserved Silage	Conserved Foggage	Grazing Spring	Grazing Summer	Grazing Autumn	Grazing Winter	Establishment Season Months
Warm coastal											
Grasses											
Acroceras macrum	Nile grass	P	S,R	2	2	1		1	1		Oct–Feb
Chloris gayana	Rhodes grass	P	S	1			2	2	2		Feb–Mar
Cynodon nlemfuensis	Star grass	P	S	3			2	2	3		Oct–Feb
Cynodon hybrid	Bermuda X	P	S	2		2	2	2	2		Oct–Feb
Panicum maximum	Guinea grass	P	T	2	2	3	2	2	2		Feb
Panicum coloratum	Small buffalo grass	P	T,S	2	2	3	2	2	2		Feb
Paspalum dilatatum	Dallis	P	T		2	2	2	2	2		Feb
Paspalum notatum	Bahia	P	S		3	2	2	1	2		Feb
Pennisetum clandestinum	Kikuyu	P	S,R	3	2	2	2	2	2		Feb
Pennisetum glaucum	Babala or pearl millet	A	T		2			2	2		Oct
Legumes											
Desmodium intortum	Greenleaf	P	S	2	2			1	2		Jan–Feb
Desmodium uncinatum	Silverleaf	P	S	2	2			1	2		Jan–Feb
Glycine max	Soybean	A	T	1	1			1	2		Nov
Lablab purpureus	Dolichos	B	S		2			1			Nov
Leucaena leucocephala	Leucaena	P	(Shrub)				1	1	1		Oct–Feb
Lotononis bainesii	Lotononis	P	S				1	1	2		Jan–Feb
Macroptilium	Siratro	P	S		2			1			Oct–Jan
Macuna pruriens	Velvet bean	A	S		2						Nov
Neontonia wightii	Glycine	B	S		2			1			Oct–Jan

Longevity[1]:   A = Annual   Growth Form[2]:   R = Rhizomatous   Usage rating: 1 = Highly suited
         B = Biennial   S = Stoloniferous   2 = Moderately suitable
         P = Perennial   T = Tufted   3 = Acceptable

Table 9.2 Forage species suited to regions which experience a summer rainfall of between 750 and 1 000 mm per annum.

| Botanic name | Common name | Growth characteristics | | Method of utilisation | | | | | | | Establishment Season |
| | | | | Conserved | | | Grazing | | | | |
		Life¹	Form²	Hay	Silage	Foggage	Spring	Summer	Autumn	Winter	Months
Grasses											
Acroceras macrum	Nile grass	P	R	2	2	1		1	1		Oct–Feb
Bromus wildenowii	Rescue grass	B	T	2		1	1		1	2	Feb–Mar
Chloris gayana	Rhodes grass	P	S	1			2	2	2		Feb–Mar
Cynodon nlemfuensis	Star grass	P	S	3			2	2	3		Oct–Feb
Cynodon hybrid	Bermuda X	P	S	2	2	2	2	2	2		Oct–Feb
Dactylis glomerata	Cockstoot	P	T	2	2	2	1	2	1		Feb–Mar
Digitaria eriantha	Smuts' fingergrass	P	T	2	2	1		2	2	3	Feb–Mar
Echinochloa crusgalli	Japanese millet	A	T	3				1			Oct
Eragrostis curvula	Weeping lovegrass	P	T	1			1	2			Feb–Mar
Eragrostis tef	Teff	A	T	1							Nov
Festuca arundinacea	Tall fescue	P	T	3	2	1	1	2	1	3	Feb–Mar
Lolium multiflorum	Italian ryegrass types	A	T	2	1		1	3	1	1	Aug, Feb–Mar
Lolium multiflorum	Westerwolds ryegrass types	A	T	2	1		1		1	1	Feb–Mar
Lolium perenne	Perennial ryegrass	B	T	2	1		1	3	1	2	Feb–Mar
Lolium boucheanum	Hybrid ryegrass	B	T	2	1	1	1	3	1	2	Feb–Mar
Panicum maximum	Guinea grass	P	T		2		2	2	2		Feb–Mar
Panicum coloratum	Small buffalo grass	P	T,S		2		2	2	2		Feb–Mar
Paspalum dilatatum	Dallis	P	T		2		2	2	2		Feb–Mar
Paspalum notatum	Bahia	P	S	3	3	2		1	2		Feb–Mar
Pennisetum clandestinum	Kikuyu	P	R		3	2	2	1	2		Feb–Mar
Pennisetum glaucum	Babala or pearl millet	A	T		1			2	2		Oct
Sorghum hybrids		A	T		1			2	2		Nov
Sorghum versicolor	Forage sorghum	A	T		1			2	2		Nov
Sorghum sudanense	Sudan grass	B	T	3	2		2	2	2		Nov
Root crops											
Beta vulgaris	Fodder beet	A									Jan–Mar
Brassica napus	Fodder rape	A	T						1	1	Jan–Feb
Brassica rapa	Fodder turnip	A	T						1	1	Jan–Mar
Raphanus sativus	Fodder radish	B	T						1	1	Jan–Mar
Legumes											
Desmodium intortum	Greenleaf	P	S	2	2		1	1	2		Jan–Feb
Desmodium uncinatum	Silverleaf	P	S	2	2		1	1	2		Jan–Feb
Glycine max	Soybean	A	T	2	1				2		Nov
Leucaena leucocephala	Leucaena	P	(Shrub)				1	1	1		Oct–Feb
Lotus corniculatus	Birdsfoot trefoil	P	S				1	1	1	2	Feb–Apr
Lotus pedunculatus	Big trefoil	P	S				1	1	1	2	Feb–Apr

Table 9.2 *Continued*

Botanic name	Common name	Growth characteristics Life[1]	Form[2]	Method of utilisation — Conserved: Hay	Silage	Foggage	Grazing: Spring	Summer	Autumn	Winter	Establishment Season Months
Medicago sativa	Lucerne – hay type	P	T	1	2		1	1	1	3	Feb–Apr
Neontonia wightii	Glycine	B	S		2			1	2		Oct–Jan
Trifolium repens	White clover	P	S		2	2	1	1	1	3	Feb–Mar
Trifolium pratense	Red clover	B	T		2	2	1	1	1	3	Feb–Mar
Trifolium versiculosum	Arrowleaf clover	A	T	2	2	2	1	1	1	3	Feb–Mar
Vicia sativa	Hairy vetch	A	S	2		2	2	3	2	3	Feb–Mar
Vicia villosa	Wolly pod vetch	A	S	2		2	2	3	2	3	Feb–Mar
Cereals											
Avena sativa	Oats	A	T	2	2				1	1	Jan–Apr
Secale cereale	Rye	A	T	2	2				1	1	Jan–Apr
Tritico secale	Triticale	A	T	2	2				1	1	Jan–Apr

Longevity[1]:   A = Annual
         B = Biennial
         P = Perennial

Growth Form[2]:   R = Rhizomatous
           S = Stoloniferous
           T = Tufted

Usage rating: 1 = Highly suited
           2 = Moderately suitable
           3 = Acceptable

Table 9.3 Forage species suited to regions which experience a summer rainfall of between 600 and 750 mm per annum.

Botanic name	Common name	Growth characteristics		Method of utilisation							Establishment Season
		Life¹	Form²	Conserved			Grazing				Months
				Hay	Silage	Foggage	Spring	Summer	Autumn	Winter	
Grasses											
Cenchrus ciliaris	Blue buffalo grass	P	T,R	2			2	2	2		Nov–Mar
Chloris gayana	Rhodes grass	P	S	1			2	2	2		Nov–Mar
Cynodon dactylon	Couch or Bermuda	P	S	2	2	2	2	1	2		Oct–Feb
Cynodon nlemfuensis	Star grass	P	S	3			2	1	3		Oct–Feb
Cynodon hybrid	Bermuda X	P	S	2	2		2	1	2		Oct–Feb
Digitaria eriantha	Smuts' fingergrass	P	T	2	2	2	2	1	2		Feb–Mar
Eragrostis curvula	Weeping lovegrass	P	T	1		1	1	2			Feb–Mar
Eragrostis tef	Teff	A	T	1							Nov
Panicum maximum	Guinea grass	P	T	3	2	2	2	1	2		Feb–Mar
Panicum coloratum	Small buffalo grass	P	T,S	3	2	2	2	1	2		Feb–Mar
Pennisetum glaucum	Babala or pearl millet	A	T		2			2	2		Oct
Sorghum hybrids		A	T		2			2	2		Nov
Sorghum versicolor	Forage sorghum	A	T		2			2	2		Nov
Sorghum sudanense	Sudan grass	B	T	3	2		2	2	2		Nov
Legumes											
Lespedeza cuneata	Poor man's lucerne	P	T	1			3	3	3		Feb–Mar
Leucaena leucocephala	Leucaena	P	(Shrub)				2	2	2		Oct–Feb
Medicago sativa	Lucerne	P	T	1	2		1	1	1	3	Feb–Apr
Trifolium versiculosum	Arrowleaf clover	A	T	1	2		1			3	Feb–Mar
Vigna unguiculata var. sinensis	Cowpea	A	T	2	1	1		2	2		Nov

Longevity¹: A = Annual
B = Biennial
P = Perennial

Growth Form²: R = Rhizomatous
S = Stoloniferous
T = Tufted

Usage rating: 1 = Highly suited
2 = Moderately suitable
3 = Acceptable

Table 9.4   Forage species suited to regions which experience a summer rainfall of between 400 and 600 mm per annum.

Botanic name	Common name	Growth characteristics		Method of utilisation							Establishment Season
		Life¹	Form²	Conserved			Grazing				Season
				Hay	Silage	Foggage	Spring	Summer	Autumn	Winter	Months
Grasses											
Anthephora pubescens	Wool grass	P	T	2			1	1	2		Nov–Jan
Brachiaria spp.	Signal grass	P	S				2	2	2		Nov–Jan
Cenchrus ciliaris	Blue buffalo grass	P	T,R	2			2	2	2		Nov–Jan
Chloris gayana	Rhodes grass	P	S	1			2	2	2		Nov–Jan
Cynodon nlemfuensis	Star grass	P	S	3			2	2	3		Oct–Feb
Cynodon hybrid	Bermuda X	P	S	2	2	3	2	1	2		Oct–Feb
Digitaria eriantha	Smuts' fingergrass	P	T	2	2	1	2	1	2		Feb–Mar
Panicum maximum	Guinea grass	P	T	3	2	2	2	1	2		Feb–Mar
Panicum coloratum	Small buffalo grass	P	T,S	3	2	2	2	1	2		Feb–Mar
Pennisetum glaucum	Babala or pearl millet	A	T		2				2		Oct
Sorghum almum	Columbus grass	B	T,R	3	2		2	2	2		Nov
Sorghum sudanense	Sudan grass	B	T	3	2		2	2	2		Nov
Sorghum hybrids	Forage sorghums	A			2			2	2		Nov
Legumes											
Stylosanthes guianensis	Finestem stylo	P	T	1		2	2	2	1		Feb–Mar
Stylosanthes humilis	Townsville lucerne	A	T	1		2	2	2	1		Feb–Mar
Vigna unguiculata var. sinensis	Cowpea	A	T	2	1			2	1		Nov

Longevity¹: A = Annual   Growth Form²: R = Rhizomatous   Usage rating: 1 = Highly suited
         B = Biennial   S = Stoloniferous   2 = Moderately suitable
         P = Perennial   T = Tufted   3 = Acceptable

Table 9.5 Forage species suited to regions which experience a summer rainfall of less than 400 mm per annum.

Botanic name	Common name	Growth characteristics		Method of utilisation							Establishment Season
				Conserved			Grazing				
		Life¹	*Form²*	*Hay*	*Silage*	*Foggage*	*Spring*	*Summer*	*Autumn*	*Winter*	*Months*
Grasses											
Anthephora pubescens	Wool grass	P	T	2			1	1	2	2	Nov–Jan
Cenchrus ciliaris	Blue buffalo grass	P	T,R	2			2	2	2	3	Nov–Jan
Trees and shrubs											
Agave mexicana	American aloe	P					2	2	2	2	Sept–Oct
Atriplex mueleri	Australian saltbush	P					2	2	2	3	Aug–Sept
Atriplex nummularia	Old Man Saltbush	P					2	2	2	3	Aug–Sept
Atriplex semibaccata	Creeping saltbush	P					2	2	2	3	Aug–Sept
Opuntia spp.	Spineless cactus	P					2	2	2	2	Sept–Oct

Longevity¹: A = Annual
  B = Biennial
  P = Perennial

Growth Form²: R = Rhizomatous
  S = Stoloniferous
  T = Tufted

Usage rating: 1 = Highly suited
  2 = Moderately suitable
  3 = Acceptable

Table 9.6   Forage species suited to regions which experience a winter rainfall of more than 500 mm per annum.

Botanic name	Common name	Growth characteristics		Method of utilisation — Conserved			Method of utilisation — Grazing				Establishment Season
		Life¹	Form²	Hay	Silage	Foggage	Spring	Summer	Autumn	Winter	Months
Grasses											
Dactylis glomerata	Cockspoot	P	T	2	2		1	3	1	2	Apr–May
Festuca arundinacea	Tall fescue	P	T	3	2		1	3	1	2	Apr–May
Festuca X Lolium	Festulolium	A	T	2	1				1	1	Apr–May
Lolium multiflorum	Italian ryegrass types	A	T	2	1		1		1	1	Apr–May
Lolium multiflorum	Westerwolds ryegrass types	A	T	2	1		1		1	1	Apr–May
Lolium perenne	Perennial ryegrass	B	T	2	1		1	2	1	2	Apr–May
Lolium rigidum	Wimmera ryegrass	A	T	2	1		1		1	1	Apr–May
Phalaris tuberosa	Harding grass	P	T	2	2		2	2	2	3	Apr–May
Cereals											
Avena sativa	Oats	A	T	2	2		1		1	1	Apr–May
Hordeum vulgare	Barley	A	T	2	2		1		1	1	Apr–May
Secale cereale	Rye	A	T	2	2		1		1	1	Apr–May
Tritico secale	Triticale	A	T	2	2		1		1	1	Apr–May
Legumes											
Lespedeza cuneata	Poor man's lucerne	P	T	1			2	2	2		Apr–May
Lotus hispidus	Boyd's clover	A	T				1	2			Apr–May
Lupinus angustifolius	Narrow leafed lupin	A	T		1		1	3	2	1	Apr–May
Lupinus luteus	Yellow lupin	A	T				1	3		1	Apr–May
Lupinus albus	White lupin	A	T		1		1	3		1	Apr–May
Medicago sativa	Lucerne	P	T	1	2		1	1	1	1	Apr–June
Medicago species	All medics	A	T,S				1	2	1	1	Apr–June
Ornithopus sativus	Seradella	A	T				1	2	1	1	Apr–June
Tagasaste	Tree lucerne	P	T				1	1	1	1	Apr–June
Trifolium repens	White clover	P	S				1		1	2	Apr–June
Trifolium pratense	Red clover	B	T				1	1	1		Apr–June
Trifolium subterraneum	Subterranean clover	A	T				1		1	1	Apr–June
Trifolium species	All annual clovers	A	T,S				1	2	1	1	Apr–June

Longevity¹: A = Annual   Growth Form²: R = Rhizomatous   Usage rating: 1 = Highly suited
        B = Biennial   S = Stoloniferous   2 = Moderately suitable
        P = Perennial   T = Tufted   3 = Acceptable

Table 9.7 Forage species suited to regions which experience a winter rainfall of less than 500 mm per annum.

Botanic name	Common name	Growth characteristics		Method of utilisation							Establishment Season
		Life¹	Form²	Conserved			Grazing				
				Hay	Silage	Foggage	Spring	Summer	Autumn	Winter	Months
Grasses											
Lolium multiflorum	Italian ryegrass types	A	T	2	1		1		1	1	Apr–June
Lolium multiflorum	Westerwolds ryegrass types	A	T	2	1		1		1	1	Apr–June
Lolium rigidum	Wimmera ryegrass	A	T	2	1		1		1	1	Apr–June
Phalaris tuberosa	Harding grass	P	T	2	2		2	2	2	3	Apr–June
Legumes											
Lupinus angustifolius	Narrow leafed lupin	A	T		1		1	3	1		Apr–June
Medicago littoralis	Harbinger medic	A	T				1	2	1		Apr–June
Medicago truncatula	Barrel medic	A	T				1	2	1		Apr–June
Ornithopus sativus	Seradella	A	T				1	2	1		Apr–June
Tagasaste	Tree lucerne	P	T				1	1	1	1	Apr–June
Trifolium subterraneum	Subterranaean clover (early cultivars)	A	T				1	2	1		Apr–June
Vicia sativa	Hairy vetch	A	S	2			2	3	2		Apr–June

Longevity¹: A = Annual
B = Biennial
P = Perennial

Growth Form²: R = Rhizomatous
S = Stoloniferous
T = Tufted

Usage rating: 1 = Highly suited
2 = Moderately suitable
3 = Acceptable

Table 9.8   Species of grasses and legumes used for special purposes.

Botanic name	Common name	Gully eroded areas	Contour banks and strips	Sand dune reclamation	Water ways	Leys	Moist vleis	Green manure	Saline sites	Open cast mines	Fine turf	Coarse turf
Acroceras macrum	Nile grass						x					
Agrostis tenuis	Bent										x	
Ammophila arenaria	Marram grass			x								
Atriplex semibaccata	Creeping saltbush								x			
Axonopus compressus	American carpet grass											x
Brachiaria latifolia	Tanner grass				x		x					
Chloris gayana	Rhodes grass		x			x				x		
Coronilla varia	Crown vetch	x										
Crotalaria	Sunhemp							x				
Cynodon spp.	Kweek or Bermuda	x			x					x		x
Dactyloctenium australe	Durban or Berea grass			x								x
Digitaria didactyla	Mauritius, MacIntosh										x	
Digitaria diversinervis	Richmond fingergrass										x	
Digitaria swazilandensis	Swaziland fingergrass										x	
Eragrostis curvula	Weeping lovegrass	x	x			x				x		
Festuca arundinacea	Tall fescue				x		x					x
Festuca ovina	Hard fescue										x	
Festuca pratense	Meadow fescue						x					
Festuca rubra	Creeping red fescue											
Hyparrhenia hirta	Common thatchgrass	x										
Lolium perenne	Perennial ryegrass					x	x			x	x	x
Lotus corniculatus	Birdsfoot trefoil											
Lupinus spp.	Lupins	x						x				
Melilotus alba	Sweet clover	x						x				
Panicum repens	Torpedograss		x							x		
Paspalum dilatatum	Dallis				x		x					
Paspalum distichum	Buffaloquick paspalum						x					
Paspalum notatum	Bahia				x		x					x
Paspalum vaginatum	Saltgrass				x		x		x			
Pennisetum clandestinum	Kikuyu	x			x							x
Pennisetum purpureum	Napier fodder or Bana											
Phragmites australis	Common reed	x		x	x		x					
Poa aquatica	Reed sweetgrass						x					
Poa pratensis	Kentucky bluegrass						x				x	
Stenotaphrum secundatum	Coastal buffalo grass											x
Trifolium fragiferum	Strawberry clover						x					
Vetiveria zizanoides	Vetiver	x	x		x					x		

Within all the regions listed above, growing conditions for pasture plants may be improved materially by irrigation. Where this is possible, productive pastures can usually be produced even in the most arid regions. Among the grasses, kikuyu (*Pennisetum clandestinum*), cocksfoot (*Dactylis glomerata*), tall fescue (*Festuca arundinacea*) and the ryegrasses (*Lolium* spp.) respond particularly well to irrigation, as do lucerne (*Medicago sativa*) and red (*Trifolium pratense*) and wild white clover (*Trifolium repens*) among the legumes.

Pasture species suited to special purposes are listed in Table 9.8. The species shown are adapted to the environmental and soil conditions which prevail at such sites, and they are well suited to achieve the most important objectives of establishing a cover in these specialised sites.

10 Establishment of pastures

CONTENTS

10.1 ESTABLISHMENT OF PASTURES IN HUMID REGIONS   *P.E. Bartholomew*

Using machinery, overseeding, fertilizers and lime, graziers have attempted to create vegetation types considered more productive than the natural vegetation which it replaces. Cultivated pastures represent the most drastic intervention in this process. Here the veld is totally replaced with selected species, usually grasses or legumes, or a mixture of both. The introduced species can be used for grazing, hay, silage and/or foggage. Between the two extremes of veld and cultivated pastures there are 'grey areas' involving different degrees of modification or replacement of veld (refer to Chapter 8). The present chapter is concerned with the establishment of the three main types of pastures (permanent, ley and annual) from both seed and vegetative material. Irrespective of the type of pasture being established, the basic principles and procedures for establishment remain essentially the same.

Cultivated pastures require reasonably good soils and favourable moisture conditions. Since many pasture species are shallow rooted, deep soil profiles are not as important as in most annual field crops, exceptions being species such as *Medicago sativa* that require deep, well drained soils. Reasonably good physical and chemical properties of soil are, however, needed for the successful production of cultivated pastures. Compared to many field crops, the permanent cover offered by pastures allows them to be grown on relatively steep slopes. Indeed, well managed pastures of *Pennisetum clandestinum* and *Cynodon* species often offer better soil protection and promote better water penetration on steep slopes than does veld.

High input costs associated with establishing cultivated pastures necessitate the careful planning and execution of planting operations. Particular attention should be given to the site, soil, aspect, slope and landscape position with respect to the species to be established (refer to Chapters 8 & 9). In addition, cognisance must be taken of legislation relating to the utilisation of various classes of land for pastures.

10.1.1 Legislation

As arable land becomes increasingly limited, there is more and more incentive for farmers to intensify by establishing cultivated pastures. However, the modification of veld, vlei, wetland and forest communities for intensive agricultural use has become an emotional issue with farmers and environmentalists. There is, however, legislation governing the use of these natural resource areas. The Conservation of Agricultural Resources Act, 1983 (Act 43 of 1983) makes provision for the protection of the natural resources. It requires that the land be adequately protected against excessive soil loss arising from '. . . erosion through the action of water . . .' and '. . . erosion through the action of wind . . .' Furthermore, legislation provides for protection '. . . against waterlogging and salination of irrigated land'. There is legislation relating to the 'utilisation and protection of vleis, marshes, water sponges and water courses'. Therefore, before any area of natural vegetation can be cultivated, permission must be obtained from the extension services of the Department of Agriculture.

Once an area has been identified as suitable for cultivated pasture, the potential of the site needs to be assessed, the appropriate species selected, the seedbed prepared and the pasture planted.

10.1.2 Site selection

Until fairly recently little attention was paid to the relative value of different sites for pastures. However, both landscape position and the soil family (series) of a site play significant roles in determining the type of pasture, the pasture species to plant and the technique of establishment most appropriate to that site. The effect of site condition assumes considerable significance where a particular species is marginal, i.e. due to restricted precipitation or extremes of temperature. Edwards & Scotney (1978) have dealt extensively with this topic. They point out that the extent of the different soil groups, their production potential for pastures and their susceptibility to erosion should form the basis for determining priorities in farm development (Edwards & Scotney 1978). They grouped the more than 500 soil series of the National Soil Classification system (MacVicar *et al.* 1977) into soil groups of similar production potential for pastures. These groups are rated under dryland conditions at optimum levels of fertilization, as well as for their erodibility. Each soil class is made up of five categories. In the first category the major soil forms are grouped into 13 categories on the basis of landscape position, diagnostic topsoil horizon, dominant colour and subsoil drainage. The other categories relate to the diagnostic topsoil horizon (five categories), intensity of leaching (three categories), soil depth (three categories) and soil texture (three categories). These five categories and their respective ratings are shown in Table 10.1 where the original classification has been modified to accommodate the more recent soil classification by the Soil Classification Working Group (1991).

The potential value of a particular site, for pastures, is derived by adding the ratings for each of the five categories shown in Table 10.1. Theoretically the potential score can range from 9 to 34. Edwards & Scotney (1978) grouped the potential ratings of a site into five classes as follows:

> 30 = very high potential for pastures
27 to 30 = high potential for pastures
22 to 26 = moderate potential for pastures
17 to 21 = low potential for pastures
< 17 = very low potential for pastures

In addition to these ratings of potential, the erosion hazard of the soil should be considered when planning land development. Based on the Soil Loss Estimator for Southern Africa (SLEMSA), Edwards & Scotney (1978) give erodibility values for the different soil families (series) ranging from 1.3 (very high) to 6.9 (very low). The values should play a role in determining the sites, techniques and plant species selected, as well as influencing the conservation measures needed (Edwards & Scotney 1978).

10.1.3   Site preparation

Before attention is given to the preparation of the seedbed, consideration needs to be given to controlling excess runoff in order to reduce soil erosion and promote the infiltration of water.

It is true that a cultivated pasture, once established, is normally very effective in preventing runoff and soil erosion. However, the seedbed invariably remains vulnerable to erosion from planting until the pasture is well established and so preventative measures usually need to be taken during this period to restrict soil loss.

10.1.3.1   Runoff and soil erosion

The slope of the land is the main determinant of runoff and thus soil erosion. If the slope is such that there is any danger of water running off the land during and following rains, and of soil being washed away, then anti-erosion measures must be taken. Besides curtailing soil loss, such measures should aim to prevent seed or seedlings of the planted pasture from being washed away or buried. There are a number of procedures that can be followed to reduce runoff and so minimise soil erosion.

(a)   *Storm water drains*
Pastures established at the foot of slopes or some distance down a slope may require mechanical structures to protect them from run-on water. Contour banks or storm water drains may be needed to divert excess runoff from the pasture area.

(b)   *Contour banks*
Contour banks are commonly used to prevent runoff water from rushing down fields or pastures by subdividing long slopes into a series of shorter slopes. Contour banks should be designed to spill excess water safely into properly constructed artificial or natural waterways. Such measures are particularly important on land planted to annual fodder species, even on relatively flat land (1% slope).

Table 10.1 Derivation of soil classes for pasture production (after Edwards & Scotney 1978) modified by Manson (pers. comm.) to incorporate the new soil classification by the Soil Classification Working Group (1991).

Group	Soil description	Main characteristics	Soil form	Rating
First category (major soil groups)				
1.	Hydromorphic vlei	Bottomland, humid to sub-humid	Champagne, Katspruit, Rensburg, Willowbrook	8
2.	Youthful colluvia and alluvia	Alluvial toe slope, dry, reasonable drainage	Dundee, Inhoek	10
3.	Melanic	Generally middle and foot slope, sub-humid to dry	Bonheim, Immerpan, Mayo, Milkwood, Steendal	6
4.	Vertisols (not listed in 1 above)	Generally middle and foot slope, sub-humid to dry	Arcadia	4
5.	Poorly drained, seasonally wet at or near surface	Foot slope and depressions, sub-humid to dry	Cartref (wet), Estcourt (wet), Fernwood (wet), Klapmuts (wet), Kroonstad, Longlands, Wasbank, Westleigh	6
6.	Shallow claypan soils lacking marked wetness	Middle slope and foot slope, sub-humid to dry	Estcourt (dry), Klapmuts (dry), Sepane, Sterkspruit, Swartland, Valsrivier	2
7.	Sands with E-horizon normally devoid of marked wetness	Usually just inland of vegetated dunes on coastal landscape	Cartref (dry), Constantia, Fernwood (dry), Kinkelbos, Vilafontes	6
8.	Podsols	Winter rain, developed under Macchia, humid to sub-humid	Concordia, Groenkop, Houwhoek, Jonkersberg, Lamotte, Pinegrove, Tsitsikamma, Witfontein	7
9.	Hydromorphic soils with moderate drainage	Generally middle or foot slope, restricted drainage in lower sub-soil	Avalon, Bainsvlei, Bloemdal, Glencoe, Pinedene, Tukulu	8
10.	Well drained non-calcareous soils	Generally crest and middle slope, not hydromorphic	Clovelly, Griffin, Hutton, Oakleaf, Shortlands	10
11.	Humic soils	Well drained, humid areas	Inanda, Kranskop, Magwa, Nomanci, Sweetwater, Lusiki	10
12.	Shallow soils (not previously included)	Generally steep slopes and crests	Coega, Dresden, Glenrosa, Knersvlakte, Mispah	2
13.	Well drained calcareous and well drained Dorbank soils	Dry areas	Addo, Askham, Augrabies, Brandvlei, Etosha, Gamoep, Garies, Kimberley, Molopo, Montagu, Oudtshoorn, Plooysburg, Prieska, Trawal	6
Second category (diagnostic top soil horizon)				
1.	Organic			3
2.	Orthic			4
3.	Melanic			2
4.	Vertic			1
5.	Humic and humic phases			5
Third category (intensity of leaching)				
1.	Dystrophic			10
2.	Mesotrophic			6
3.	Eutrophic			2
Fourth category (soil depth)				
1.	Deep (>50 cm)			5
2.	Shallow (50 to 20 cm)			3
3.	Very shallow (<20 cm)			1
Fifth category (soil texture)				
1.	Sand (<15% clay in the B-horizon)			3
2.	Loam (15% to 35% clay in the B-horizon)			5
3.	Clay (>35% clay in the B-horizon)			3

The length of the slope, soil erodibility, natural drainage, soil type and rainfall (intensity and amount) dictate the need for and spacing of contour banks on land planted to pastures. They can be broad based so that the pasture can be planted on and over the banks. This is particularly important for irrigated pastures, as it allows for productive utilisation of the whole area. However, while contour banks are effective in controlling erosion on slopes of up to 15%, steeper slopes should be terraced unless planted to sod-forming pasture species.

(c) *Terracing*
Terraces are more effective than contour banks in reducing runoff, improving infiltration and controlling erosion. Well designed and well constructed terraces can be effective in controlling erosion on slopes from 15% to 50% (Anon. undated (b)). They are, however, very costly and can be recommended only where land is limiting or where it has a very high value.

Not all areas that are planted to pastures need to be contoured or terraced to contain erosion. In many instances grass strips could adequately contain soil erosion resulting from runoff.

(d) *Grass strips*
Broad strips of sod-forming grasses can be planted on the contour to contain excess runoff and promote infiltration. However, they should be used only on gentle slopes, where the soil has a high infiltration capacity and where it is unlikely that runoff volumes will be large. While such strips of grass check the rate of flow of water and, in so doing, reduce erosion and promote infiltration, excess water is not easily diverted from the land by means of these strips. They are, therefore, not generally recommended as the sole means of controlling runoff. However, the higher the infiltration capacity of the soil, the more effective such strips are likely to be.

(e) *Strip establishment*
On steep slopes that are to be established to sod-forming grasses the pasture can be established in 2 m to 3 m wide strips along the contour. Strips of veld (2 m to 6 m wide, depending on slope and soil erodibility) are left intact between the planted strips. Once the planted strips have become well established the remaining intact strips of veld are planted. Provided the planted strips are not too wide and provided the intact strips of veld are wide enough to contain erosion, there should be no need for mechanical protection when strip planting sod-forming grasses.

Excess runoff can also result from, or be aggravated by, poor infiltration or poor drainage within lower soil horizons. Attention should thus be given to alleviating problems arising from poor drainage within the soil profile.

10.1.3.2 *Impeded soil drainage*

Restricted percolation of water to lower soil horizons may result from inherent drainage problems associated with the particular soil type, from a plough-sole or plough-pan. In either case it is necessary to rectify potential waterlogging problems if pastures are to be established and if erosion is to be controlled. As a general rule, sites subject to waterlogging occur in vleis and wetlands. It is not generally recommended that these areas be drained and used for crop or pasture production.

(a) *Establishing a pasture without draining*

An alternative to draining vleis is to use a moisture-loving grass, e.g. *Acroceras macrum, Hemarthria altissima, Paspalum dilatatum* or *Festuca arundinacea* on the undrained site. Such pastures are, however, difficult to manage. To be successful, establishment practices require correct timing. Hydromorphic weeds such as sedges and rushes readily invade such pastures, while waterlogged soil makes for inefficient use of applied fertilizer and so the productivity from these areas is usually low. These wet sites are usually cold and so the growth period is restricted. In addition, poor drainage imposes limitations on utilisation. From an animal health point of view, liver fluke in particular, and stomach worms in general, can cause problems in undrained areas. Foot-rot can also cause problems in water-logged pastures. For these reasons undrained waterlogged areas are seldom used for pastures.

(b) *Natural impedance to soil water drainage*

If waterlogging is caused by a heavy clay subsoil, then the insertion of drains may be worthwhile. Either enclosed drains (such as tile drains, perforated pipe drains or mole drains) or open type drains (ditches or open furrows) may be used (Hill *et al.* 1981). The former tend to be very expensive and so open drainage systems are generally used. Of these, the 'ridge and furrow system' is well suited to pastures (Hill *et al.* 1981).

(c) *Artificial impedance to soil water drainage*

In addition to the inherent drainage problems, soils may develop waterlogging problems resulting from the formation of impervious layers due to continuous cultivation – the plough-pan or plough-sole which often results from annual ploughing to the same depth. Ripping to a depth several centimetres below the plough-sole will break the impervious layer, promote drainage and encourage the development of a good, deep, root system.

10.1.4 Preparing the seedbed

In all soils, under all conditions and for all kinds of pastures, one of the essentials for obtaining a good stand of the planted species is a properly prepared seedbed.

The methods used to prepare seedbeds vary considerably. They depend on climate, soil type, method of planting, time available from the first mechanical operation to planting and the machinery available. In spite of these variables the ultimate aim is everywhere the same – to work the soil to such a condition that the sown seed will germinate readily so that plants will establish and develop as rapidly as possible. This condition can be described as well-drained, weed free, fine, firm, moist and level (White 1973). More specifically, the various operations aim to provide:

(a) adequate moisture in the surface soil and in the subsoil. This may entail the need for moisture conservation practices (White 1973) and it may also be necessary to rip to encourage water percolation to lower soil horizons;

(b) a soil surface which is reasonably smooth and even. This facilitates even placement and covering of the seed, improved efficiency of operations such as fertilization, fencing, irrigation (uniform water penetration) and mowing;

(c) a firm, well consolidated seedbed. This promotes an even depth of planting, ensures good soil/seed contact (promoting rapid and even germination) and promotes the upward movement of soil moisture;

(d)   a good soil structure. A powdery or pulverised 'structure' can result in the development of a crust after heavy rains (White 1973). This would restrict soil aeration and the penetration of water into the soil;

(e)   the soil which is firm throughout the ploughed layer to ensure adequate support to the developing seedling and to promote close contact between the roots and the soil solution;

(f)   the ploughed layer which is in 'direct' contact with the lower soil layers to allow for the uninterrupted upward movement of soil moisture (the soil should therefore be allowed to settle before planting);

(g)   a surface which provides good conditions for seeding and other mechanical operations;

(h)   a topsoil as free of weed seeds, insect pests and diseases as possible; and

(i)   an adequate concentration of nutrients for the pasture species being planted. Of particular significance is the incorporation of the 'immobile' nutrients, P and calcium, and the correction of soil acidity (pH) before planting (see Chapter 11).

To achieve these requirements, the general sequence of operations would be to remove any overburden, rip, plough, disc, roll, harrow, roll, plant and again roll.

(a)   *Removing existing vegetation*
      Excessive amounts of overburden can be removed by grazing, burning, mowing and removing or burning the material, or, if the material is green, spraying with a herbicide (e.g. paraquat) and burning when dry.

(b)   *Ripping*
      If a ripping treatment is needed, this should be done before the area is ploughed. Ripping is usually to a depth of 30 cm to 50 cm and at a spacing of 100 cm in dry loam and clay soils and 70 cm in moist and sandy soils. Soils with a high clay content should be ripped when dry. The area should then be ploughed.

(c)   *Ploughing*
      The soil should be ploughed some time before the pasture is seeded. This allows the topsoil to consolidate, it allows time for the decomposition of plant material from the previous crop and time for weed control. The latter should extend over several weeks to reduce the weed seed population as effectively as possible and it must coincide with suitable moisture conditions which promote germination of weed seeds. Early ploughing also allows for early lime incorporation. This facilitates the mineralisation of organic matter.

      For spring planting, an autumn ploughing is advocated, provided there is adequate soil moisture, and for autumn planting land preparation should commence in the spring to early summer.

      Normal ploughing depth for most pastures is regarded as 15 cm to 25 cm but care needs to be taken not to lift subsoil to the surface during ploughing. On sandy soils an off-set disc can be used in preference to a plough.

(d)   *Discing*
      If the area needs to be limed, the lime should be incorporated with an off-set disc. Better mixing is achieved with this implement than with the plough. When a rotary plough is used and the lime is applied before ploughing, the discing operation may not be necessary.

The discing operation should produce a relatively smooth surface with only small clods remaining. A Cambridge roller should then be used to encourage the germination of as many weed seeds as possible (MacDonald 1991a).

(e) *Harrowing*

Harrowing has four objectives, viz. to control weeds, to smooth the surface, to break down clods and to facilitate the even mixing of lime and applied fertilizer with the soil. These objectives may be variously achieved using either disc type or tine (tiller) type harrows. To control weeds, these implements should be used when the weed seedlings are most vulnerable (2 to 3 weeks old or 3 cm to 7 cm tall). Two or more crops of weeds should be controlled by harrowing before planting. It is important to apply weed control measures when conditions are hot and sunny.

Harrowing will also assist in smoothing the surface but the surface should not be overworked. Too many operations can destroy the structure of the soil. Therefore herbicides can be used if continued harrowing is likely to impact too severely on soil structure. Their use is, however, costly.

(f) *Final seedbed preparation*

Final seedbed preparation involves ensuring that the soil has been adequately fertilized for the species in question, and that the seedbed is in an acceptable condition for planting.

1) *Soil fertility*

The fertilization of pastures is discussed in Chapter 11. However, since certain fertilizers, and particularly those which are immobile, should be applied as basal dressings before or during seeding, these will be briefly mentioned in relation to seedbed preparation.

If lime is required, it is best applied well before seeding. It should be well mixed into the topsoil and should therefore be applied before the land is ploughed or disced for the first time.

Some P should always be incorporated throughout the topsoil (10 cm to 15 cm) at planting irrespective of the inherent soil P level. This is generally done by discing or harrowing in a P containing fertilizer immediately before the final seedbed operation or when planting.

Potassium moves relatively freely through the soil profile and may be applied with the pre-planting phosphatic fertilizer. Since it does not become fixed in an unavailable form it can also be top-dressed following seedling establishment.

Nitrogen is readily soluble and is easily leached into and through the soil. For this reason N fertilizers should be applied as top-dressings only once the seedlings have become established (5 cm to 6 cm tall, or 2 to 4 weeks after planting). This also discourages the growth of the often faster germinating weed seedlings when N fertilizer is applied at planting.

2) *Rolling*

If the surface of the seedbed is loosely packed (fluffy), depth control at planting will be difficult and the contact between the seed and soil is likely to be poor. Such seedbeds also dry out rapidly. The final operation in preparing the seedbed is therefore to roll the land. This should provide a fine, firm, compact seedbed which facilitates a uniform planting depth and so improves the economy of seeding rates; ensures close contact between soil and seed (especially if rolled again after seeding); provides a good foundation for the seedling roots; and promotes the conservation of soil moisture. The importance of rolling is aptly summed up by Frame (1992) who commented that the final consolidation is the key element, and here the Cambridge roller with its rigid compressing rings is the preferred choice whether just prior to or following drilling or broadcasting.

10.1.5   Seeding or planting the pasture

Methods used to establish pastures from seed or vegetative material are many and varied. Only the more commonly used techniques will be discussed here.

10.1.5.1   *Seeded pastures*

Seeding requires the uniform distribution of seed over the whole area to be planted. There are two basic methods of 'sowing' the seed – broadcasting the seed evenly over the area or planting the seed in closely spaced rows (drilling).

(a) *Broadcast planting*
Broadcast procedures range from hand broadcasting to the use of machines to distribute the seed.

1) *Hand broadcasting*
This is the simplest and oldest method of sowing seed but requires skill and experience if the seed is to be sown uniformly over the area to be planted. To facilitate the even spread of seed, (i) large areas of land should be divided into smaller 'workable areas'; (ii) the amount of seed to be sown on each workable unit of land should be divided into two equal lots and each spread separately over the entire field, moving in a different direction with each sowing operation; (iii) small seed should be mixed with sand to give a more bulky parcel to work with; (iv) in the case of mixed pastures, each type of seed should be sown separately and split into two lots as outlined in (ii) above.

2) *Cyclone seeders*
These are small, hand operated seeders. With experience, uniform distribution of seed is possible using the suggestions (i) to (iv) outlined in 1).

3) *Barrow seeders*
These manually operated seeders use a seed-box mounted on a wheelbarrow type structure. The 'barrow' is pushed manually so the seedbed needs to be relatively even, firm and smooth.

4) *Fertilizer distributors*
Here the seed and fertilizer (usually phosphatic) are mixed together and both are distributed in the same operation. Since any delay following mixing could result in the seed being 'burnt' by the fertilizer, the seed needs to be sown immediately after the two have been mixed.

5) *Brillion or cultipacker seeders*

These seeders have two rollers in tandem with a broadcasting seed box mounted between them. They produce excellent results, since the seedbed is compacted before and again after planting (Decker *et al.* 1982). However, because the rollers are light it is usually necessary to roll with a Cambridge roller following the seeding operation.

6) *Thatching*

This is an old, yet simple method of establishing certain grass pastures. A stand of the pasture species to be established is cut when the plants are in full seed. The harvested material is then spread evenly over the site prepared for the pasture. The procedure is only recommended where seed of a particular species is not commercially available.

(b) *Drilling*

Here the seed is planted in rows of predetermined widths and at predetermined depths. Good drilling equipment is available today for all but fluffy seeded species, e.g. *Cenchrus ciliaris*. However, such seeds can be pelleted, and then planted with conventional machinery.

There is little doubt that for most purposes drilling is superior to broadcasting, particularly on sandy soils and in dry areas. The seed can be placed at the correct depth, there is good contact with soil moisture and the seeding rate need be only about two-thirds of that required for broadcast planting (Frame 1992). Furthermore, many drills are able to apply fertilizer below, next to, or with the seed. However, the machinery is expensive and the seeding operation slower than are some of the broadcast methods.

(c) *Rolling*

Whether broadcasting or drilling, seeding should be followed by rolling to ensure good contact between seed and soil, except where seed-drills have press wheels which provide adequate compaction, though they may not always do so. Rolling is particularly important when seed has been broadcast. It is often advisable to roll with two Cambridge rollers in tandem, particularly on sandy soils. This would eliminate the need for harrowing to cover broadcast seed. As a general guide to the effectiveness of compaction, footprints on the rolled area should leave only shallow indentations (3 mm to 4 mm) in the seedbed.

10.1.5.2 *Vegetative planting*

Many pasture species are not suited to establishment from seed and must be propagated vegetatively. Appropriate planting material includes:

(a) rooted plants or groups of tillers;
(b) plant material 'started' in speedling trays;
(c) rooted or unrooted stem cuttings; and
(d) runners (stolons or rhizomes).

Rooted plants or groups of tillers need to be planted by hand. Such planting is labour intensive, costly and slow, but establishment is usually good.

Several pasture species, e.g. coastcross II, are sold as 'rooted tillers' in speedling trays. The plants are expensive and planting is labour intensive. Speedling plants are therefore often planted out in a nursery where they are bulked up to provide the material for larger mechanically assisted plantings. Establishment with speedling plants is usually extremely successful.

Napier fodder (*Pennisetum purpureum*) provides a good example of the use of stem cuttings for establishing a pasture. Stems are cut into three to four noded lengths and each length is layered in the soil with one node left above ground for the development of shoots. The soil needs to be compacted around each cutting. Stem cuttings can also be layered in a plough furrow (see later).

The use of runners as planting material requires less exacting planting procedures than is required for tillers, speedling plants or stem cuttings. For example, kikuyu (*P. clandestinum*) and Nile grass (*A. macrum*) rhizomes and stolons can be distributed over the seedbed and lightly disced or rotovated into the soil (to a depth of 5 cm to 8 cm). Here it is essential to roll after discing or rotovating. Alternatively, runners can be planted behind a single furrow plough by leaning the material against the turned furrow slice, with some material protruding above the turned sod. The next plough slice covers the material, leaving about 5 cm uncovered (MacDonald 1991a). A double furrow plough or a potato ridger can also be used. It is not advisable to harrow to smooth the surface following planting, as the runners could be pulled out.

Successful establishment of kikuyu, Nile grass and *Cynodon dactylon* can also be achieved by laying the runners along the bottom of a plough (single or double furrow plough or potato ridger) furrow. The following plough slice covers the runners. Since they are completely buried, the area can be harrowed and rolled to obtain a smooth, firm surface. Total burial of the runners is not, however, recommended for all species. For example, coastcross II, Star grass and Napier fodder require at least one node to be exposed above the soil following planting.

Approximately 75 bags of runners, each weighing about 50 kg, are required to plant one hectare at a 1 m row spacing and one hectare of a sod-forming grass pasture should provide sufficient material to plant 10 ha of pasture.

A variety of implements is available to mechanically plant vegetative material but many are specific to certain types of material only.

Where machinery and implements cannot be used (as in steep and/rocky situations), sod-forming grasses can be successfully established in the veld by spot planting vegetative material. The veld overburden should be removed by burning after the first reasonable spring rains. Indentations are then made in the soil, small quantities of phosphatic fertilizer spread evenly across the bottom of the indentations and the fertilizer covered with a thin layer of soil. A divot of the sod-forming grass (with roots) is placed in the indentation and covered with the soil removed while making the indentation and is then compacted with the foot (MacDonald 1991a). This establishment procedure is recommended only for high rainfall areas (>750 mm).

10.1.5.3   *Depth of sowing*

The seeding depth used is usually a compromise between (a) the capacity of food reserves contained in the seed to support seedling growth (the larger the seed the larger the reserves); and (b) the better anchorage, moisture supply and nutrition associated with an increasing depth of seed placement. There is a general tendency for farmers to plant pasture seeds too deep.

While some species, such as *Digitaria eriantha*, will germinate successfully when the seed is sown directly onto the soil surface provided the soil is then compacted (as with the wheel of a tractor), the seed of most pasture species should be planted into a firm seedbed to a depth of between 0.5 cm and 1.5 cm. As a general guide seed should not be planted deeper than 2.5 times its width (MacDonald 1991b) so that the smaller the seed, the shallower it should be planted. Actual depth of seeding is also influenced by the method of seeding and by soil conditions. The looser, fluffier (less compact) and sandier the seedbed, the deeper seed placement should be. Seedlings from seed planted too deep are slow to emerge and are considerably weaker than those from shallower plantings.

With broadcasting, the depth of seeding is controlled by the compactness of the seedbed at sowing, the depth of the after-seeding harrowing and the degree of rolling following seeding. With drilling, the depth of seeding can be regulated by adjusting the furrow opening device.

10.1.5.4 Time of sowing

In considering the best time to sow, cognisance must be taken of the local climatic conditions (rainfall, temperature), soil moisture, the species to be seeded (annual, perennial, temperate, tropical), the time when the forage is required (especially if an annual) and the weed status of the land.

Where irrigation is not possible soil moisture status is of paramount importance in determining when to plant. Seeding dates should therefore be based, in the first instance, on the dependability of moisture at the time of, or immediately after, planting. Thus pastures should be seeded during the period of most reliable rainfall. Within such periods, best results are achieved when the final seedbed preparation and seeding procedures are performed on an already moist soil, with rain falling immediately after seeding. This is possible under irrigation and should be an objective for irrigated pastures. For rain-fed conditions where there can be little control of moisture conditions, the best option is to plant following rainfall sufficient to have replenished the subsoil moisture (soil profile should be moist to about 30 cm). Here seeding should commence as soon as the topsoil has dried out sufficiently to permit the use of seeders and rollers without causing soil to pack on the implements.

Besides soil moisture, suitable temperature conditions are a requirement for optimum seed germination and seedling growth. Most tropical and subtropical species require a soil temperature of at least 10°C for effective seed germination and seedling establishment. Temperate species will germinate and establish at soil temperatures of 5°C (Carpenter *et al.* 1990). High temperatures will decrease the effectiveness of establishment in temperate species, while very low temperatures can be critical to the establishment of tropical and subtropical species (seedlings are more easily killed by frost than are older plants). Suitable temperatures for germination and seedling growth of both temperate and tropical species are experienced in spring and autumn which, in most pasture producing areas, are also the periods of most reliable rainfall. Thus the most suitable time for sowing pastures is either in spring or in autumn. However, weed seeds tend to germinate best in spring and while this problem can be partially overcome by delaying planting until later in the summer, at least for perennial tropical species, the most opportune time to plant pastures tends to be the autumn.

Tropical and subtropical annual species must be planted in spring or early summer when, unfortunately, weed problems are at their worst. Fortunately, most of these species (millet, sorghum, teff) have vigorous seedlings which can compete strongly with weed seedlings.

Summer establishment of temperate species (annual and perennial) is possible under irrigation in much of South Africa, at least where summer temperatures are reasonably moderate. Thus, for example, *Lolium multiflorum* can be seeded in January in the higher altitude inland summer rainfall areas provided that the area can be irrigated with 5 mm to 10 mm of water every two to three days (unless natural rainfall is sufficiently regular to support seedling growth) and provided weeds are not a problem in the area. Frequent light irrigations would be necessary to cool the soil surface and prevent the seedlings from being burnt off at soil level.

Grasses can be established vegetatively from early spring through to autumn. However, spring or early summer planting following good rains is generally advocated. This ensures that the plants are well established before the first frosts occur. Vegetative material should not be planted within about six weeks of the first expected autumn/ winter frosts. Where supplementary irrigation can be applied, it is advisable to plant vegetative material as soon after the danger of spring frost has passed as possible.

10.1.6   Seeding rate

The literature abounds with 'recommended seeding rates' for different pasture species. Table 10.2 provides an example of such recommendations. These data were derived from local experience, Metcalf (1973), Fair (1986), Carpenter *et al.* (1990) and MacDonald (1991b). While such tables provide an indication of appropriate seeding rates, actual seeding rates depend on a number of factors, including the following:

(a)   purity of seed – the number of pasture seeds in a sample expressed as a percentage of the total sample. Purity provides an indication of contamination, i.e. the percentage of foreign material present;

(b)   germination capacity of the seed – the percentage of seeds in the sample which will germinate;

(c)   size and weight of the seed. Although seeding rate is expressed on a weight basis, it is actually the number of viable seeds per kilogram of seed that is important. Table 10.2 gives an indication of the expected numbers of seeds per kilogram of seed;

(d)   moisture supply – the amount of moisture in the soil at the time of sowing and the rainfall of the area. The less moisture in the soil and the lower the rainfall of the area, the lower the seeding rate should be;

(e)   soil fertility – the less fertile the soil, the lower the seeding rate should be;

(f)   effectiveness of seedbed preparation – the poorer the seedbed, the higher the seeding rate should be because of the likely failure of some seeds to germinate or of some seedlings to survive;

(g)   method of sowing – the less efficient this is, the higher the seeding rate must be. Broadcast, drilled or spaced rows each, in turn, require lower seeding rates;

(h)   purpose of the pasture – pastures to be used for hay or silage are usually seeded more heavily than those used for grazing so as to enforce an upright growth habit; and

(i)   annual or perennial – relative to perennial species, annual species have a limited tillering capacity and thus require higher seeding rates.

Table 10.2 Average number of seeds per kilogram and generally recommended seeding rates for selected pasture crops when row planted (drilled). For broadcast planting, increase seeding rates by 30%.

Botanical name	Common name	No of seeds per kg seed	Seeding rate (kg/ha)
Grasses			
Acroceras macrum	Nile grass	373 000	10
Anthephora pubescens	Wool grass	900 000	7
Avena sativa	Oats	28 600	40–60
Bromus wildenowii	Rescue grass	136 400	25
Cenchrus ciliaris	Blue buffalo grass	368 000	3–4
Chloris gayana	Rhodes grass	4 714 600	4–6
Cynodon dactylon	Kweek	4 000 000	5–7
Dactylis glomerata	Cocksfoot	950 000	15
Digitatia eriantha	Smuts' fingergrass	2 000 000	6–12
Echinochloa crusgalli	Japanese millet, Barnyard millet	340 000	15–20
Eragrostis curvula	Weeping lovegrass	3 218 000	2–3
Eragrostis tef	Teff	4 224 400	10–12
Festuca arundinacea	Tall fescue	500 000	20
Hordeum vulgare	Barley	27 000	40–50
Lolium multiflorum	Diploid cultivars	454 500	20
Lolium multiflorum	Tetraploid cultivars	250 000	25
Lolium perenne	Diploid cultivars	500 000	20
Lolium perenne	Tetraploid cultivars	256 400	25
Lolium rigidum	Wimmera ryegrass	460 000	20–30
Lolium hybrids	Hybrid ryegrass	500 000	15–20
Panicum maximum	Guinea grass, Green panic	2 000 000	7
Paspalum dilatatum	Common paspalum	660 000	30–40
Paspalum notatum	Bahia grass	280 000	20
Pennisetum clandestinum	Kikuyu	400 000	2–3
Pennisetum typhoides	Babala, Pearl millet	100 000	25–30
Phalaris tuberosa	Harding grass	781 000	12–15
Phleum pratense	Timothy	2 500 000	9–12
Secale cereale	Rye	45 000	45–50
Setaria sphacelata	Golden millet, pigeon grass	1 600 000	6–7
Sorghum almum	Columbus grass	117 000	15–20
Sorghum bicolor *(S. vulgare)*	Sudan grass, forage sorghum	120 000	15–20
Triticum x Secale	Triticale	27 000	40–45
Legumes			
Coronilla varia	Crown vetch	240 000	5–10
Desmodium intortum	Greenleaf desmodium	620 000	2–3
Desmodium uncinatum	Silverleaf desmodium	200 000	3–4
Glycine max	Soybean	11 000	50–65
Lotononis bainesii	Lotononis	3 300 000	0.5–1
Lotus corniculatus	Birdsfoot trefoil	850 000	4–6
Macroptilium atropurpureum	Siratro	76 000	3–4
Medicago littoralis	Harbinger medic, Strand medic	380 000	6–10
Medicago sativa	Lucerne	440 000	15–20
Medicago truncatula	Barrel medic	231 000	10–15
Neonotonia wightii/Glycine wightii	Glycine	120 000	3–5
Onobrychis viciifolia	Sainfoin	62 000	20–25
Ornithopus compressus	Yellow seradella	193 000	15–20
Ornithopus sativus	Pink seradelia	350 000	10–15
Stylosanthes guianensis	Stylo	300 000	3–5
Stylosanthes humilis	Townsville stylo	455 000	3–5
Trifolium alexandrinum	Berseem clover	380 000	12–15
Trifolium balansae	Balansa clover	1 400 000	2–3
Trifolium fragiferum	Strawberry clover	700 000	6–7
Trifolium pratense	Red clover	600 000	5
Trifolium repens	White clover	1 760 000	2–3
Trifolium resupinatum	Shaftal or Persian clover	1 450 000	3–4
Trifolium semipilosum	Kenya white clover	850 000	2–4
Trifolium subterraneum	Subterranean clover	143 000	12–15
Trifolium versiculosum	Arrowleaf clover	850 000	10–12
Vicia villosa	Hairy vetch	44 000	25–35
Vigna unguiculata var. *sinensis*	Cowpea	6 600	25–30

Under intensive conditions (high rainfall, irrigation) a young pasture requires 100 seedlings per square metre or 30 seedlings per metre of row length. This represents 1 million seedlings per hectare for broadcast planting or pastures established in 30 cm rows. Because purity and germination percentages are seldom, if ever, 100%, it is necessary to first determine the percentage of live pure seeds in the seed sample. Purity (%) multiplied by germination (%) and divided by 100 will give the percentage of live pure seeds. The 'quality' of the seedbed, climatic and soil conditions and the purpose of the pasture (see points (d), (f), (h) and (i) before) will also influence the seeding rate, as will the weed status of the land. As an example, *F. arundinacea* seed with no impurities, 100% germination and no seedling mortality would need to be seeded at 2 kg/ha (500 000 seeds per kg of seed, cf. Table 10.2) to provide the required million seedlings per hectare. With 80% purity and 70% germination, there would be 56% live pure seeds in the seed to be planted. Thus a seeding rate of 3.6 kg/ha would be required. Furthermore, if, as a result of poor seedbed preparation and poor soil moisture, the seedling mortality was 50%, the required seeding rate would be 7.2 kg seed/ha. With poor weather conditions following seeding, and/or a strong weed component, there could be a further 60% seedling mortality. The seeding rate for *F. arundinacea* would then need to be 18 kg seed/ha to obtain the desired million seedlings per hectare.

10.1.7   Re-establishing without reseeding

It is appropriate to briefly mention a relatively common practice adopted by many farmers in South Africa of re-establishing annual pastures such as *L. multiflorum* (in KwaZulu-Natal in particular) and annual medics (in the Western Cape where they often form part of a crop rotation) without sowing additional seed. The following procedure is normally adopted in re-establishing *L. multiflorum* pastures in KwaZulu-Natal: at the end of the season (from October to November) the pasture is withdrawn from utilisation and allowed to set seed. It is then 'fallowed' until January/February when it is disced lightly, sometimes harrowed if there is not too much overburden, and rolled. These mechanical operations incorporate the overburden and the seed, and compact the seedbed. Such a procedure reduces establishment costs and the erosion hazard and results in the return of a large quantity of organic matter to the soil. However, the following disadvantages tend to far outweigh the advantages. It is difficult to incorporate lime and P adequately and effectively with only a single shallow discing; there may in time be a gradual shift in the genetic composition of the species; the effective seeding rate is not known; weed problems may become very severe and the period of withdrawal to allow seed to be produced results in a substantial wastage of forage. Consequently, self-seeding to re-establish *L. multiflorum* pastures is not recommended.

However, should self-seeding, for the re-establishment of this species, be practised, the following points provide some guidelines for this management option:
(a)   weed control must be effective from the start;
(b)   self-seeding should not continue for more than two years in succession;
(c)   should the seed crop be too heavy, the crop should be ploughed or disced deeply to bury as much seed as possible. Should there be a meagre seed yield, additional seed should be applied before discing, and discing should be light;

(d) cultivars that are made up of different components (such as Italian and Wester-wolds ryegrass) should not be self-seeded;

(e) self-seeding re-establishment should be restricted to those cultivars that are not protected by plant breeders' rights; and

(f) the pastures should not be withdrawn from utilisation too late in the season or seed yields will be low and seed quality poor.

In an effort to ensure that annual medic pastures re-establish following a year of cropping in the Western Cape, care should be taken to retain the seed produced by the pasture crop in the surface 50 mm or so during the cropping phase. The soil should therefore be no more than scarified (i.e. a minimum tillage system should be adopted) when preparing it for the crop. Experience has shown that a tined implement retains more medic seed in the topsoil than does discing or ploughing (Kotze 1998). Being hard seeded, a large proportion of the annual temperate legume seed remains viable in the soil during the cropping year. However, where wheat is grown in the cropping phase it is important to remove or bury the wheat straw before the seeds begin to germinate because of its allelopathic effect on the germination of medic seed and the emergence of medic seedlings.

10.1.8 Nurse/cover crops

Several commonly grown pasture species, e.g. *D. eriantha, F. arundinacea, E. curvula*, are relatively slow to establish. This has led to the practice of sowing a rapidly establishing species with the slower establishing pasture species. The main aim here is to provide additional forage while the pasture species is establishing but the nurse crop may also provide 'protection' to the desired slower establishing species, and thus may improve establishment of the pasture.

The terms 'cover crop', 'nurse crop' and 'companion crop' are all used to describe the rapidly establishing species sown with the desired, slow to establish, pasture species. Crops such as wheat, barley and oats, linseed, peas or the brassicas, turnips and rape are often used worldwide (Santhirasegaram & Black 1965; White 1973), with the pasture species and the cover crop usually being sown at the same time (Santhirasegaram & Black 1965). In South Africa the commonly used cover crops are *Eragrostis tef* (sown with *E. curvula*), *Chloris gayana* (sown with *D. eriantha*), *Sorghum* species (sown with *E. curvula, D. eriantha* and *P. clandestinum*) and *L. multiflorum* and *Avena sativa* (sown with *Medicago sativa, F. arundinacea* and *P. clandestinum*).

In effect, however, a nurse crop probably seldom improves the establishment of a pasture species (Santhirasegaram & Black 1965), largely because of the competition it affords to the slower growing pasture species (White 1973). Careful thought should therefore be given to their use in pasture establishment although there may be situations where their use could be expedient. These situations would include:

(a) where there is an immediate urgent need for forage;

(b) where wind blown sand adversely affects the establishment of the pasture species; and

(c) where pastures are established on steep land susceptible to erosion. Where they are used, their seeding rates need to be relatively low and their growth needs to be kept in check by, for example, frequent mowing, so that they do not offer too much competition to the establishing pasture species.

10.2 SEMI-ARID REGIONS   *R.H. Drewes*

In the semi-arid summer rainfall areas of South Africa pastures are now largely being established on soils which are marginal for cash-cropping and which have been withdrawn from annual cultivation. Many of these soils should not have been ploughed in the first place. However, many such soils are able to support very productive perennial pastures. This applies particularly to the vertic and melanic soils which, due to their high base status, not only produce palatable forage, but in wet years are able to produce high yields with little or no fertilizer addition. Species such as *D. eriantha* (Smuts' fingergrass) can be established very successfully even on shallow, stony Mispah and Glenrosa soils.

In the semi-arid cropping region of the summer rainfall area the normal practice therefore is to use the high potential soils for annual cash crops, and to plant perennial pastures on soils ranging from fair to poor. Exceptions will occur, as when the forage requirements of dairy cows or of ewes or fat lambs demand that pastures be available near the homestead or dairy. Here high potential soils may be used for pastures.

Establishment procedures for perennial pastures can range widely from high input to low input, depending on the local circumstances. The ultimate purpose and use of the pasture will play an important role in this decision. Pasture to be used for extensive production will normally be established using low input cost procedures. Conversely, where pasture is being planted for grazing by dairy cows, establishment procedures will normally be more intensive and more costly. Also, the more urgent the need, the more costly establishment procedures can be justified.

In the semi-arid summer cropping areas, pastures mostly comprise perennial tropical/subtropical tufted grasses. *Medicago sativa* (lucerne) is the most important forage legume. In this chapter emphasis will be given only to the establishment of these pasture types.

10.2.1 Recommended planting season

A general recommendation is to delay planting grass pastures until the soil contains adequate reserves of moisture. In practice, planting is therefore often delayed until well into the season. Every effort should, however, be made to plant early where the growing season is short but planting can be delayed until later in the season in areas with a long frost-free season. In all areas early establishment sometimes has the advantage of providing useful forage during the establishment season but late planted pastures are known to remain green later into the autumn and so provide high quality forage in the late autumn and early winter. More commonly, however, pastures provide little useful grazing during their first summer and must of necessity be reserved for winter foggage. Summer grazing will normally be available only during the second season.

On arable lands subject to wind erosion it is advisable to make use of the protection provided by a widely spaced row crop such as maize for the establishment of both grass and lucerne pastures. This means that planting must be delayed until at least February, when the mature maize is intersown with the pasture crop. This procedure should, however, be adopted only during above average rainfall seasons and in areas which do not experience early frosts.

Lucerne should preferably be planted from March through to June on land subjected to moisture conservation measures through the summer season. Here weeds should have been controlled by regular shallow cultivation. Establishment early in March increases the probability of sufficient rain falling after planting to stimulate germination, but it does introduce the risk of severe insect damage, particularly from the lucerne caterpillar. By delaying planting until the autumn or winter, the risk of insect damage is reduced, but so is the likelihood of receiving sufficient rain to stimulate germination. Follow-up rains are particularly important on clay soils, whereas germination on sandy soils will often readily take place even without any further rain.

10.2.2   Seedbed preparation

The ideal seedbed is one in which the surface soil is fine, firm, level and weed free (refer to section 10.1). However, the extent to which the pasture is likely to contribute to the livestock feeding programme, and therefore its importance in the livestock system, will normally determine the extent to which it would be wise to aim for the ideal situation. If the end product of the production system has the capacity to bring in high returns, then a great deal of effort can be spent on preparing the seedbed. If not, then such effort is unlikely to be worthwhile.

The procedures which should be used in preparing a pasture seedbed in the semi-arid areas are no different from those outlined above for the humid areas (section 10.1) except that here attention needs to be paid to conserving as much moisture as possible in the soil prior to planting.

Where pastures are established between the rows of an existing crop such as maize, cultural operations will normally be restricted to no more than light harrowing between the maize rows. Because it will not normally be possible to use the roller in such situations, the tractor wheels will need to be used to provide the necessary compaction.

10.2.3   Fertilizer application

Arable lands which are converted to pastures in the summer rainfall cropping areas generally have a relatively high P status. Since many of the soils of the area are either eutrophic or mesotrophic, the calcium and magnesium status of the soils is usually also relatively high. Potassium is also seldom deficient. For these reasons it is seldom necessary to apply fertilizer during the establishment phase of a pasture.

10.2.4   Seed and seed mixtures

Not only are the seeds of grasses typically used in these areas very small (e.g. *D. eriantha* – Smuts' fingergrass, *Panicum maximum* – Guinea grass and *C. gayana* – Rhodes grass), but they require, to a greater or lesser extent, an after-ripening period before they will germinate. Because of their size they contain only relatively small reserves of energy with which to support the germinating seedling. Seedling survival may therefore be expected to be poor when environmental conditions do not suit establishment. The requirement for an after-ripening period means that the germination of the seeds is normally erratic, but this has the advantage of increasing the probability that germination of at least some seed will be delayed until conditions favour successful establishment.

Because of the risks associated with the small seeds of these species and those brought about by seed dormancy, it is generally considered advisable to sow mixtures of different species in the hope that at least one of them will establish successfully. A mixture of, for example, species such as Smuts' fingergass and one or more of Guinea grass, *Anthephora pubescens* (Anthephora) and Rhodes grass, has been used successfully in such situations. These four grasses are all palatable and have similar growth patterns. Even if one or more of them should fail to establish or to persist, the pasture will still remain productive.

Lucerne, planted in alternate rows with any of the above species, also makes an extremely useful pasture not only because of the N it supplies but also because, mixed with the grass in this way, the danger of bloat is greatly reduced.

10.2.5   Method of establishment

Under rain-fed conditions there is no doubt that row planting should be resorted to. It leads to a saving of seed, allows the soil between the rows to be worked if this should become necessary, and it allows different species to be established in alternate rows. Row planted stands of both lucerne and grasses tolerate drought conditions better than do broadcast stands.

The above species can all be planted successfully with the maize planter (fitted with seed-plates) or wheat drill. The seed can be fed through the seed-bin or, as in the case of Anthephora, the seed may be mixed with fertilizer or sifted manure and planted through the fertilizer bin. A number of locally designed planters have also been produced. All these implements deposit the seed in the compacted soil produced by the tractor wheel or the seed is compressed into the soil by the press-wheel of the wheat drill or maize planter. Where a procedure does not give sufficient soil coverage to the seed, a double folded chain should be dragged behind the planter.

Where the tractor wheel is used to compact the seed into the soil, prior soil preparation should be designed to produce a firm seedbed. Ideally, the wheel tracks should not be deeper than 70 mm. Also, the maize or wheat planter should be adjusted so that the seed is not covered by more than about 10 mm of soil.

It would appear that the success which can be achieved with any of the above methods can largely be attributed to the extent to which:
(a)   the soil onto which the seed is sown is compacted;
(b)   the soil is forced in from the sides of the furrows to cover the seed; and
(c)   the furrows which are formed accumulate moisture so that the soil surrounding the seed remains relatively moist for a longer period of time after rain.
It is obvious that heavy rains immediately after planting can cause severe wash and, in so doing, bury the seeds too deeply.

The vigour of seedlings of the tropical perennial grasses is usually relatively poor. At least a part of this problem stems from the seeds exhibiting a degree of dormancy and because they require after-ripening before they will germinate. As a general rule the seed of many of these tropical grasses tends to germinate over an extended period of time, sometimes up to two to three years. For this reason judgement on the success of establishment should be delayed until at least the season following planting. A hasty decision to plough up newly established pastures has undoubtedly brought an untimely end to many a potentially productive pasture.

10.2.6   Seeding rates

Seeding rates for species commonly grown in the semi-arid regions under rain-fed conditions are presented in Table 10.3.

Table 10.3. Recommended seeding rates for species commonly planted in rows in the semi-arid regions.

Species	Seeding rate (kg/ha)
Anthephora pubescens (Anthephora or Wool grass)	8
Chloris gayana (Rhodes grass)	3
Digitaria eriantha (Smuts' fingergrass)	3
Medicago sativa (Lucerne)	3
Panicum maximum (Guinea grass)	4
Sorghum spp. (Forage sorghums)	15

10.2.7 Weed control

In pastures in the semi-arid regions many of the weeds are well grazed by animals and some may even be preferred to the planted species. Therefore the extent to which it is necessary to control weeds will largely depend on the ultimate purpose of the pasture and the urgency with which the pasture needs to be brought into full production. Where the invading weeds are acceptable to the grazing animals or where the degree of infestation is light, it may not be necessary to implement any control measures. However, where infestation is so heavy or weeds are so tall that the pasture plants are subjected to excessive shading, or where conditions are so dry that the weeds compete strongly for moisture, weed control is likely to be essential for successful establishment.

10.2.8 Utilisation of the pasture

The decision on whether or not to utilise a pasture during its establishment year may often be dictated by the severity of the weed problem. It may be necessary either to graze or mow a pasture in its establishment year to reduce the competitive effect of a dense weed population. However, where this is not so and the decision on when to utilise a pasture is based on the condition of the pasture plants, then it is normal to delay utilisation at least until the flowering stage. It has also become apparent that moderate grazing, even at a very early stage, does little harm to some pasture species (including Smuts' fingergrass) provided they are allowed a reasonable recovery period. However, where the initial stand is very sparse, pastures comprised of species which are able to reseed themselves should be rested out fully in order to maximise the amount of seed produced by the established pasture plants.

10.3 ARID REGIONS

N.M. Tainton

Reference will be made here only to the establishment of the three main drought fodder crops used in arid areas. Acknowledgement is made to Jordaan (undated) for information presented in this section.

10.3.1 *Opuntia* spp. (Spineless cactus)

Spineless cactus should preferably be planted on northern or north-eastern aspects because of its intolerance of temperatures less than 10°C. Only phosphatic fertilizer is normally needed, and then only where it has been shown to be deficient.

Plantations of this drought fodder plant should normally be planted between September and October. Only 'leaf' segments older than one year should be used as planting material. When planted into grassveld, some mechanical disturbance of the soil to reduce the competitive effect of the grass is recommended (De Kock 1980a).

Various establishment procedures can be used. Segments can be laid flat on the soil surface and anchored with a stone or a spade-full of soil. This procedure is particularly suited to sites where the soil cannot be mechanically disturbed. It requires little labour, but establishment is slow. Where it is possible to draw furrows through the site (with, for example, a single-furrow plough), segments can be placed along one side of the furrow. Here the cut surface of the segment should remain exposed to prevent fungal attack and rotting. This procedure is more labour intensive than the procedure described above, but results in more rapid establishment.

Establishment is most rapid when double-segments are planted and this is advisable where planting material is freely available (Aucamp & De Kock 1970; De Kock 1980a).

Spineless cactus segments are normally planted 1.5 m apart in rows 3 m to 4.5 m apart. Such plantings would require between about 1 500 and 2 300 segments of planting material per hectare.

10.3.2 *Atriplex nummularia* (Old Man Saltbush)

Seed of Old Man Saltbush should be soaked in water for 2 to 3 days before planting in order to remove as much of the salt from the seed as possible. The water should be changed two to three times daily or, better still, seeds should be immersed in running water. They should then be planted immediately on removal from the water.

Seed should be planted into seedbeds or preferably into moveable containers (tins or plastic bags) in August or September and lightly covered with soil. They should then be watered regularly.

Seedlings should be transplanted when 150 mm to 200 mm tall. If transplanting is delayed beyond this, the seedlings should be trimmed back to this height at planting. Planting should be into previously ploughed land where this is possible, or, at worst, into furrows drawn across the site. Where neither of these operations is possible, seedlings can be planted into individually prepared micro-seedbeds. Transplanting should normally take place in February or March.

Seedlings should be planted in 2 m rows, with a spacement of 1 m to 2 m within the rows. Between 2 500 and 5 000 seedlings would be required per hectare (note that between 20 000 and 30 000 seedlings can be produced from 1 kg of seed). Soil should be firmed around the roots and water applied soon after planting. The seedlings should again be watered three to four weeks after planting, and again three to four weeks later. According to De Kock (1980a), establishment is normally satisfactory when 1 litre of water is applied to each seedling when transplanted and a further 2 litres 10 days and again 20 days later.

Old Man Saltbush can also be established from cuttings. Young stems at least 6 mm thick should be cut into lengths of approximately 250 mm and the upper pair of leaves left attached to the stem segments. Approximately half the length of the cutting should then be buried in a moist sandy soil medium during early spring or autumn and kept moist. The cuttings should begin to root after about six weeks and can normally be planted out about four weeks later (De Kock 1980a).

10.3.3 *Agave mexicana* (American aloe)

Two year old slips or seedlings should be planted in furrows 3 m to 5 m apart, with plants spaced 1 m to 2 m apart in the row.

10.4 LEGUME NODULATION AND INOCULATION

V.D. Wassermann

10.4.1 Precautionary measures for effective nodulation

Inoculation of legume seed is often essential to ensure that appropriate bacteria are present when the seed germinates. However, additional precautionary measures are generally needed to ensure that effective nodulation takes place. A sufficient number of bacteria must be available in the vicinity of the seed at germination to ensure that effective entry is achieved into the first formed root hairs. Because the root hairs are confined to the region immediately behind the root tip, there is a tendency for them to develop out of reach of the bacteria which are applied with the seed at inoculation. Incomplete nodulation may therefore take place. Also, while rhizobium bacteria can survive for long periods in soil under favourable conditions (Date & Brockwell 1978), they are able to multiply only in the presence of root exudates (Rovira 1961; Van Egeraat 1975). Since inoculated seed is often sown into soils which do not have the appropriate bacteria, it is important for germination conditions to be as near ideal as possible so that it can proceed rapidly and before a large proportion of the bacteria die off.

There are a number of factors which determine the rate at which the bacteria will die:

(a) moisture content of the soil: because germination is normally slow in dry soils, a large proportion of the bacteria may die before a sufficient number of root hairs have developed;

(b) temperature: high summer soil temperatures reduce the survival of the rhizobium bacteria (Chatel & Parker 1973). Different strains of bacteria differ considerably, however, in their tolerance of high temperatures (Date & Brockwell 1978). There is also a strong interaction between soil temperature and soil moisture content; the bacteria are less tolerant of high temperatures when the soils are wet than when they are dry (Date & Brockwell 1978);

(c) soil texture: bacteria survive better in heavy soils containing high levels of organic matter and clay colloids than in light sandy soils (Hely & Brockwell 1962; Marshall 1964). The reason for this is that a high colloidal fraction reduces the rate of moisture loss from the soil (Marshall 1975);

(d) soil acidity: rhizobium bacteria survive best when the soil pH is near neutral (pH in water of 6.5 to 7.0). Survival in very acid soils (pH 4.5 to 5.0) is very poor (Date 1970). Different species differ, however, in their tolerance of low pH. So, for example, clover bacteria (*R. trifolii*) survive relatively well at a pH (water) of 4.7 to 4.8, while those of lucerne and other *Medicago* species (eg. *R. meliloti*) require a minimum pH of 5.8 to 5.9 (Date 1970). In general, therefore, liming to increase the pH of acid soils promotes the survival of the bacteria. At the same time, the calcium contained in the lime stimulates nodulation (Date 1970);

(e) seed coat toxicity: the seed coats of some legume seeds contain hydrophobic compounds which suppress rhizobium bacteria and can lead to the failure of inoculation (Bowen 1961; Thompson 1961). Special procedures may therefore be needed for effective nodulation in these seeds; and

(f)   other toxicities: any contact between seed and acid fertilizers such as superphosphate (pH of approximately 2.0) will severely reduce the survival of the bacteria (Date 1970). The application of molybdenum as sodium molybdate during inoculation can have deleterious effects if inoculated seed is not sown immediately (Gault & Brockwell 1980). However, ammonium molybdate and molybdenum trioxide can be safely used. Chemicals must be of good quality. Also, seed treatment with the majority of fungicides, insecticides, eel worm remedies and herbicides is likely to kill the bacteria (Date 1970). There are, however, some chemicals which can be safely used.

A further factor which affects nodulation is the antagonism between the rhizobium bacteria and other micro-organisms found in soils. A number of these are antagonistic towards rhizobium bacteria (Date & Brockwell 1978). For example, they may limit nodulation by effective strains, particularly if the latter are not present in sufficient numbers to effectively nodulate the root hairs as they develop (Date & Brockwell 1978). In the final analysis, the success of inoculation may largely depend on the ability of the rhizobium bacteria to compete with other soil micro-organisms.

From the previous discussion it should be obvious that there are a number of measures which need to be taken into account to ensure successful inoculation. The first golden rule is that conditions should be as favourable for the germination of the crop as possible. Also, if soil conditions are likely to be unfavourable for bacterial survival, bacterial dosage levels should be high. According to Date (1970), as few as 100 bacteria per seed will ensure good nodulation under favourable conditions, but as many as 100 000 may be needed to ensure effective nodulation when soil conditions do not favour the survival of the bacteria, or when other organisms are likely to compete with these bacteria.

10.4.2   Inoculation procedures

The inoculant most commonly used on legumes is produced from fine, sterilised peat, to which a large number of bacteria are added. The quality of inoculant available in South Africa is normally very good (because of the strict controls applied) but it is difficult to control the handling of inoculants once they leave the producer. Those responsible for their distribution need to appreciate that the material contains living organisms which will die if exposed to unfavourable conditions, and particularly to high temperatures. This can lead to the death of the bacteria long before the advertised expiry date of the inoculant.

Different circumstances demand different inoculation procedures. Where conditions are favourable for germination of the crop and for the survival of the bacteria, the seed should be moistened with water containing a suitable sticker and well mixed with the inoculant. The sticker helps to hold the inoculant to the seed. Of the different stickers available, a 2% solution of methyl cellulose is usually recommended. It has proved to be effective, is relatively inexpensive and does not harm the bacteria. Sugar is not as effective as methyl cellulose, and the use of milk can no longer be recommended.

Where the conditions for bacterial survival are less satisfactory, more complicated inoculation procedures must be adopted. Here the seed may need to be pelleted, i.e. the inoculated seed is covered with a layer of lime or some other suitable material to protect it from desiccation or from unfavourable soil conditions. This procedure has proved to be extremely successful (Brockwell & Phillips 1970) but it does require that the seed, sticker and pelleting material be mixed in the correct ratio (Strijdom & Wassermann 1980). Here the type of sticker used is extremely important. Either gum arabic (40% solution) or methyl cellulose (4%) can be used, although Australian data suggest that gum arabic is the more effective of the two (Date 1970). Apparently it protects the bacteria against the action of toxic substances in the seed coat of some legumes (Vincent *et al.* 1962).

Lime should not be used as the pelleting medium in all situations. For example, it may suppress certain rhizobia of tropical and subtropical legumes (Norris 1967). It also kills the rhizobia of lupins and serradella, both of which are temperate legumes. Nonalkaline pelleting material, such as bentonite, should be used for these legumes.

Since bacteria have a limited life span, inoculated seed should always be planted as soon as possible after inoculation.

Good results have also been achieved when applying the bacteria in a suspension in water to the seeded row after planting, as well as with the use of dry pelleted inoculant which is sown together with the seed (Hely *et al.* 1976; Brockwell *et al.* 1980). These procedures are, however, considerably more expensive than conventional procedures and are unlikely to gain general acceptance.

Direct application of liquid inoculum to seed at planting is often better than conventional seed inoculation. For large-scale sowings, this requies an inoculant tank, a pump, a manifold and capillary tubes to deliver liquid into the seedbed beside and beneath the seed (Brockwell *et al.* 1995). The inoculant may comprise a suspension of peat culture in water or a specially prepared formulation.

Nutrition of planted pastures

N. Miles & A.D. Manson

CONTENTS

Pasture ecosystems, like natural ecosystems, are characterised by energy flow chains and nutrient cycles (Booysen 1981). Intensive pastures are by objective high growth rate and highly productive systems which require high rates of energy and nutrient flow. High rates of energy flow are achieved by using photosynthetically efficient plants, while high rates of nutrient flow are achieved through nutrient (principally fertilizer) additions to the system. The development of economically and environmentally sound fertilizer programmes requires an adequate understanding of nutrient cycling.

A diagrammatic representation of the rudimentary pools and flows of nutrients in a grazed pasture is shown in Fig. 11.1. The soil, plants and animals constitute the major nutrient pools in the system. Additions of nutrients to the system are mainly through the deposition of dung and urine derived from imported animal feeds and through the application of fertilizer and lime. Major pathways of nutrient loss are through leaching, the removal of harvested plant material and animal products and excretal deposition off the pasture.

The 'available' pool in the soil constitutes the reservoir of nutrients available for plant production. An objective of pasture management is to maintain the quantities of all plant nutrients in the available pool at levels that will not limit plant growth. There has, as will be discussed in more detail later, been some success in estimating the size of the available pools of phosphorus (P), potassium (K), calcium (Ca), magnesium (Mg) and zinc (Zn), and in relating their size to levels of plant production. However, the flow of nitrogen (N) between the available pool and the soil organic matter, together with the tendency for nitrate to move freely with the soil water, largely precludes the development of a relation between plant yield and the available soil N.

Pasture utilisation options have a major bearing on the magnitude of nutrient movements through the various compartments of the pasture system (Fig. 11.1). Losses are most severe where harvested material is removed from the pasture (refer to Table 11.1 for an indication of the amounts of nutrients that 10 tons of material would contain). High fertilizer inputs, in particular of K and N in grasses, are therefore needed to sustain pasture productivity under a removal strategy. Where, however, forage is grazed *in situ* by animals which remain permanently on the pasture, most of the ingested nutrients are recycled through the excreta, resulting in low long-term fertilizer requirements.

Intermediate between these extremes is utilisation by intermittent grazing, as in most dairying operations. Here, appreciable nutrient losses result from excretal deposition off the pasture. Although supplementary feeds may result in the addition of nutrients to the pasture system, these additions are unlikely to match the above losses which would then need to be made good by fertilizer application.

In subsequent discussions, accumulations, transfers and losses of individual nutrients are dealt with in greater detail. It should be pointed out that although much of the research discussed is based on the authors' own work in KwaZulu-Natal, the principles discussed apply universally.

11.1 NITROGEN

Nitrogen is quantitatively the most important element taken up by pasture plants from the soil. Its availability, together with temperature

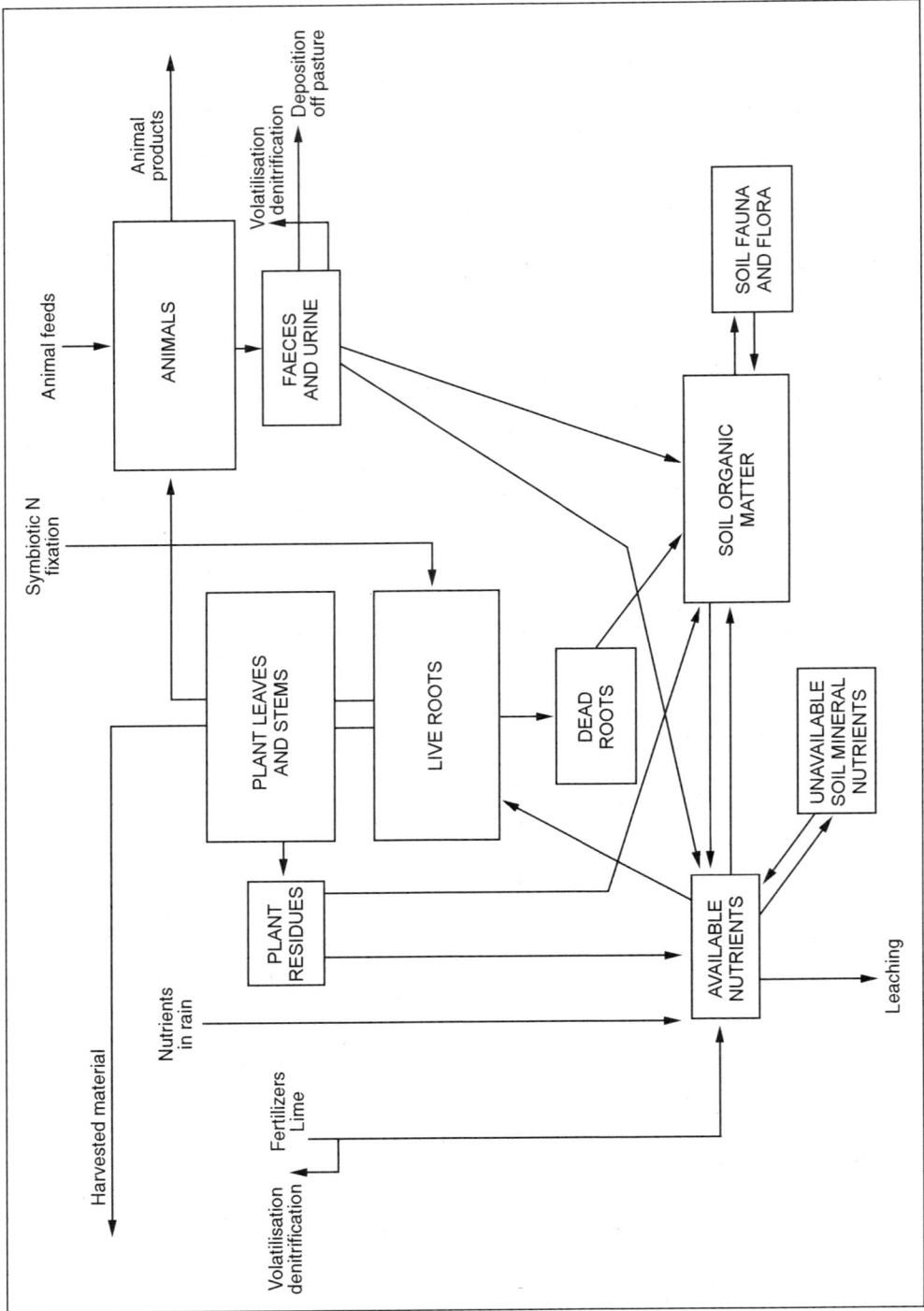

Figure 11.1　Cycling of nutrients in a grazed pasture system.

Table 11.1 Typical amounts (in kg) of the major nutrients in 10 t dry matter in each of three pasture grasses and a legume.

Pasture species	N	P	K	Ca	Mg
Italian ryegrass	450	28	400	55	22
Kikuyu	350	34	350	28	32
Eragrostis curvula	160	15	140	25	14
White clover	480	38	180	150	30

and moisture supply, are usually the major factors determining the productivity of pastures.

It is supplied to pastures in various ways. Major pathways of supply are through additions of fertilizers, symbiotic N fixation by legumes and through the excreta of animals being fed supplements. Rainwater provides a small but generally insignificant amount.

11.1.1 Nitrogen in pasture soils

11.1.1.1 *Total nitrogen in the topsoil*

The topsoils of pastures, and particularly of perennial pastures, contain large amounts of N. In the rooting zone of pastures in the United Kingdom these levels range between 5 t/ha and 15 t/ha (Whitehead *et al.* 1986a). In KwaZulu-Natal the total N in the top 100 mm of soil under pastures is usually within the range 1 t/ha to 5 t/ha. As much as 98% of this N is bound in soil organic matter and so is not immediately available to plants.

11.1.1.2 *Mineralisation/immobilisation of nitrogen*

As indicated above, most soils, and in particular those under perennial pasture, contain appreciable organic N reserves which in the short term are not available to plants. However, through the action of certain soil micro-organisms in a process known as mineralisation, organic N may be converted to ammonium nitrogen (NH_4^+). The ammonium nitrogen thus formed can be changed further, again by microbial activity in a process called nitrification, into nitrate (NO_3^-). Both the ammonium and nitrate forms are taken up by grasses and legumes.

Micro-organisms responsible for the decomposition of organic matter in the soil can satisfy their own nutritional needs for growth from two sources: from nutrients already present in the soil in an available form, and from nutrients contained in the organic matter. When N in the organic matter is in short supply, N from soil reserves is assimilated into the cell material of micro-organisms. This process of N depletion through conversion to an organic non-available form, is termed immobilisation.

Many factors affect the activities of the soil microbia and hence the mineralisation and immobilisation of N. The most important of these are the chemical composition of decomposing litter, environmental factors (particularly moisture and temperature), soil chemical factors and soil disturbance (tillage).

The ratio of the percentage of carbon (C) to that of N (the C:N ratio) of organic matter added to the soil provides useful guidelines on N release patterns in soils. The C:N ratios of some organic materials are reported in Table 11.2. The low C:N ratios of leguminous crop residues and the very high C:N ratios of straw and sawdust are noteworthy. Incorporating organic matter having a C:N ratio less than 25:1 (i.e. relatively rich in N) into the soil results in rapid mineralisation (release) of N (Whitehead 1986). Between 25:1 and 30:1, immobilisation of N dominates for the first few days or weeks of decomposition, after which there is net mineralisation. With C:N ratios greater than 30:1, prolonged immobilisation occurs. This gives rise to a temporary depletion of inorganic N reserves (the negative period). In pastures receiving N fertilizer in the range 350–450 kg N/ha/yr, grass roots have C:N ratios around 25–30:1 and dead herbage ratios of <25:1. Clover roots and herbage usually have ratios between 12:1 and 20:1. Decomposing residues from pastures receiving high rates of N are unlikely, therefore, to bring about appreciable immobilisation of N.

Table 11.2   Approximate carbon:nitrogen (C:N) ratios of a selection
of organic materials (adapted from Tisdale *et al.* 1985;
Haynes 1986; Whitehead 1986).

Organic substance	C:N ratio
Clover roots and herbage	12–20:1
Grass roots (fertilized with 350 kg N/ha/yr)	25–30:1
Grass litter (fertilized with 350 kg N/ha/yr)	<25:1
Grass roots and herbage (low fertilizer rate)	40–50:1
Barnyard manure (rotted)	20:1
Maize stover	60:1
Wheat straw	80: l
Sawdust	400:1

The combined effects of high temperature and high moisture favour microbial growth and thus mineralisation of soil N where the C:N ratio of the organic substrate is sufficiently low. Mineralisation has been found to increase sharply as temperature increases over the range 5°C to 35°C (Tisdale *et al.* 1985). Below 5°C and above 40°C the rate of N mineralisation declines markedly. As a consequence of moisture and temperature effects on mineralisation, the availability of N from soil reserves usually peaks in the first half of the summer, particularly in summer rainfall climates (Sanchez 1976).

Soil pH also affects N mineralisation. Low pH inhibits mineralisation, probably through some combination of hydrogen and aluminium toxicities and Ca deficiency (Adams & Martin 1984). The application of lime to acid soils promotes decomposition and results in rapid increases in soil inorganic N levels (for example, from 67 kg/ha inorganic N in the top 150 mm of a Balmoral soil where no lime was applied, to 96 kg/ha and 114 kg/ha one month after lime, as calcium hydroxide, was applied at rates of 2 t/ha and 6 t/ha respectively). Therefore observed yield responses to lime, particularly of acid tolerant grasses, may result from an improved N supply rather than from the effect of lime *per se.*

Cultivation, by promoting soil aeration and exposing organic matter to microorganisms, enhances processes such as oxidation of organic matter, mineralisation of organic N, and nitrification of ammonium N. Cultivation, therefore, usually markedly increases the mineralisation of N (Haynes 1986; Whitehead *et al.* 1990), with the amount of N mineralised being related to the composition of incorporated residues, the amount and composition of organic matter in the soil and past management. A large increase in mineralisation generally accompanies the ploughing of fertilized grass swards. In the United Kingdom, for example, the amount of N mineralised in the first year following the ploughing of three- and eight-year-old grazed swards was 201 kg and 306 kg N/ha, respectively (Whitehead *et al.* 1990). Such high levels are likely to lead to the leaching of nitrate.

11.1.1.3 Factors influencing the accumulation of nitrogen in pasture topsoils

Following the establishment of a permanent pasture on a previously cropped soil, soil organic matter and total N increase gradually over many years. Additions of organic matter and N are through the death and decay of plant residues (both shoots and roots) and, in grazed swards, through the return of animal excreta, although the latter will add little organic matter to the soil.

Following the establishment of a permanent pasture, soil N usually increases at an annual rate of between 50 kg and 150 kg N/ha (Ryden 1984; Ball & Field 1987), depending on organic matter input. The C:N ratio in soils of well managed pastures remains relatively constant at about 10 (Table 11.3). The extent of accumulation of N in the soil organic matter is largely dependent on the amount of carbon added to the soil through the death and decay of roots and unutilised herbage rather than on N input. Since the level of herbage utilisation controls organic matter input, it also controls N accumulation in pasture topsoils. Under intensive utilisation, the rate of oxidation of resident organic matter may exceed the input of fresh organic residues, resulting in a net decline in total N. Ball & Field (1987) predict that, for New Zealand pastures, utilisation by animals of more than 70% of the measured herbage depletes soil N reserves. Annual pasture and cropping systems, because they enhance organic matter oxidation, also restrict the accumulation of organic matter, as illustrated in Table 11.3.

Table 11.3 Soil organic carbon and nitrogen in topsoils (0 mm to 100 mm) under various management regimes.

Management regime	Yrs[1]	Clay (%)	C (%)	N (%)	C:N
Cedara					
Maize silage/grain	15	33	1.91	0.15	12.7
Annual Italian ryegrass	11	34	1.76	0.17	10.4
Annual Italian ryegrass	3	33	2.60	0.24	10.8
Kikuyu	15	59	5.70	0.56	10.2
Tabamhlope					
Maize silage	12	33	2.72	0.22	12.4
Eragrostis curvula hay	10	33	3.89	0.32	12.2
Kikuyu	12	30	4.12	0.42	9.8
Swartberg (East Griqualand)					
Annual Italian ryegrass	>10	17	0.64	0.08	8.0
Perennial ryegrass	10	16	1.07	0.14	7.6
Veld	–	22	2.41	0.23	10.5
Tsitsikama					
Annual pasture crops	5	8	0.87	nd	–
Perennial tropical pasture	5	8	1.20	nd	–

[1] = years under indicated management system. Prior management uncertain
nd = not determined

11.1.2   Nitrogen and plant growth

11.1.2.1   Plant uptake of nitrogen

Plant dry matter normally contains between 1% and 5% N. Roots absorb N as nitrate (NO_3^-) and ammonium (NH_4^+) ions, and as urea ($CO(NH_2)_2$), although the rate of absorption of the latter is generally slow (Mengel & Kirkby 1987). Once absorbed, nitrate is converted (reduced) to ammonia which is incorporated into various organic compounds to form proteins.

The high N requirement of pasture grasses, coupled with their pervasive root systems, results in efficient absorption of N from the soil. In all-grass swards, about 50–70% of applied fertilizer N is normally taken up, although this decreases at very high N levels. Inorganic N therefore does not accumulate in significant amounts in pasture topsoils except in excretal, and particularly urine, patches.

In pastures adequately supplied with other growth requirements, an abundant supply of N is associated with vigorous growth and a dark green colour. Where plants are deficient in N, they become stunted and yellow. In grasses the lower leaves usually 'fire' or turn brown, beginning at the leaf-tip and progressing along the midrib until the entire leaf is dead. The young upper leaves tend to remain green due to the translocation of N from older to younger plant tissues.

11.1.2.2   Yield response to nitrogen and factors affecting it

A characteristic response curve for the relation between dry matter yield of grass and N supply is presented in Fig. 11.2. When N is applied there is usually an initial near linear response (A in Fig. 11.2), a phase of sharply diminishing response (B) and a point (C) beyond which N has little or no effect on yield. The amount of dry matter produced for each kilogram of N applied within zone A depends largely on the species under consideration, the frequency of defoliation and growth conditions.

Tropical grasses generally produce more dry matter per unit of N than do temperate grasses (Andrew & Johansen 1978). *Eragrostis curvula*, for example, has produced up to 60 kg dry material per kg N applied (Tainton *et al.* 1981), but irrigated Italian ryegrass only between 25 kg and 34 kg dry material per kg N applied (Eckard 1989). Perennial ryegrass produces an average of 23 kg DM/kg N over an N application range of 0–300 kg N/ha (Morrison *et al.* 1980). Data such as these are averages over the season and conceal wide variations in response efficiency within the season. For example, in perennial ryegrass the spring response is two to three times greater than at other times.

Pasture management and environmental factors both impact on the response to fertilizer N. Data from Bartholomew & Miles (1984) illustrated the characteristic interaction between cutting frequency and fertilizer N rate (Fig. 11.3). Here the longer the interval between harvests, the higher the yield. Response to N was, however, greatest at the 4 week cutting interval. Similar results have been reported by Morrison & Russell (1980) and Frame *et al.* (1989).

Climate and soil profoundly influence the N response of grasses. In general, N responses increase with increasing rainfall, temperature, day length and solar radiation, but may be low when grasses are producing seed (Whitehead 1970). Water is required to move topdressed fertilizer N into the rooting zone and, thereafter, to the root surface. Where water is often in short supply, as in the relatively drier central and western areas of southern Africa, there is generally little or no benefit in applying more than

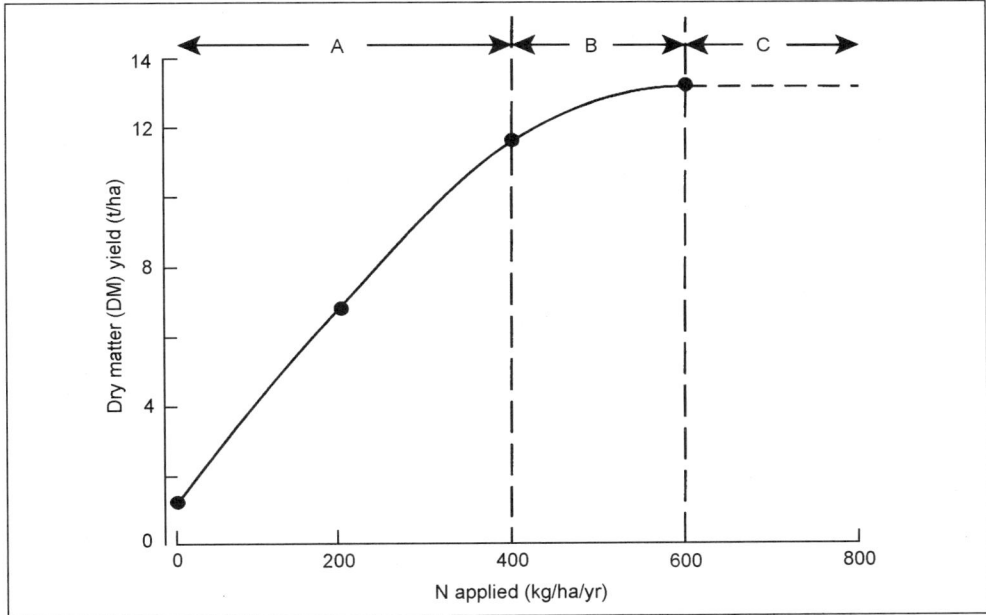

Figure 11.2 Response of kikuyu (cut and removed) to applied nitrogen (N) in the Highland Sourveld of KwaZulu-Natal (the extrapolation above 600 kg/ha is to facilitate discussion in the text).

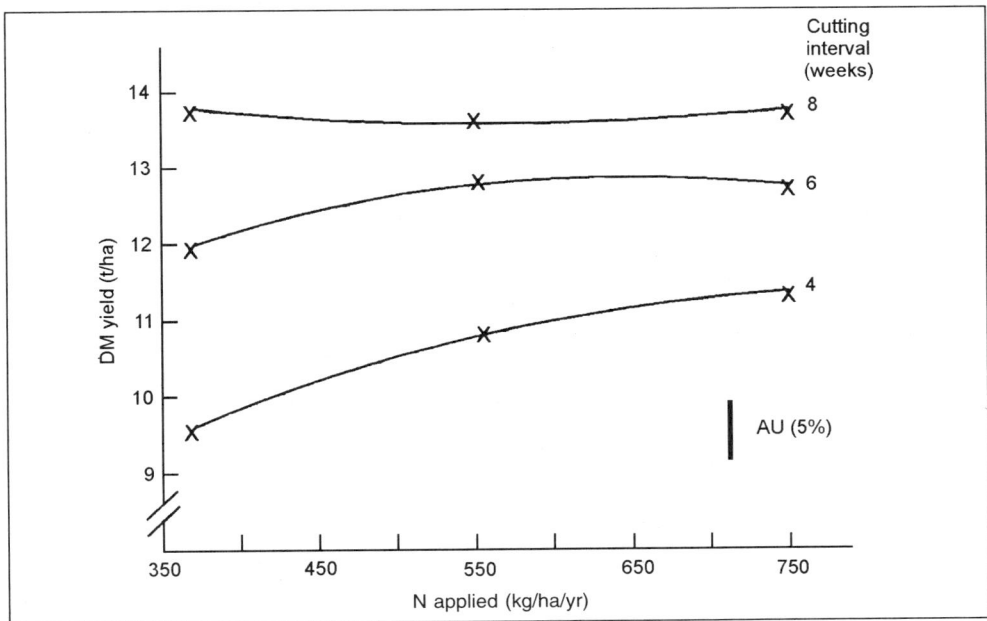

Figure 11.3 The influence of nitrogen (N) fertilizer and cutting interval on the total seasonal dry matter production of irrigated Italian ryegrass (Bartholomew & Miles 1984).

150–200 kg N/ha/yr. In the higher rainfall eastern areas, however, yield responses may continue to levels of 350 kg N/ha/yr or more, particularly in the warm eastern coastal areas where winters are mild.

Soil factors have an important influence on the pattern of response of pastures to fertilizer N. Where available soil N reserves are high, as in virgin soils and those previously under long-term pasture or where unfavourable factors (waterlogging, high soil acidity or mineral deficiencies) inhibit plant growth, responses are normally poor. Also, factors which promote the mineralisation of organic N, such as cultivation or liming, reduce the fertilizer N requirement.

11.1.2.3   Nitrogen effects on plant chemical composition

Nitrogen fertilizers may profoundly influence the chemical composition of herbage. In general, both dry matter and soluble carbohydrate content are depressed at high rates of N, while crude protein and non-protein N are increased (Anon. 1982), but these responses will be affected by stage of regrowth (Wilman 1975a, 1975b) (Table 11.4). Here crude protein and nitrate contents increased with N rate and decreased sharply with

Table 11.4   Yield and variations in chemical constituents in the regrowth of Italian ryegrass (data means over three seasons of experimentation in the UK; adapted from Wilman 1975a, 1975b).

N rate	Harvest interval (weeks)				
	1	2	4	8	12
(kg/ha)	DM yield (kg/ha)				
28	56	347	1 893	5 029	6 160
84	67	515	2 498	6 070	8 221
140	56	482	2 878	7 269	8 355
196	45	504	2 923	7 269	9 061
(kg/ha)	DM content of herbage (%)				
28	15.6	10.2	13.7	17.7	30.4
84	15.0	9.6	11.2	14.0	28.6
140	15.9	9.4	10.3	13.8	27.2
196	16.5	9.6	9.4	12.7	28.0
(kg/ha)	Digestible organic matter in herbage (*in vitro* % in DM)				
28	65.9	63.6	68.8	60.3	52.5
84	68.1	65.9	66.8	57.1	51.7
140	66.8	64.5	66.5	57.7	50.3
196	67.7	64.9	64.0	58.1	51.5
(kg/ha)	Crude protein content of herbage (% in DM)				
28	19.5	25.0	12.8	6.6	5.8
84	22.1	31.5	15.9	8.6	5.8
140	24.3	34.8	19.4	9.8	7.3
196	23.6	35.6	23.6	11.2	7.4
(kg/ha)	Nitrate-N content of herbage (% in DM)				
28	0.11	0.05	0.01	0.01	0.01
84	0.27	0.34	0.03	0.01	0.01
140	0.30	0.63	0.10	0.02	0.01
196	0.32	0.77	0.37	0.05	0.01

increasing age of the herbage beyond two weeks. However, the level of N applied had little or no influence on the digestibility of the ryegrass. Possible implications for animal health of the high nitrate levels in material harvested bi-weekly are discussed in section 11.9.1.

Levels of P, K, Ca and Mg in the plant may be increased or decreased by N applications, depending on the form in which N is applied and on the availability of these elements in the soil. In general, where nutrient levels in the soil are adequate, concentrations in herbage increase with increasing N supply. If, on the other hand, their levels are low, herbage concentrations of that element tend to decline with increasing N supply through dilution resulting from increased yields.

11.1.3 Symbiotic fixation of nitrogen

Nitrogen, which makes up 78% of the earth's atmosphere, can be converted into plant-utilisable forms by certain micro-organisms, and in particular by bacteria of the genus *Rhizobium* which associate with the nodules which develop on the roots of legumes. Throughout the world, use is made of the N-fixing capabilities of legumes to meet the N requirements of pastures. Their use does, however, have both advantages and disadvantages. Major advantages are improved herbage quality and substantial savings on fertilizer costs. Disadvantages include the possibility of bloat in animals and the often poor predictability of legume growth, and hence the amount of N fixed under local conditions.

11.1.3.1 Extent of nitrogen fixation

The amount of N fixed by legumes in mixed grass/clover swards is influenced by the soil, the climate and by management and is reported to vary from virtually zero to over 500 kg N/ha/yr (Whitehead *et al.* 1986b; Simpson 1987). Data of Smith (1987) suggest that, in the Mpumalanga Highveld, red or white clover grown with Italian ryegrass produce in the region of 170 kg N/ha/yr even where the soil has a relatively high residual N level.

Soil chemical and physical conditions greatly affect the N fixation processes. Poor legume performance and reduced N fixation may frequently be traced to excessive soil acidity and/or nutrient deficiencies, in particular of P, sulphur (S) and molybdenum (Mo). Limiting or excessive soil moisture may also seriously impair the N fixation process (Ryden 1984; Simpson 1987; Salette 1988). These considerations underline the importance of both sound irrigation scheduling and soil fertility management in legume-based pastures.

11.1.3.2 Transfer of nitrogen from legume to grass

There are several known mechanisms of N transfer from legumes to grasses. Direct transfer involves an exchange of N between legume plants and grass roots, either concurrently with fixation or, more importantly, after a phase of nodule decomposition in the soil (Simpson 1987). Nodule senescence is reportedly stimulated by defoliation and adverse environmental conditions (Chu & Robertson 1974) and is characteristically more rapid than that of the remaining root material. Interestingly, Herriott & Wells (1960) have reported N uptake by grass to be more closely related to the N yield of the clover in the previous than in the current season.

Other important mechanisms of N transfer from legume to grass are through the decomposition of ungrazed, senescent, or trampled leaf and stem components of legumes at the soil surface, and through animal ingestion of herbage and redistribution in dung and urine (Ball & Field 1987; Simpson 1987). The latter is, however, a relatively inefficient process because of the losses incurred.

11.1.3.3 *Grass–legume swards and fertilizer nitrogen*

In grass–legume pastures, legumes supply insufficient N for maximum dry matter production by associated grasses. Additional N therefore generally needs to be provided by strategic applications of low levels of fertilizer N. This, however, creates certain management difficulties, since the additional N reduces fixation by the legume (for example by 50 kg N/ha for every 100 kg fertilizer N applied in the range 0 kg to 300 kg N/ha (Ryden 1984)) and it also improves the relative competitive ability of the grass. However, spring applications to grass–clover swards may be highly profitable because both fixation and mineralisation of soil N are extremely slow at this time (Ryden 1984; Simpson 1987).

11.1.4   Nitrogen losses from pastures

Nitrogen may be lost from pastures by the removal of animal products, excretal deposition off the pasture (most of it being excreted in the urine), and by leaching, volatilisation and denitrification. The amounts removed in animal products are generally small. Of that consumed, the proportion utilised for the production of milk or meat is 15–25% for dairy cows, 5–10% for beef cattle and 10–15% for sheep and lambs (Whitehead *et al.* 1986a). Ball & Field (1987) concluded that an input of only 40 kg N/ha would compensate for all the N abstracted in milk and meat.

The amount of N lost through excretal deposition off the pasture is related to the system of animal management and may reach levels of 100 kg to 200 kg N/ha/yr where animals spend only a part of the day on the pasture, as they do in typical dairy operations in South Africa. Leaching, denitrification and volatilisation may, together or singly, also greatly affect the N economy of pasture systems. Fertilizer N, excretal N and that mineralised from soil organic matter may be lost by these pathways. Nitrate, because of its negative charge, is not retained on clay colloids but is free to move with the soil water. It is therefore the principal form in which N is lost by leaching.

Denitrification is a process whereby, under conditions of excessive soil moisture, the N in nitrate is converted to gaseous nitrogen and nitrous oxide which may then be lost to the atmosphere. This is carried out by certain soil bacteria. Its rate increases as soil temperature increases above 5°C and so denitrification tends to be least pronounced in winter. In general, leaching losses are likely to exceed losses by denitrification on well drained soils but on poorly drained soils the reverse is likely.

Volatilisation is the process whereby gaseous ammonia located at or near the soil surface is lost to the atmosphere. Topdressed N fertilizers, in particular urea and ammonium forms, and the N contained in urine are susceptible to being lost in this way. Such losses increase with increasing soil pH and increasing fertilizer application rate and are most pronounced when N is applied to an initially wet soil which then undergoes drying.

Ball & Ryden (1984) and Ball & Field (1987) have drawn attention to the major N losses which take place from ruminant excretal patches. The bulk of the N ingested by ruminants is excreted in the urine as urea. Amounts of N in urine patches equate to between 300 kg and 1 000 kg N/ha. These quantities greatly exceed the short-term N requirements of plants and major losses through leaching, volatilisation and denitrification are therefore inevitable, as shown in Table 11.5 (Ryden 1984). Grazing animals can therefore be viewed as causing two distinct high intensity processes in pastures: localised high N availability and rapid N uptake by plants, and localised accelerated N losses.

Table 11.5 Inputs, outputs and storage of N in cut and grazed ryegrass swards in the United Kingdom. Nitrogen fertilizer source was ammonium nitrate (after Ryden 1984).

	Cut ryegrass	*Grazed ryegrass*
	kg N/ha/yr	
Inputs		
Fertilizer	420	420
N fixation	0	0
Atmosphere	15	15
Outputs		
Herbage or liveweight	300	29
Leaching	33	160
Denitrification	20	40
Ammonia volatilisation	0	120
Storage		
In soil organic matter	70	100

11.1.5   Nitrogen fertilization in practice

11.1.5.1   *Application rates and schedules*

No soil N test suited to routine use is at present available, and so pasture N recommendations are based largely on the results of field fertilizer trials. Optimum N rates for pastures are dependent on the desired level of production. Advisory services, in making N recommendations for grass pastures, generally cater for three production levels, viz. low, medium and high. Expected yields of various pasture species at these production levels are listed in Table 11.6.

Fertilizer N requirements for the levels of production reported in Table 11.6 are as follows:

low production   = 60–150 kg N/ha/yr
medium production = 175–250 kg N/ha/yr
high production   = 275–375 kg N/ha/yr

The above represent general guidelines on N usage. High rates of N are pointless if growth rates will be limited by factors such as water stress, low temperatures or other nutritional disorders, and it clearly becomes uneconomical if there is no real need for the additional material which is produced.

Due to the tendency for applied N to be lost from the rooting zone and for heavy N dressings to promote dangerously high protein and nitrate levels in herbage, fertilizer N

Table 11.6   Dry matter production levels for various pasture species.

Species	Level of production		
	Low	Medium	High*
	t/ha		
Eragrostis curvula (lovegrass)	5–8	9–13	14–16
Pennisetum clandestinum (kikuyu)	5–8	9–13	14–16
Cynodon spp. (coastcross, star)	5–8	9–13	14–16
Paspalum dilatatum (Dallis grass)	4–6	7–12	13–16
Digitaria eriantha (Smuts')	4–6	7–12	13–16
Dactylis glomerata (cocksfoot)	4–6	7–11	12–15
Festuca arundinacea (fescue)	4–6	7–11	12–15
Lolium multiflorum (Italian rye)	4–6	7–11	12–16
Lolium perenne (perennial rye)	4–6	7–11	12–14

* Production at this level will be achieved only under good management and favourable climatic and soil conditions for the species in question.

is normally applied to pastures in split dressings throughout the growing season. It is best applied immediately after grazing or cutting. The amount applied in each dressing depends on the level of production required and on the length of the intervals between applications. Topdressings of from 40 kg to 75 kg N/ha generally give near maximum yields in each regrowth period. In colder, high lying areas, N applications should be reduced or withheld completely over the mid-winter period when growth is restricted by low temperatures.

Similar levels to the above are generally used at establishment, when the N may be included with the basal fertilizers prior to planting or soon after seedling emergence.

In grass/legume pastures where heavy N applications may suppress clover N fixation, recommended N levels are generally 25% to 50% lower than those for pure grass pastures. As noted earlier, the bulk of the N should be applied in spring and autumn to allow the grasses to express their full growth potential. In perennial ryegrass/clover pastures, N topdressings in early spring are particularly important since ryegrass exhibits a distinct growth-rate peak at this time.

11.1.5.2   Nitrogen fertilizers

Nitrogen fertilizers used in southern Africa are based on urea, ammonium nitrate and ammonium sulphate. These fertilizers do not differ greatly in price (per kg N) and, in practice, product availability, transport costs and a possible requirement for sulphur frequently determine the source used.

The typical compositions of some common sources of fertilizer N are reported in Table 11.7. Mixtures of N and P and NPK (not listed in Table 11.7) are widely used at establishment, while topdressings are usually made with 'straight' N fertilizers.

There appears to be little difference in the efficiency of the different forms of N fertilizer (Hyam & Clayton 1968; Nash & Tainton 1975; Rethman 1987; Hefer 1989; Eckard 1990). There have been reports, however, of high volatilisation losses accompanying the use of urea on sandy soils or on those with a near-neutral or alkaline pH, or on damp soils under drying conditions (Black *et al.* 1985, 1987; Whitehead & Raistrick 1990). These losses increase markedly with increasing application rates so that frequent light N topdressings are preferable to less frequent heavier applications. However, where urea is washed into the soil shortly after application, little or no N is lost by volatilisation.

Table 11.7 Typical compositions of some nitrogen fertilizer sources.

Source	N	P	S
		%	
Limestone ammonium nitrate (LAN)	28	–	–
Ammonium sulphate	21	–	24
Ammonium sulphate nitrate (ASN)	27	–	12.8
Urea	46	–	–
Ammoniated superphosphate	2.8	12.2	9.8
Monoammonium phosphate	11	22	1
Diammonium phosphate	18	20	–
Urea – ammonium nitrate (solution)	32	–	–
Ammonium nitrate (solution)	19	–	–

On poorly drained soils, and particularly in warm weather, limestone ammonium nitrate (LAN) has been shown to be a relatively poor N source due to high denitrification losses from the nitrate component of this fertilizer (Ball & Ryden 1984). Under such conditions, ammonium sulphate or urea should be used in preference to LAN.

In recent years there has been an increase in the use of liquid N fertilizers (urea ammonium nitrate (UAN) and ammonium nitrate solutions) on pastures. UAN has been found to be as efficient as LAN (Hefer 1989). A practical advantage of using liquid N is its low labour requirement since the solution is transported in tanks and transferred by pumping. When applied with a tractor mounted boom-spray, application is usually more accurate and uniform than with solid N fertilizers.

The soil acidifying potential of the various N fertilizers should be taken into account in pasture fertilizer programmes. The acidity generated by the 'straight' solid N fertilizers decreases in the order: ammonium sulphate > ammonium sulphate nitrate > urea > limestone ammonium nitrate. The continuous application of sulphur-based N fertilizers may promote rapid soil acidification (Table 11.8). Although urea, LAN and ammonium sulphate all increase exchangeable acidity and decrease pH, ammonium sulphate has twice the effect of the other two (Eckard 1989). Leaching losses of Ca and Mg may also be more pronounced where ammonium sulphate is applied. Persistent use of ammonium sulphate would, therefore, necessitate greatly increased use of lime.

Table 11.8 Variations in soil chemical properties (0–100 mm) with fertilizer nitrogen sources following the application of 360 kg N/ha in split dressings to irrigated Italian ryegrass over a ten-month period. Metz soil series; 37% clay (Eckard 1989).

	Fertilizer source			
	Zero N	Urea	LAN*	AS†
Exchangeable Ca (mg/ℓ)	872	749	758	569
Exchangeable Mg (mg/ℓ)	164	118	151	83
Exchangeable Al + H (cmol$_c$L)	0.48	0.85	0.80	1.94
Acid Saturation (%)	7.1	14.2	12.8	34.1
pH (KCl)	4.45	4.28	4.37	4.03

* Limestone ammonium nitrate
† Ammonium sulphate

The use of slurries and manures from milking parlours and holding areas can cut costs by reducing the need for inorganic fertilizers, but it is then not possible to control the ratio of nutrients being applied, and special management and equipment may be required to restrict losses of N. A further advantage in their use is that they improve soil physical properties such as infiltration rate, and are therefore particularly useful for annual pastures. In some situations, the use of organic fertilizers such as chicken litter may be justified because of their effect on soil physical properties. However, the N contained in these organic fertilizers is released only slowly as the organic matter is mineralised.

11.2 PHOSPHORUS

In stark contrast to the situation in many intensively-farmed overseas countries, P is severely deficient in the majority of virgin soils in southern Africa, and particularly in the high potential dystrophic and mesotrophic soils of the humid eastern areas. Plant response to applied P on these soils is often spectacular, and correcting P deficiency presents a formidable economic obstacle in pasture and cropping programmes.

11.2.1 Forms of soil phosphorus

The total P content of soils varies greatly, depending on the parent material and intensity of weathering. Phosphorus exists in both the organic and inorganic form. The inorganic fraction occurs in numerous combinations with iron, aluminium, calcium, fluorine and other elements, many of which are very insoluble. Phosphates may also be bonded to the surfaces of clays and other inorganic soil constituents. The organic fraction, on the other hand, is found in humus and in plant, animal and microbial residues. Organic P relations in soils are in many respects analogous to those of organic C and N. Management practices which promote the build-up of soil organic matter result in an accompanying increase in organic P levels. Furthermore, like N, organic P is subject to mineralisation/immobilisation reactions. However, little is known of the role organic P plays in regulating P supply to plants.

11.2.2 Phosphorus sorption

The availability of P applied to soils is reduced through reactions with soil constituents. These reactions (referred to as phosphorus sorption, immobilisation or fixation) are extremely complex and are influenced by many factors, the most important of which are the type and amount of clay. Phosphorus sorption is particularly intense in medium to fine textured highly weathered soils which typically have high levels of kaolinite, oxides and hydroxides of iron and aluminium and organic matter–aluminium complexes. The less crystalline the oxides and hydroxides, the larger their P fixation capacity because of their large surface area. In these soils, which occur extensively in the relatively moist eastern parts of southern Africa, correction of P deficiency poses a major economic constraint.

Contrary to popular belief, soil acidity is not the reason for strong sorption of P and liming generally has little effect on the ability of a soil to immobilise P.

An indication of both the intensity and variability of P sorption in soils is provided by the data presented in Table 11.9. Here the amounts of P (kg P/ha) required to increase the soil test for individual soils by a single unit are indicated. In the absence of sorption, the P requirement for raising the soil test by a single unit in the top 150 mm soil layer would be 1.5 kg P/ha. Table 11.9 provides evidence of considerable sorptions, particularly on the heavy clay dystrophic soils. In virgin soils of this type, P applications of over 100 kg P/ha may be required for successful pasture establishment.

Table 11.9 Phosphorus required to raise the soil test P level by a single unit in a number of KwaZulu-Natal soils (NH_4HCO_3/EDTA/NH_4F soil test; 0–150 mm soil layer).

Soil	Locality	Clay %	pH (KCl)	P requirement kg P/ha
Avalon	Dundee	11	4.5	3.4
Katspruit	Kokstad	30	4.5	6.3
Hutton	Cedara	53	4.0	16.4
Clovelly	Rietviei	47	4.2	13.2
Griffin	Nottingham Road	61	4.0	15.8

11.2.3 Role of phosphorus in the nutrition of pasture plants

11.2.3.1 Function, uptake and phosphorus content of plants

Phosphorus is an essential macronutrient for all living organisms. It is involved in a number of physiological functions within the plant including energy transformations in the form of adenosine triphosphate (ATP) and carbohydrate transformations such as the conversion of starch to sugar. It is also a constituent of the genetic material in the cell nucleus (Beard 1973). Phosphorus is absorbed by plant roots mainly as the monovalent di-hydrogen phosphate ion $H_2PO_4^-$.

The total P content of pasture plants varies between species and is also dependent on growth stage. Plants adequately supplied with P generally have P contents in the range 0.25% to 0.4% in the dry matter. Thus plant P contents are roughly one-tenth those of N or K.

11.2.3.2 Phosphorus and pasture establishment

Plants require more P during early growth than at maturity (Baylor 1974; Kerridge 1978; Mays *et al.* 1980). For example, Fox *et al.* (1974) found that, whereas 0.2 mg/L P was required in the soil solution for establishment of *Desmodium aparines*, only 0.01 mg/L was required for regrowth after the first harvest. Data from KwaZulu-Natal reflect the same need for P following autumn establishment (Fig. 11.4). Adequate P at establishment promotes root growth and branching (Beard 1973), thereby minimising the effects of weed competition.

11.2.3.3 Seasonal patterns in response to phosphorus

Pasture plants, and particularly those of temperate species, are most responsive to P when they are under some form of stress, as during dry (Miles *et al.* 1985a; Saunders *et al.* 1987) or cold (Miles *et al.* 1991) periods. Kikuyu is unusual among tropical species in that it has a greater P requirement in spring/early summer than when growth is at its peak in mid-summer (Miles 1986). The effects of temperature on the availability of soil

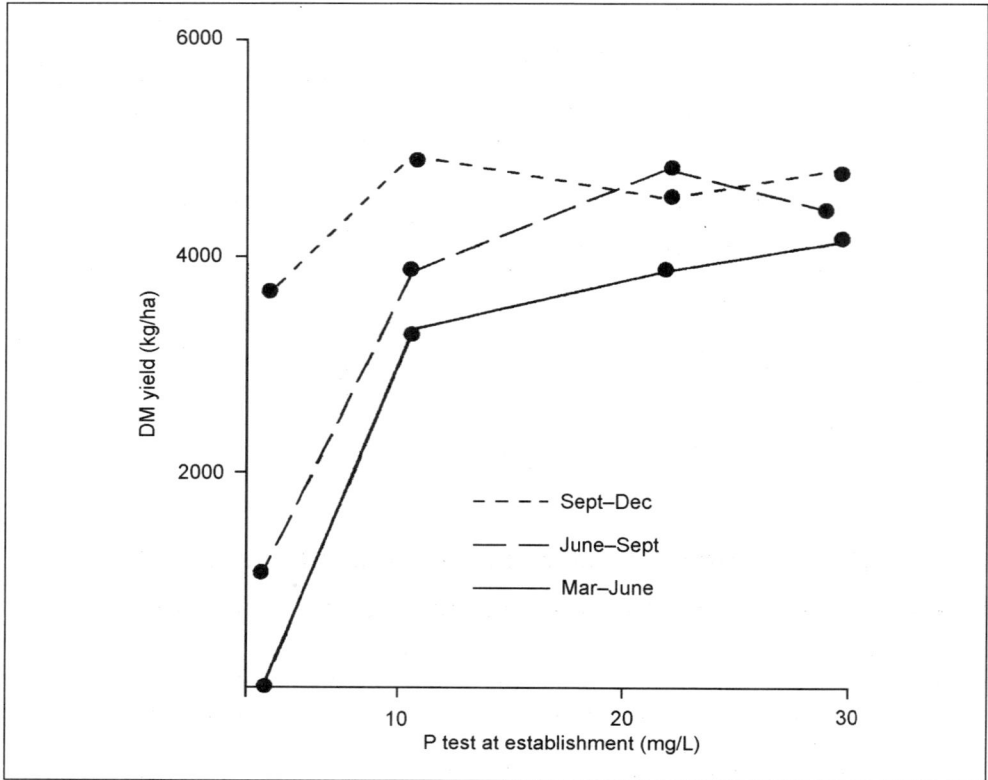

Figure 11.4   Effects of applied phosphorus (P) on the dry matter production of irrigated Italian
ryegrass (cv. Midmar) over three periods on a Cleveland soil in the Highland
Sourveld of KwaZulu-Natal.

P presumably contribute to these seasonal variations in P response; a decrease in soil P
availability with decreasing soil temperatures has been widely reported (Sutton 1969;
Barber 1980).

The fact that the response to P is often greatest during slow growth periods is of
considerable practical significance. Where pasturage is the sole feed source, stocking
rate will normally be governed by the growth rate of the pasture during such periods. In
order to increase stocking rate, therefore, low soil P should not be allowed to restrict
growth at these times.

11.2.3.4   Species differences

Some important differences exist between pasture species in both their ability to extract
P from the soil and in their overall P requirements for growth. In general, legumes utilise soil P less efficiently than do grasses and therefore have a higher soil P requirement.
This has been ascribed largely to the more extensive root network of grasses and so
their more extensive exploration of the soil mass.

Significant differences exist among the grasses in P utilisation and P requirement. In
general, tropical grasses utilise P more efficiently than temperate grasses (Andrew &

Johansen 1978) but noteworthy species differences have emerged even among the former. As an example, on virgin Mistbelt soils (>50% clay), the average yields of the tropical grasses kikuyu and *E. curvula* in the absence of applied P were 34% and 57% of their respective maximum yields (Miles 1988). Also, the soil P test threshold for maximum yield was higher in kikuyu than in *E. curvula*, while the required herbage P content for maximum yield was 2.2 g/kg to 2.6 g/kg for kikuyu and only 1.5 g/kg to 1.6 g/kg for *E. curvula*. The latter species therefore has both a lower internal (plant) and external (soil) P requirement than kikuyu. Further evidence is provided by the data of Fig. 11.5. Here kikuyu responded poorly to the addition of P; clover responded relatively strongly, with a yield maximum at 100 kg P/ha; and ryegrass responded very strongly, to as much as 150 kg P/ha and a yield increase of 268%. The greater requirement of ryegrass than of clover for P is at variance with reports that legumes are more responsive to P than grasses. However, as discussed earlier, lower soil temperatures during the main growth period of ryegrass (autumn through to spring) may contribute to it responding more than the summer-growing clover.

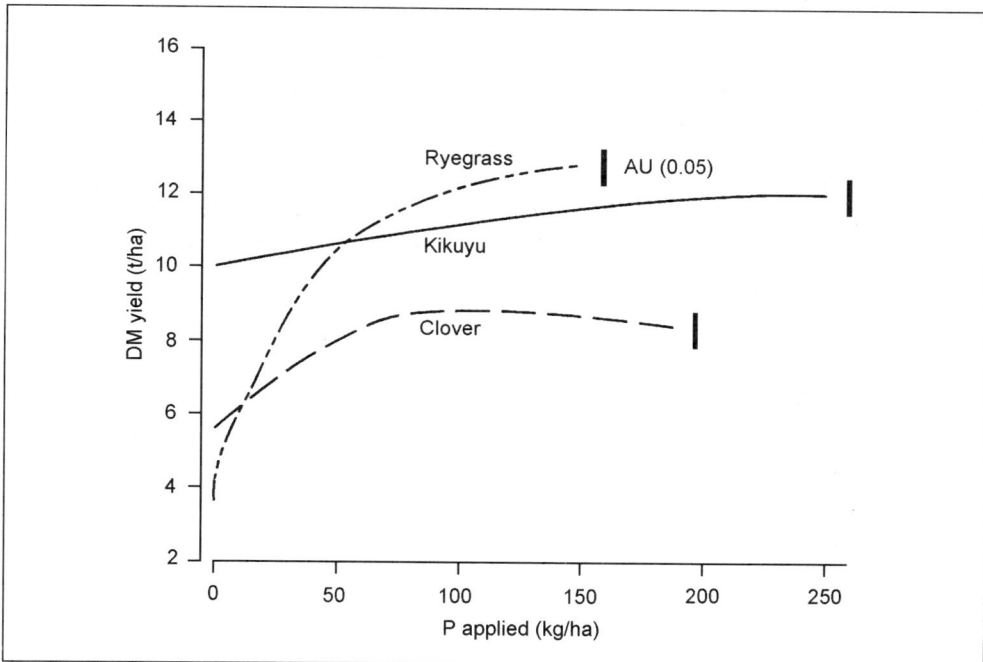

Figure 11.5 Effects of applied phosphorus (P) on the dry matter production (seasonal totals) of kikuyu, white clover (cv. Ladino) and irrigated Italian ryegrass (cv. Midmar) on a Cleveland soil in the Highland Sourveld of KwaZulu-Natal.

11.2.3.5 *Phosphorus and symbiotic nitrogen fixation*

The overriding importance of P for the survival of, and N fixation by, legumes in a pasture has been widely recognised. Phosphorus plays a major role in nodule development and in the activity of the associated rhizobia (Andrew & Johansen 1978; Crowder & Chheda 1982) to the extent that N concentrations in tops of legumes increase with

P supply beyond the P rates required for maximum yields (Andrew & Robins 1969; Ozanne 1980). This implies that maximum N fixation requires a higher P supply than does maximum dry matter production. Interestingly, because of the effect of P on N fixation in legumes, grass grown with clover in the absence of applied N responds to much higher levels of P than when grown alone and supplied with N fertilizer (Ozanne 1980). Here the grass tends to respond to the level of applied P which gives rise to maximum N fixation, rather than to that required by the grass for its own needs. Not surprisingly, therefore, animal performance on legume-based pastures is often closely related to inputs of P fertilizer.

11.2.3.6 Interaction of phosphorus with nitrogen

Two or more growth factors may interact when their individual effects are modified by the presence of one or more of the others. When a combination of factors results in a growth response that is greater than the sum of their individual effects, the interaction is positive, and when the combined effects are less, the interaction is negative. Additivity indicates the absence of interaction.

The importance of interactions to future advances in farm profitability have been highlighted in excellent reviews by Wagner (1981) and Sumner & Farina (1986). Phosphorus interactions with both N and lime are frequently of major economic significance in intensive pasture operations. Phosphorus–lime interactions are dealt with in section 11.4.4.3.

Typically, P–N interactions are positive. Yield responses of kikuyu on a virgin Balmoral soil (clay = 54%) in the Mistbelt of KwaZulu-Natal provide an example of this type of interaction (Fig. 11.6). Here, responses to both N and P were small or negligible

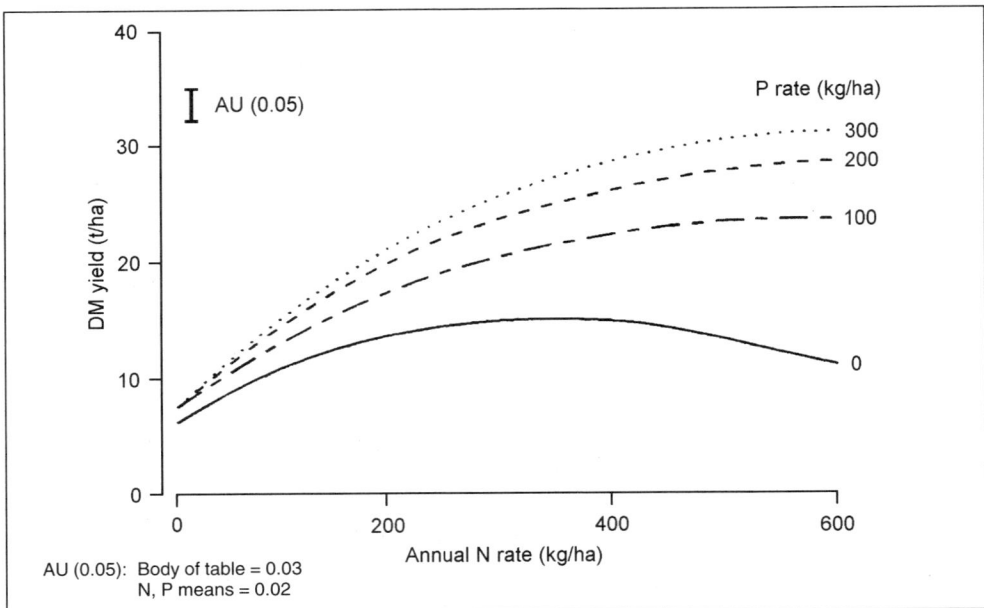

Figure 11.6 Effects of phosphorus (P) (applied at establishment) and nitrogen (N) on the total dry matter production of kikuyu over four seasons on a Balmoral soil.

at deficient levels of the other nutrient. However, yield increases in response to combinations of high P and N rates were substantial, indicating an increasing P requirement with increasing N rate. These interactive effects are thought to be due mainly to N-induced P absorption by the plant (Adams 1980; Sumner & Farina 1986). Data presented in Table 11.10 provide evidence of this. On this severely P deficient soil, applied N had no effect on plant P concentrations in the absence of applied P. With increasing P applications, however, N additions markedly increased P concentrations.

Banding N and P fertilizers together has often been shown to promote root development and increase P uptake (Adams 1980). In this respect ammonium (NH_4^+) fertilizers have a greater stimulatory effect on P absorption than do nitrates (NO_3^-). This probably results, at least in part, from increased solubility of soil P in response to the acidity generated by nitrification of the NH_4^+.

Table 11.10 Variations in kikuyu P concentration (sampled in mid-January) with applied N and P on a virgin mistbelt soil.

Establishment P	Annual N rate (kg/ha)				Mean
	0	*200*	*400*	*600*	
kg/ha	%				
0	0.21	0.21	0.21	0.21	0.21
100	0.23	0.23	0.24	0.25	0.24
200	0.26	0.28	0.29	0.30	0.28
300	0.28	0.35	0.37	0.39	0.34
Mean	0.24	0.26	0.28	0.29	

11.2.4 Phosphorus fertilizers and their properties

11.2.4.1 Sources

Commercially available inorganic P fertilizers may be divided into two groups, depending on their solubilities. 'Readily-available' or 'water-soluble' sources include the superphosphates, ammoniated superphosphate, and mono- and di-ammonium phosphates. These products are formed by a reaction of ground rock phosphate with sulphuric or phosphoric acids, and when N is a component, by inclusion of ammonia in the manufacturing process. All the P in these products is essentially water-soluble and readily available for plant uptake. The chemical compositions of these P sources are listed in Table 11.11. In addition to use as 'straight' P and NP fertilizers, these products are used in the manufacture of granulated and 'bulk blend' NPK mixtures.

Table 11.11 Typical compositions of water-soluble P fertilizers (%).

Type of fertilizer	P	N	Ca	S
Superphosphate (single supers)	8.3–10.5	–	20–22	10–12
Double superphosphate	19.6	–	16	3
Ammoniated superphosphate (AMP)	12.2	3.8	17.1	9.8
Monoammonium phosphate (MAP)	22.0	11.0	0.7	1.8
Diammonium phosphate (DAP)	20.0	18.0	0.4	–

The second group of inorganic P fertilizers are the 'sparingly soluble' forms. These include products such as the rock phosphates, partially acidulated rock phosphates and so-called thermophosphates. Rock phosphates differ widely in their fertilizer value depending on their origin. Hard crystalline apatites are very insoluble and of little fertilizer value. In some rock phosphate deposits, however, the solubility of apatite is increased by a substitution within the crystal lattice of phosphate by carbonate, and fluoride by hydroxyl ions (Le Mare 1991). Partially acidulated rock phosphates appear to be assuming increasing importance in overseas countries. These products are prepared by treating rock phosphates with less (usually 20–50%) than the amount of sulphuric or phosphoric acid necessary to produce completely water-soluble sources such as superphosphate. Thermophosphates comprise rock phosphates which have been thermally treated with silicates and carbonates (Sanchez 1976). The reactivities of both partially acidulated rock phosphates and thermophosphates tend to be intermediate between those of water-soluble P fertilizers and rock phosphates.

Organic fertilizers, when used, contribute to the organic P fraction, about which little is known. However, it is known that organic P in manure is fairly easily mineralised, and it may be assumed that 50% of the P in manure will become available to plants in the season after incorporation.

11.2.4.2 *Reactions in soil and plant availability*

The relative merits of water soluble and sparingly soluble P sources continue to be the subject of much debate. The incentive to use sparingly soluble P sources stems from their appreciably lower cost per unit of total P. However, the low efficiency of these sources under certain conditions frequently outweighs any cost benefits.

Water soluble P sources rapidly dissolve in moist soil, irrespective of the soil's properties. The released P is available for plant uptake and for reaction with soil components (sorption), the latter reducing its solubility. The initial reaction products between soil and P are relatively soluble in soils with a high capacity for P sorption. However, much of the applied P is eventually converted into plant-unavailable forms.

The solubility and hence plant availability of rock phosphates and similar products are governed by a number of factors, including the nature of the product itself and the soil to which it is applied. The finer the rock phosphate is ground, the greater its reactivity, and grinding to ensure that 90% of the rock has a particle size of less than 0.15 mm is generally recommended (Le Mare 1991). Solubility increases with increasing acidity (decreasing pH) and decreasing exchangeable Ca. In addition, uptake by plants is greater in coarse (sandy) than in fine-textured soils. In the light of the above considerations and since the immediate availability of P from rock phosphates is low, their use should normally be restricted to perennial pastures and crops on inherently acid soils. An adequate supply of soluble P, such as from superphosphate, is essential for pasture establishment on P deficient soils. This is well illustrated by the fescue yield data presented in Fig. 11.7.

11.2.5 Phosphorus fertilization: practical considerations

11.2.5.1 *Predicting requirements*

Soil testing provides the only reliable means of estimating the P requirement. For a soil test to be of practical value it must have been calibrated with yield responses for the particular species and soils under consideration. Examples of calibrations for Italian

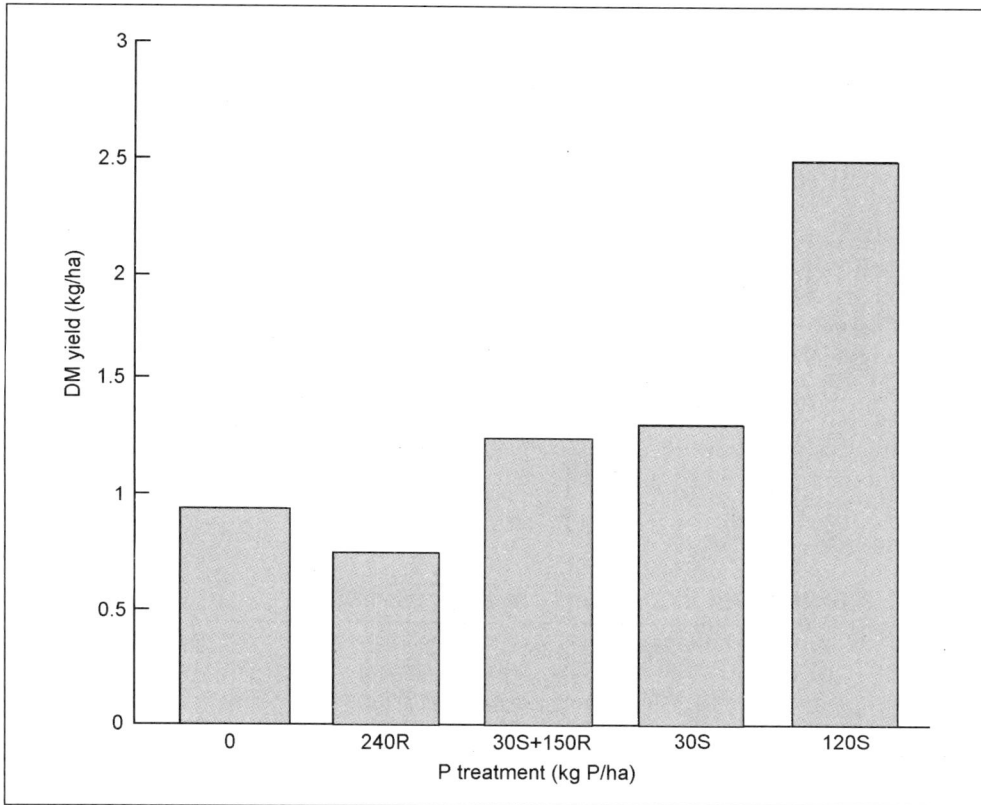

Figure 11.7 The influence of phosphorus (P) fertilizers on the first harvest yield of irrigated tall fescue established on a Hutton soil in the Mistbelt of KwaZulu-Natal (R = rock phosphate; S = double superphosphate).

ryegrass and *E. curvula* are presented in Fig. 11.8. The dashed horizontal and vertical lines in these graphs conveniently separate the data points into responsive and non-responsive populations, and thereby provide indications of threshold P-test values (Cate & Nelson 1971). These values vary between species (Fig. 11.8), and also between soils differing texturally. Threshold (optimum) P test values are higher on sandy soils than on clays, while the amount of fertilizer P required to raise the soil test by a single unit increases with increasing clay content. Where fertilizers are to be incorporated into the top 10–15 cm of topsoil, typical amounts of P required for unit increase in soil test are 10 kg P/ha for a clay soil, 7 kg P/ha for a loam soil and 3 kg P/ha for a sand soil. If the P is to be applied as a topdressing, the factor for loams and clays is 5 kg P/ha, and 3 kg P/ha for sands. Unfortunately, relatively few P test calibrations currently exist for pastures in southern Africa.

It should be borne in mind that extractants used to determine soil nutrient levels frequently differ from one laboratory to another, and these extractants vary in the proportions of available nutrients they extract. This means that the threshold soil test value for a particular nutrient may vary widely, depending on the extractant used. Since

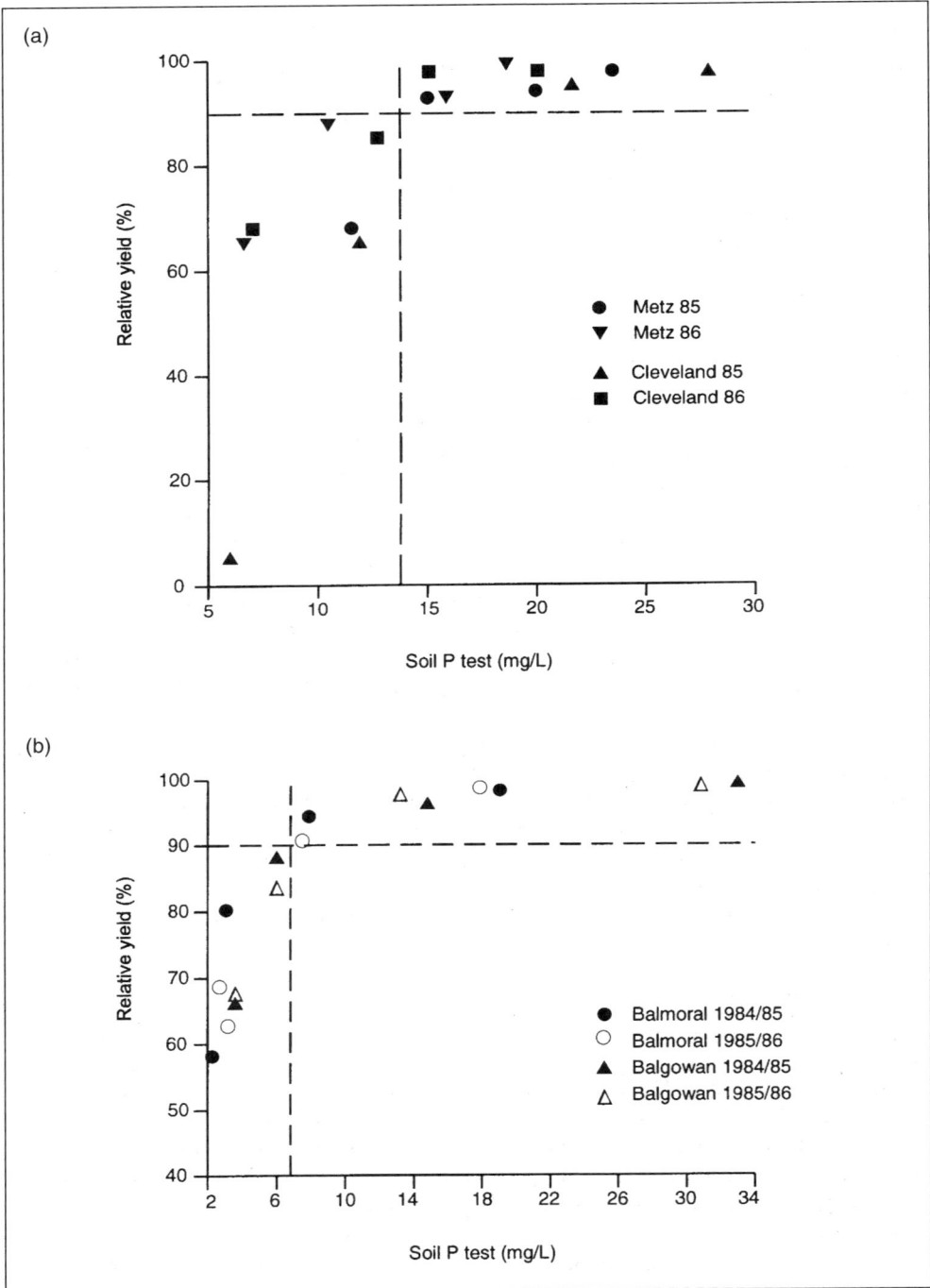

(a)

(b)

Figure 11.8   The relation between soil phosphorus (P) test (NH_4HCO_3/EDTA/NH_4F) and the relative yields of (a) Italian ryegrass in autumn (March–June) on two soils and (b) *Eragrostis curvula* (total yields over two seasons) on two soils.

relations between soil test values obtained with different extractants are generally poor (in particular for P), fertilizer recommendations must be based on calibrations developed for the extractant in question.

The use of herbage P concentrations for evaluating fertilizer P needs has met with mixed success. Species differ markedly in their 'critical' P levels, while P concentrations in plants may be greatly influenced by stage of growth, time of year and moisture supply. For example, in the annual Italian ryegrass higher critical P values apply in autumn than in spring. The summer-growing perennial, kikuyu, on the other hand, appears to have a higher internal P requirement in spring than at other times. Tissue testing cannot, therefore, be seen as a reliable method of establishing the P needs of plants (Gartrell & Bolland 1987).

11.2.5.2 *Timing and placement*

Since the P requirements of pastures are greatest during the establishment phase, adequate amounts of water-soluble P must be supplied at planting. Establishment P is usually broadcast and incorporated to a depth of 50 mm to 150 mm immediately before planting. With row planting, band placement of P directly underneath the seeded row has been found to greatly enhance seedling growth on P deficient soils.

Correcting P deficiencies in established pastures generally implies topdressing, although no-till planters do permit the sub-surface banding of P. The effectiveness of surface applications of P to pastures has often been questioned because of the immobility of P but evidence suggests that it is indeed effective (Mays *et al.* 1980) since the actively absorbing roots of most pasture species are concentrated in the top 50 mm to 100 mm of soil where they have access to surface P (Ozanne *et al.* 1965; Jackman & Mouat 1972; Mays *et al.* 1980). Even deep-rooted lucerne is reported to make efficient use of surface-applied P. An obvious disadvantage of concentrating P and other nutrients at the soil surface is, however, that nutrient uptake will be restricted under dry, hot conditions. The response to topdressed P may therefore be poor where the soil surface frequently dries out, and it may be the consequent restriction of nutrient uptake, rather than moisture stress, which is largely responsible for slow growth at such times (Ozanne & Sewell 1980). In irrigated pasture, frequent light water applications are preferable to less frequent heavy applications to ensure efficient utilisation of immobile topdressed nutrients.

11.3 POTASSIUM

11.3.1 Potassium in the soil

Soils generally contain large amounts of K relative to plant requirements, but much of it is usually unavailable to plants. Soil K can be grouped into three fractions according to its availability: non-exchangeable mineral K, exchangeable K and solution K. Non-exchangeable K is found in the crystal lattice of relatively unweathered minerals or in the interlayers of clay minerals such as illite and vermiculite. It is available to plants only after weathering to the exchangeable form. Exchangeable K^+ is held by electrostatic forces to negatively charged clay and organic matter surfaces. Solution K is that minute fraction which is present in the soil solution. An equilibrium exists between the above three forms. That between the former two fractions develops only slowly, while the development of an equilibrium between the exchangeable and solution forms is rapid. In most soils, short term plant K supplies are derived essentially from the exchangeable and solution forms. Both mineral lattice K and interlayer (non-exchangeable) K are released gradually into the plant available forms, but in soils containing appreciable amounts of illites, vermiculites and chlorites, the reverse reaction may occur. Here fertilizer K may be trapped as interlayer K, thus reducing its availability to plants.

11.3.2 Role of potassium in pasture plants

The primary role of K in plants is metabolic – there are numerous plant physiological and biochemical functions for which K is indispensable (Mengel & Kirkby 1987). Of particular importance is that K deficiency reduces tolerance to drought and frost and increases susceptibility to fungal attack and lodging.

11.3.2.1 Plant uptake

Potassium is the nutrient cation generally taken up in the largest amounts by pasture plants. The K concentration in above-ground plant parts varies widely between species, with temperate grasses, in particular, requiring high K contents for optimum growth. Critical levels for grasses and legumes usually fall within the range 1.5 % to 3.0 % of the dry matter. Potassium concentrations decline rapidly as plants mature. Because it is highly mobile in plants, under conditions of deficiency it is transported from older to younger tissues. Hence deficiency symptoms, usually in the form of browning or 'firing' of leaf tips and edges, first appear in older leaves.

The dense, fibrous root systems of grasses makes them highly efficient utilisers of soil K. This allows grasses to compete successfully with legumes for limited soil K reserves. An adequate K supply is, therefore, essential for the maintenance of a vigorous legume component in mixed swards. Data presented in Fig. 11.9 show that white clover (a legume) was more severely affected by a limited supply of K than the grasses. Also of interest in these data is that the tropical kikuyu utilised K much more efficiently than the temperate Italian ryegrass.

11.3.2.2 Interactions with other elements

Relations of considerable practical significance exist between K and N, Ca, Mg and sodium (Na).

The N status of plants has been widely shown to have a considerable influence on requirements for K (Kresge & Younts 1963; Prins & Den Boer 1985; Smith *et al.* 1985). Where K supplies are adequate, its uptake increases with increasing N supply, and the required K concentration (critical level) for maximum yield has been found to increase with increasing N concentration. Data from Prins *et al.* (1985) show, for example, that recommended herbage K concentrations in ryegrass increase steadily from 19.7 g/kg DM at a herbage N concentration of 8 g/kg DM to 38.7 g/kg DM at a herbage N concentration of 48 g/kg DM. Talibudeen *et al.* (1976) concluded that plants limit K uptake in the presence of an N deficiency and that they also limit N uptake in the presence of a K deficiency.

A strong antagonism in nutrient uptake exists between K and the cations Ca, Mg and Na. Excessive K supply may, as a consequence, lead to low herbage Ca, Mg and Na levels. Although this depressed uptake is generally not sufficiently severe to influence

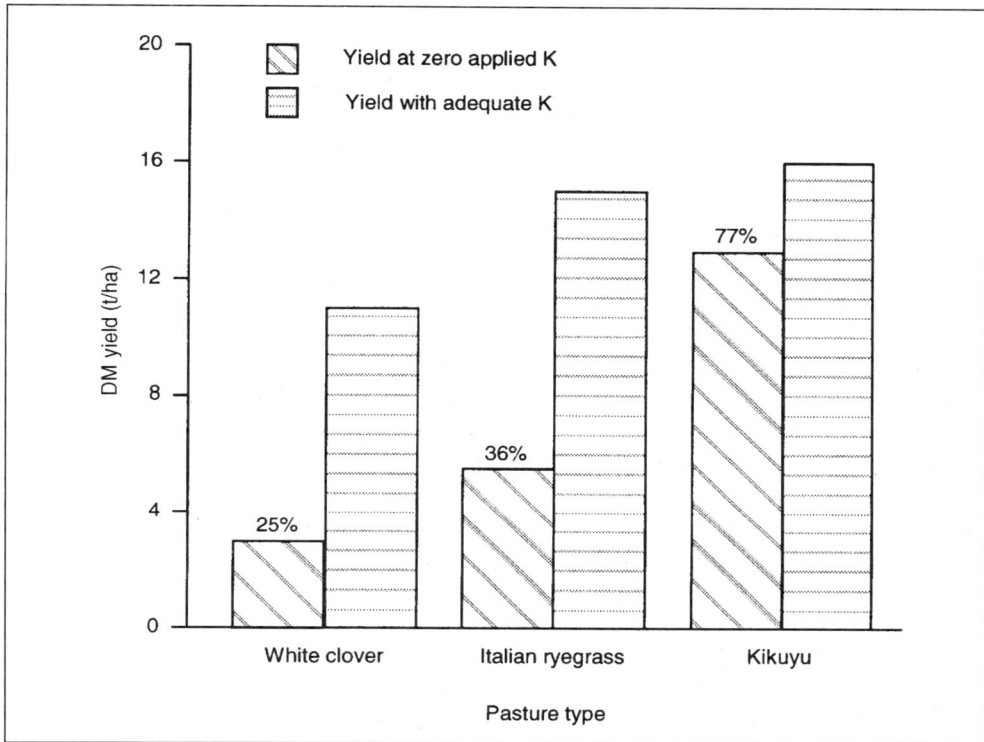

Figure 11.9 Yield responses (under cutting) of three pasture species to applied potassium (K) on a K deficient soil in the Highland Sourveld of KwaZulu-Natal (numbers presented on top of the zero K columns are the percentage yields at zero K relative to the respective yields with adequate K).

plant growth, the resultant altered mineral balance may be detrimental to the health of grazing animals.

11.3.3   Potassium fertilization: practical considerations

In their virgin state, the majority of soils in southern Africa contain sufficient plant-available K to sustain short-term pasture production. However, non-exchangeable K reserves, especially in high potential well drained soils, are extremely low. Therefore, where large amounts of K are removed from the pasture, fertilizer K addition is needed for sustained pasture production.

11.3.3.1   Leaching and fixation of potassium

Leaching of K is restricted to sandy soils (<15% clay). In loam and clay soils, including highly weathered soils which have had their cation exchange capacities increased by liming, K is essentially immobile.

Fixation of K may greatly affect fertilizer K requirements. It is negligible in highly weathered soils whose clay mineral suites are dominated by 1:1 clays and oxides and hydroxides of iron and aluminium. However, where clay minerals such as illite and vermiculite abound, K fixation may appreciably increase fertilizer K requirements. In KwaZulu-Natal, for example, the amount of fertilizer K required to increase the soil test in the top 150 mm is generally less than 2 kg K/ha for highly weathered soils (irrespective of clay content) and up to 8.8 kg K/ha for structured, high base status soils (Johnston pers. comm.).

11.3.3.2   Potassium fertilizers

Potassium may be applied in the inorganic form as a 'straight' fertilizer or in NPK compounds or blends. By far the most widely used and cheapest inorganic K fertilizer is potassium chloride (muriate of potash) which contains 50% K. Other 'straight' K fertilizers include the more expensive potassium sulphate and potassium nitrate which are generally used on specialist crops such as tobacco, fruits and vegetables.

All the K contained in organic fertilizers is plant-available. Manures and slurries, although often neglected, are potentially valuable on-farm sources of K.

11.3.3.3   Impact of pasture management on potassium balance

Pasture management options, such as the amount of N fertilizer used and the method of pasture utilisation, have major impacts on K fertilizer requirements. As indicated earlier, a close relation exists between plant N and K contents. As yields increase in response to increasing N supply, K requirements and resulting K responses also increase. An example of this is provided by the coastcross II data presented in Table 11.12. In the first season (1976/77), 8 t to 9 t DM/ha were produced in the absence of applied K, irrespective of N rate. However, when soil reserves had been depleted by the fourth season (1979/80), a satisfactory response to applied N was forthcoming only where K had been topdressed annually.

Table 11.12   The effects of applied N and K on the dry matter production (cutting-and-removal) of coastcross II on a Hutton soil in the KwaZulu-Natal Mistbelt (Miles *et al.* 1985b).

Annual N (kg/ha)	Annual K (kg/ha)		
	0	200	400
	DM (kg/ha)		
		1976/77	
100	8 400	9 928	10 884
300	9 348	10 429	10 478
500	8 567	10 095	10 185
AU (0.05)	1 158		
		1979/80	
100	2 851	3 997	3 830
300	3 993	10 548	9 958
500	4 893	13 138	12 891
AU (0.05)	1 163		

Not surprisingly, K requirements of pastures are determined, to a large extent, by the method of pasture utilisation. Where the herbage is cut and removed, soil K levels are rapidly depleted and fertilizer K requirements, particularly under high N fertilizer regimes, are substantial (Table 11.12). Under grazing, where the bulk of the ingested K is returned to the pasture, additional K requirements are generally low or negligible. Indeed, K may reach alarming levels under permanent pasture. Winter feeding of stock on pasture, together with the injudicious use of fertilizers and slurry, are largely responsible for this. An important distinction therefore needs to be made between systems in which animals remain permanently on the pasture and those in which animals spend only a part of each day on the pasture.

11.3.3.4 Predicting pasture potassium requirements

In general, both the system of pasture management as well as soil test data are used to establish fertilizer K recommendations for pastures. At establishment, any deficiency of K should be rectified on the basis of a soil test. Thereafter, fertilizer K topdressings are based on estimated K losses resulting from the method of pasture utilisation. In perennial pastures, the soil should be tested at least once every second year to monitor the balance of K and other nutrients.

Optimum exchangeable K levels for pasture species, established in field calibration trials in KwaZulu-Natal and the Mpumalanga highveld, generally fall within the range 60 mg to 120 mg K/L soil. Where K fertilizer is incorporated into the topsoil, 2.5 kg K/ha is required to raise the soil test by 1 mg K/L. Where K is topdressed on pastures, 1.5 kg K/ha is required for a one unit increase in soil test value. As intimated earlier, K competes strongly in cation uptake and exchangeable K optima appear to be largely independent of wide variations in the composition of the soil cation suite. Data presented in Table 11.13 serve to illustrate this. Here a doubling of the soil Ca level did not significantly influence uptake of K by ryegrass. Potassium requirements should not, therefore, be related to exchangeable cation ratios (e.g. K:Ca+Mg) (Edmeades pers. comm.; Farina *et al.* 1992).

Table 11.13   The effects of varying soil (0–100 mm) K and Ca levels on K concentrations in Italian ryegrass.

Ca soil test (mg/L)	K soil test (mg/L)			
	66	122	192	374
	% K			
818	2.83	4.40	5.38	6.19
1 445	2.69	4.16	5.04	5.97
1 782	2.73	4.01	5.36	6.08
AU (0.05) = 0.55				

As with P, establishing K sufficiency from plant analysis is fraught with difficulties. Critical K levels vary appreciably between species and there are few reliable values for pastures. Also, adequate K concentrations for optimum growth vary enormously with the stage of plant maturity and N supply, while topsoil moisture status affects K uptake (Farina *et al.* 1992). Considerable circumspection is therefore needed in the interpretation of leaf K data.

11.4 SOIL ACIDITY AND LIMING

Acidity is a widespread limitation to pasture and crop production in the higher rainfall areas of southern Africa. Control of soil acidity is facilitated by liming. In this section, the nature of soil acidity, the function of lime in pasture systems and practical aspects relating to the use of lime on pastures are discussed.

11.4.1 The nature of soil acidity

The most widely used, but not the most useful, chemical measurement of soil acidity is pH. This reflects the amount of active hydrogen (H^+) present. The lower the pH, the more hydrogen is present. The pH (measured in KCl) of soils having excessive acidity (from a plant growth point of view) usually lies between 3.9 and 4.5 (pH values measured in water are generally 0.7 to 1.0 units higher than KCl values). Hydrogen itself is, however, seldom harmful to plant roots. Rather it is the increase, as pH decreases, in the solubility of highly toxic aluminium (Al) which is the major cause of acid soil infertility. Aluminium restricts root growth of acid sensitive plants and thereby limits their ability to extract water and nutrients.

Laboratory measurements of soluble Al levels in soils in most cases automatically include H^+, and the combined Al + H^+ thus determined is referred to as 'extractable acidity'. Since H^+ represents only a small fraction (usually <5%) of the total extractable acidity, the latter essentially reflects the amount of soluble (or 'exchangeable') Al in the soil. An example of the relation between soil pH and extractable acidity in KwaZulu-Natal soils is shown in Fig. 11.10a. Notable here is the evidence that for a specific pH value within the range 3.8 to 4.4, the level of exchangeable acidity varies considerably in the two types of soil. Therefore pH alone is a poor indicator of the severity of the soil acidity hazard. A number of factors account for this poor relation, with soil texture being one of the most important. Exchangeable acidity is generally lower in sandy than in clayey soils.

Although the amount of exchangeable acidity in the soil provides a useful indication of the extent of the acidity problem, high levels of calcium (Ca) and magnesium (Mg) in the soil have been found to reduce the harmful effects of Al. For this reason the percentage acid saturation, rather than exchangeable acidity, has proved to be a more reliable criterion for determining whether Al toxicity is a problem. Its value is calculated by dividing the exchangeable acidity by the sum of exchangeable cations:

$$\text{Acid sat. \%} = \frac{\text{Al} + \text{H}}{\text{Al} + \text{H} + \text{Ca} + \text{Mg} + \text{K}} \times 100$$

where concentrations are expressed in equivalents ($cmol_c$/L)

The relation between acid saturation and pH is generally similar for sands and clays (Fig. 11.10b). A consequence of this is that, although a sandy soil may have a much lower level of exchangeable acidity than a clayey soil, the levels of acid saturation and the risk of Al toxicity to plants in the two soils may be similar.

11.4.2 Origin of soil acidity

Refer to the review of Fey *et al.* (1990) for a detailed account of the processes responsible for the acidification of soils. Naturally acid soils are widespread in southern Africa,

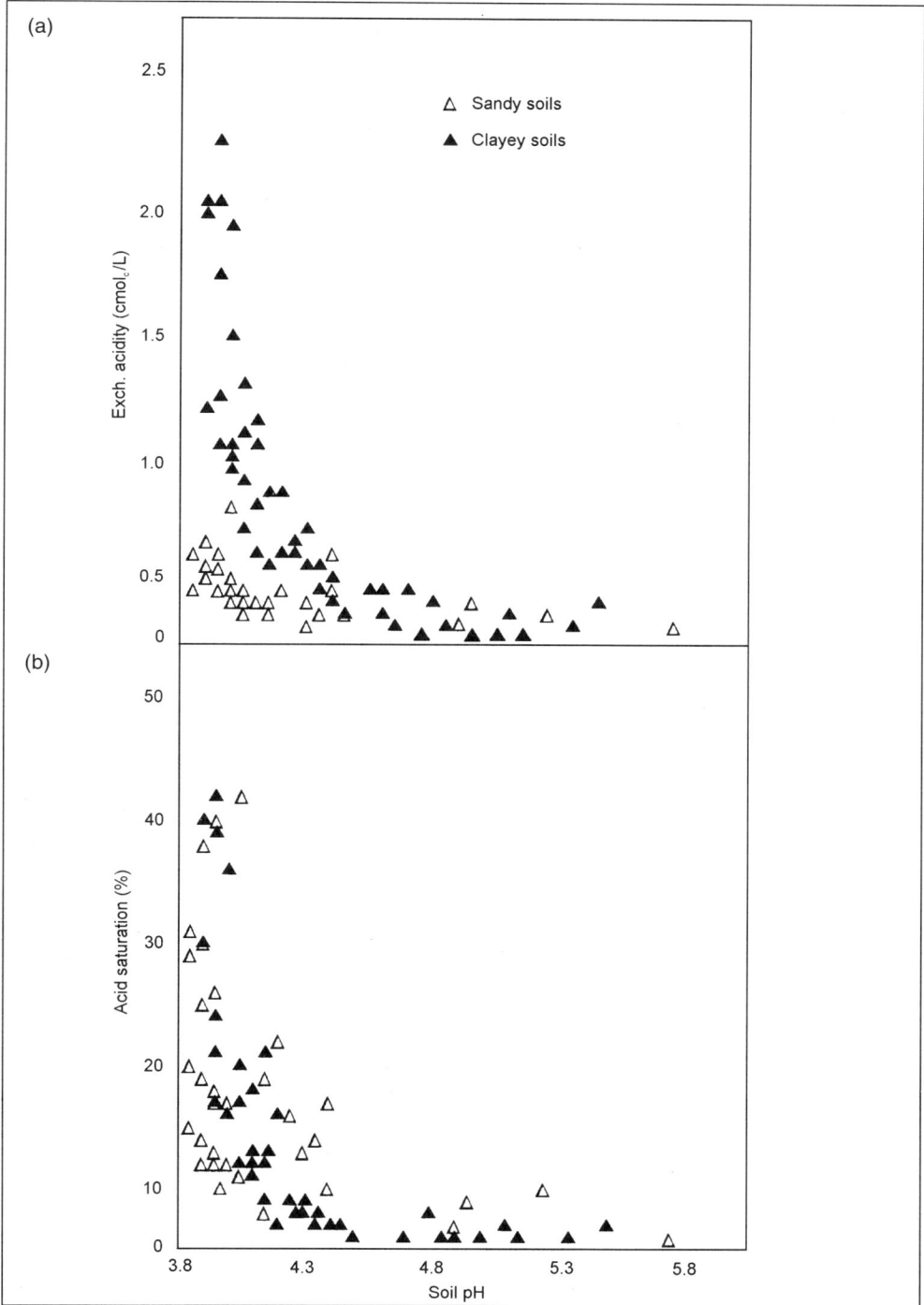

Figure 11.10 The relation between (a) exchangeable acidity (Al + H) and pH (KCl) and between (b) acid saturation and pH for sandy and clayey soils.

and indeed throughout the world, with their distribution being closely correlated with mean annual rainfall. Impoverishment of soil bases through leaching is the major factor in natural acidification, but soil acidification may be greatly accelerated by human intervention. Afforestation and acid rain have recently been identified as major causes of soil acidification. In agricultural systems, numerous processes accelerate acidification, including:

(a) cultivation, which enhances aeration and a more rapid bacterial oxidation of N-containing organic compounds and thereby a net production of acid;
(b) nitrification of ammoniacal fertilizers;
(c) the subsequent leaching of NO_3^- with Ca and Mg;
(d) the export of basic nutrients (K, Ca and Mg) in the harvest; and
(e) symbiotic N fixation of leguminous plants.

The potential magnitude of soil acidification resulting from the use of ammoniacal fertilizers was illustrated earlier (Table 11.8). The tendency for legumes to gradually acidify soils has been demonstrated by researchers in Australia and New Zealand. Nitrogen cycle effects are largely responsible for such acidification (Haynes 1983).

11.4.3 Effect of lime on soil properties

The dominant effects of lime on soil properties are:

(a) an increase in soil pH;
(b) a decrease in exchangeable acidity (Al + H);
(c) an increase in Ca and Mg levels;
(d) an increase in total cations; and
(e) a decrease in acid saturation (Table 11.14).

Marked increases in total cations with liming reflect lime-induced increases in the soil's negative charge (effective cation exchange capacity) and thereby its ability to retain nutrient cations against leaching. This may be important in soils with low cation exchange capacity. Lime may also influence the availability of micronutrients to the plant and promote mineralisation of organic N, P and S. These aspects are considered in greater detail later in this chapter.

Table 11.14 Effect of dolomitic lime on selected chemical properties of a Hutton soil (43% clay) at the Cedara Research Station.

Lime rate (t/ha)	Soil test data						
	Ca (mg/L)	Mg (mg/L)	Al + H (cmol$_c$/L)	Total cations (cmol$_c$/L)	Acid sat. (%)	pH (KCl)	Zn (mg/L)
0	380	70	1.01	5.18	19.5	4.27	4.0
4	776	122	0.17	6.77	2.5	4.78	3.2

Lime requirement is closely related to soil buffering capacity, which in turn is largely dependent on texture and organic matter content. This is illustrated by the curves presented in Fig. 11.11. Much more lime is needed to increase the pH by one unit on highly buffered clay soils than on poorly buffered sandy soils. However, re-acidification is more rapid on sandy soils than on clays. Thus control of soil acidity in poorly buffered sands requires relatively frequent small lime additions.

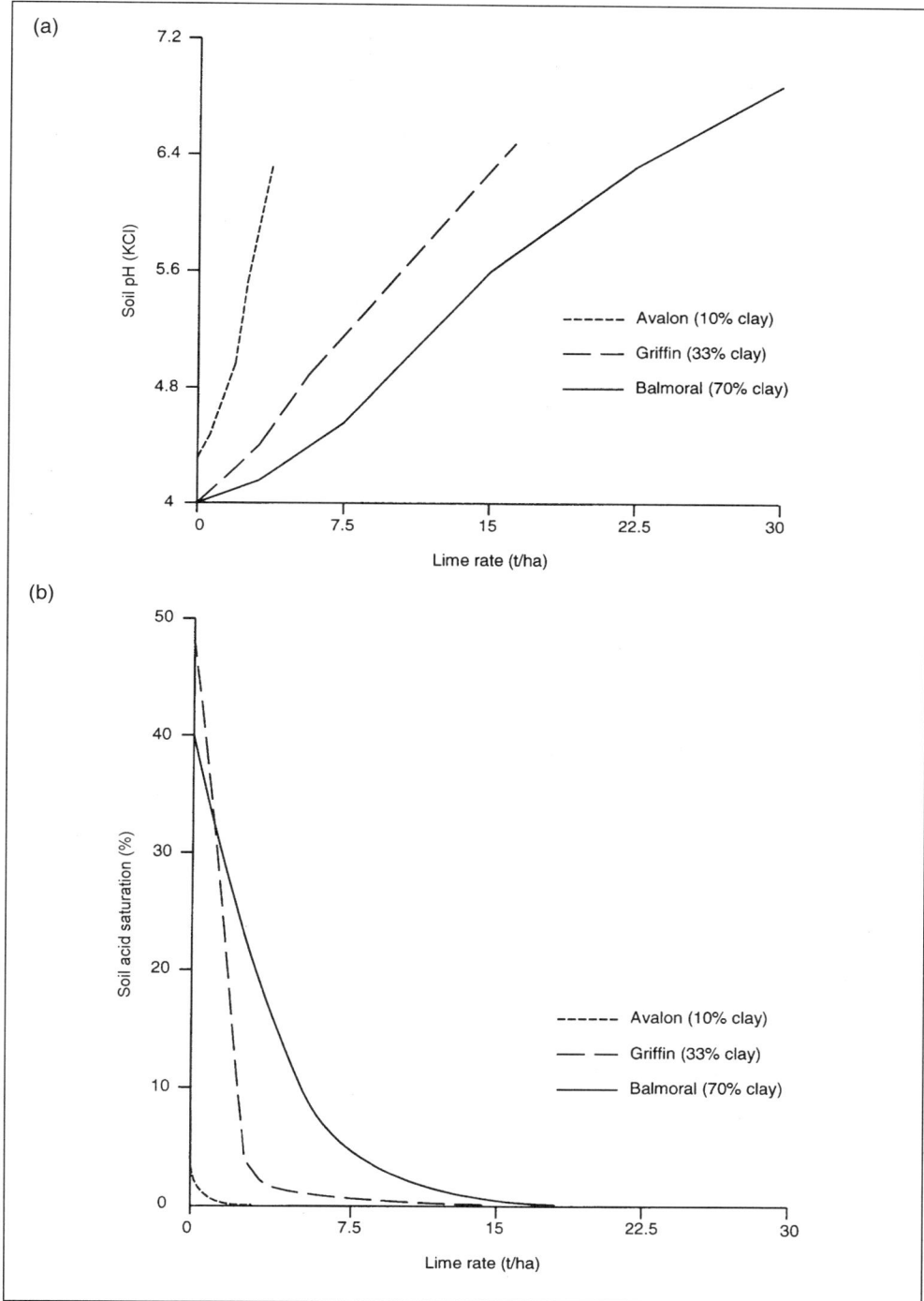

Figure 11.11 Relation between lime rate and (a) soil pH (in KCl) and (b) soil acid saturation in three soils of varying clay content in KwaZulu-Natal.

11.4.4   Plant responses to liming

11.4.4.1   Aluminium effects on plant growth

Soluble Al is the most prominent growth limiting factor in acid soils. Toxic levels disrupt physiological and biochemical processes in plant cells, with the impact being most severe on the root systems of acid-sensitive plants: both root elongation and branching are inhibited.

Visual symptoms of Al toxicity in pasture plants are not easily identified. In severe Al toxicity, foliar symptoms may resemble P deficiency – an overall stunting, small dark green leaves and a purpling of stems and leaves. In deep-rooted, acid-sensitive species such as lucerne, impaired root development resulting from Al toxicity is more easily identified.

11.4.4.2   Differences in tolerance to aluminium

Plants differ widely in their tolerance of acid soil conditions. Tropical grasses and legumes are, in general, relatively tolerant of soluble Al, whereas temperate species are not (Sanchez & Salinas 1981; Crowder & Chheda 1982; Miles 1986; Helyar 1991).

Tentative guidelines on the tolerance characteristics of some important pasture species are listed in Table 11.15. Among these data, those for *E. curvula*, kikuyu, Italian ryegrass, white clover and lucerne should be reasonably reliable. An example of the differences in yield response to lime of three of these species is shown in Table 11.16.

Table 11.15   Tolerance categories for pasture species to soil acidity.

General tolerance class	Species	Permissible acid saturation
Very highly tolerant	*Eragrostis curvula*	50
Highly tolerant	kikuyu, star grass, coastcross, Rhodes, oats, Smuts', fescue	40
Tolerant	Italian ryegrass, cocksfoot	25
Mildly sensitive	perennial ryegrass, barley	10
Sensitive	white clover, red clover, lucerne	1

Table 11.16   Dry matter yield responses to lime of three pasture species on a Griffin soil (acid saturation = 75%; pH (KCl) = 4.0) in the Natal Highland Sourveld.

Species	Season DM yield (kg/ha)		Per cent yield response to lime
	No Lime	+ Lime	
Kikuyu*	11.9	11.9	0
Italian ryegrass	10.7	13.0	21.5
White clover*	5.2	9.9	90.4

* Yield data relate to the first season following establishment

Within a species, significant varietal differences in tolerance to soil acidity may also occur. In a recent study, Edmeades *et al.* (1991) found that cultivars of *Lolium perenne* of either European or Australian origin were generally more tolerant of Al than New

Zealand cultivars. Preliminary results suggest variations in tolerance to soil acidity as well as in yield potential of lucerne cultivars (Fig. 11.12). Here a newly bred acid tolerant line far outperformed commonly used cultivars at both a low and a high pH. It is perhaps noteworthy that the cultivar which responded most to liming was SA Standard, which originates from the drier irrigated areas of the Eastern Cape where soils are generally neutral to alkaline.

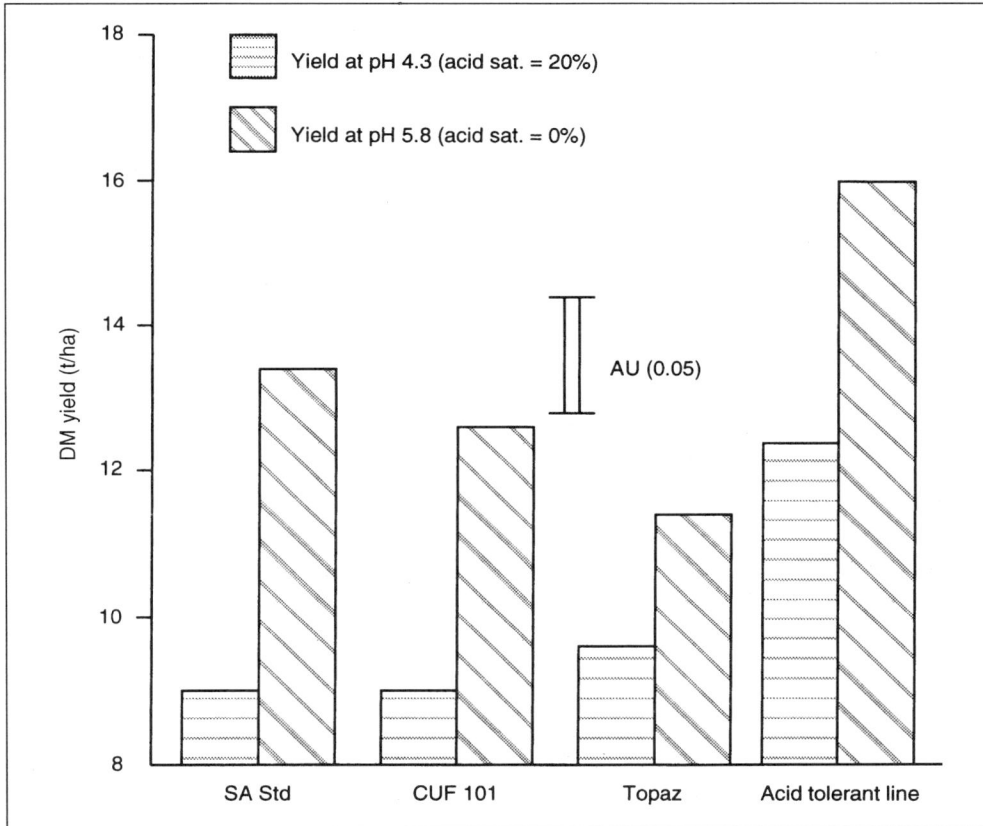

Figure 11.12 Dry matter yields over a 12-month period following establishment of four lucerne cultivars in a field trial in the Mistbelt of KwaZulu-Natal (the AU relates to comparisons both within and between cultivars).

While acid sensitive temperate species such as white clover and perennial ryegrass may perform well in the short term under mildly acid conditions, they will lack persistence. This is demonstrated in Fig. 11.13 for clover. This means that a higher level of soil acidity is permissible in annual than in perennial pastures.

11.4.4.3 *Effects of liming on nutrient availability*

Lime is able to improve N, P and S supply by way of accelerated mineralisation of organic matter by soil micro-organisms. From one-half to two-thirds of the total P in most mineral soils is in the organic form (Kamprath & Foy 1985) so that the effect of lime-

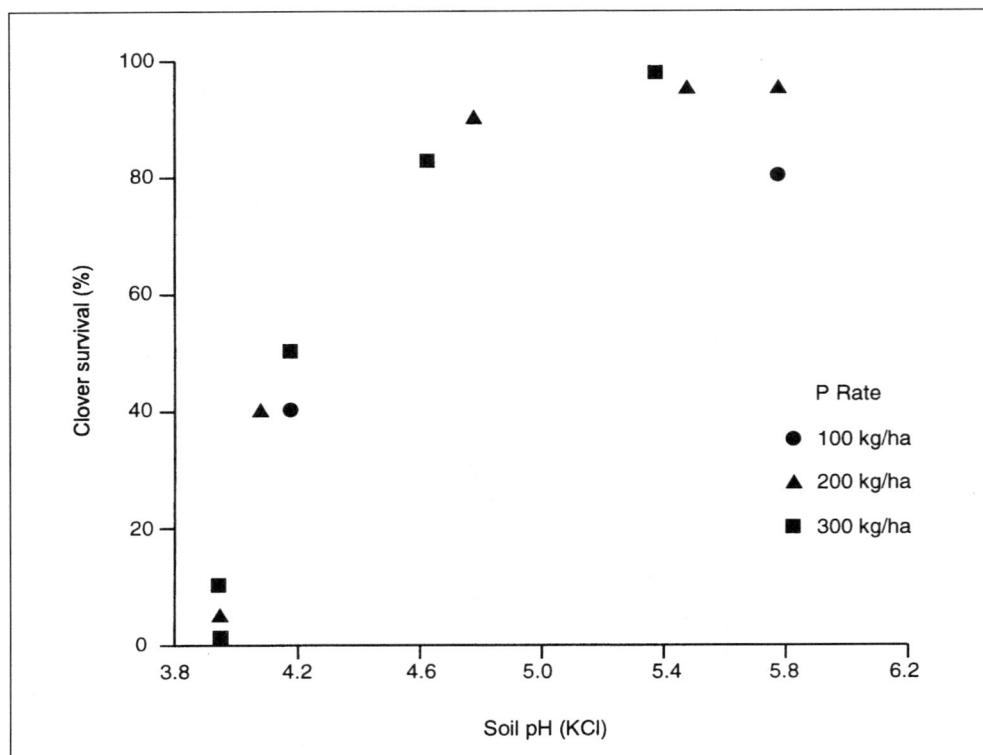

Figure 11.13 Effects of applied phosphorus (P) and soil pH on white clover survival four-and-a-half years after establishment on an acid soil (P applied at establishment; soil pH measured at the time of evaluation of clover survival).

induced mineralisation on P availability is substantial in certain soils. Phosphorus availability to plants may also be influenced by several other lime related reactions (see reviews by Haynes 1984; Sumner & Farina 1986 and Edmeades *et al.* 1990). However, the processes involved are extremely complex and lime–P interactions remain relatively poorly understood.

A typical lime–P interaction on lucerne growth in the KwaZulu-Natal Midlands is shown in Fig. 11.14. Here the yield increased continuously with increasing P soil test from 6 mg P/L to 16 mg P/L at a soil acid saturation of 25%. Where acid saturation had been reduced to 4% by liming, yields were substantially higher and did not respond to a P test level greater than 11 mg/L. Such interactions have been widely observed and have been responsible for the generalisation that 'lime increases P availability'. The implication essentially is that the plant availability of P is improved through reduced P fixation. However, attempts to elucidate the effects of lime on P fixation have generally yielded contradictory results. Indications are that the majority of lime–P interactions result from increased mineralisation of organic P and/or an improved ability of plants to utilise already plant-available P through improved root systems. Contributory processes may include the elimination of Al toxicity, the mineralisation of organic N and an increased availability of the essential trace element molybdenum (Mo).

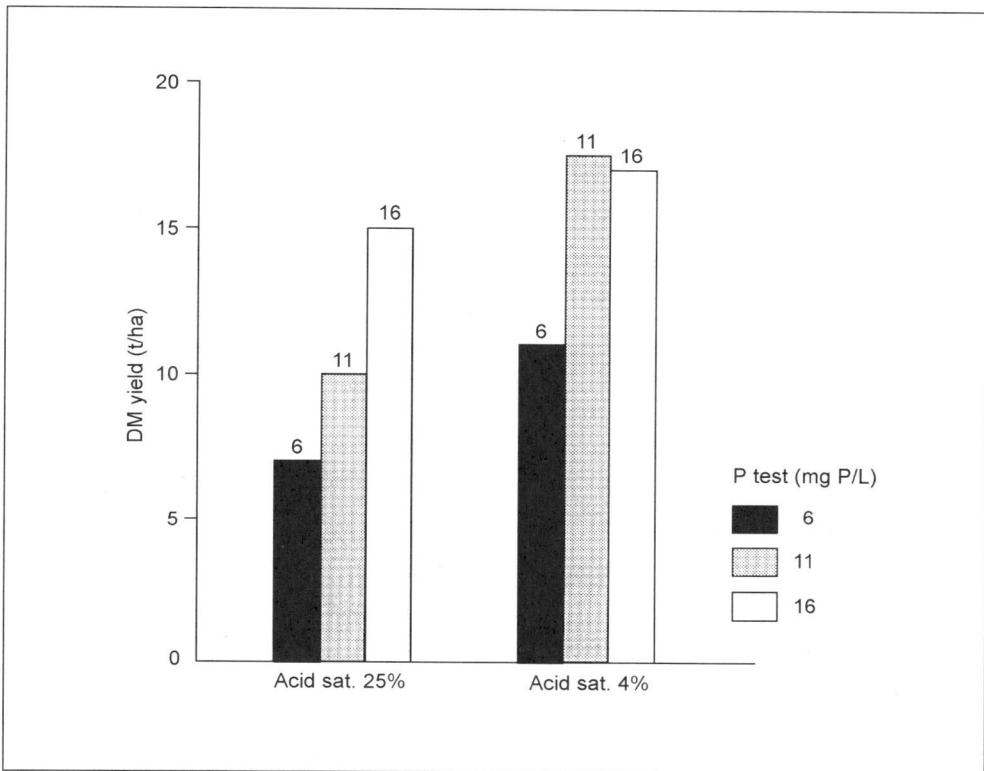

Figure 11.14 Effects of varying soil acid saturation and phosphorus (P) test (NH$_4$HCO$_3$/EDTA/ NH$_4$F) on the yield of lucerne (cv. Baronet) over two seasons on an acid KwaZulu-Natal soil.

The addition of lime also augments soil Ca and Mg levels. Although these minerals seldom restrict plant growth in southern Africa, an increase in their levels may improve animal performance.

Excessively high rates of lime have been known to negatively affect plant growth. Many of these effects, especially those observed on naturally acid soils, have not been fully explained, but deficiencies of the micronutrients manganese (Mn), copper (Cu), zinc (Zn) and boron (B) are known to occur with overliming.

11.4.4.4 Lime and earthworms

Recent research has provided clear evidence of the impact of earthworms on soil fertility (Bradshaw 1980; Cossens 1984; Syers & Springett 1984). They have been found to stimulate plant growth by:
(a) incorporating organic material into the soil;
(b) releasing metabolic products into the soil; and
(c) improving soil aeration, water relations and penetrability.
Most of the beneficial earthworm species appear to be sensitive to soil acidity, and liming frequently results in a rapid increase in earthworm populations.

11.4.5   Practical considerations in the use of lime

11.4.5.1 Lime type and quality

In the agricultural context, 'lime' refers to any product whose Ca or Ca and Mg compounds are able to neutralise soil acidity. Carbonates of Ca and Mg are the most widely used for this purpose. Dolomitic lime contains 16% or more magnesium carbonate, and calcitic lime somewhat less than this. In addition to natural carbonates, various by-products of industrial processes are frequently used as liming materials: these include calcium oxide (burnt lime), calcium hydroxide (slaked lime) and calcium silicate (slag).

The effectiveness of various liming materials varies widely, with the following factors being particularly important in this regard (Farina 1983):

(a)   Chemical purity – the presence or otherwise of non-reactive materials such as sand and clay.
(b)   Chemical composition – the nature of the calcium and magnesium compounds present.
(c)   Fineness – lime particles larger than 0.84 mm are of little value. Very coarse liming materials are completely ineffective.
(d)   Hardness – the solubility, and hence neutralising value, of lime depends on whether it is derived from hard crystalline material or from softer relatively unconsolidated material.

Where uncertainty exists as to the quality of a particular liming material, a sample should be submitted for analysis.

11.4.5.2   Estimating lime requirements

Lime recommendations for pastures are essentially based on three criteria:

(a)   the acidity level in the soil;
(b)   the tolerance of the species concerned to soil acidity;
(c)   the efficiency of the liming reaction.

Soil testing is essential for the derivation of lime requirements. Because acidity problems typically occur irregularly (Farina & Johnston 1987; Rowell 1988), a soil sample from a large area may frequently fail to indicate a problem, as good areas mask the weak areas. Weak areas, as reflected by poor growth, should therefore be sampled separately. Control of sampling depth (<150 mm) is also essential.

Pasture species differ widely in their ability to tolerate soil acidity (Table 11.15) and this has a major bearing on lime requirement. In KwaZulu-Natal the 'permissible acid saturation' (PAS) thresholds listed in Table 11.15 are used to determine the need for lime. When the soil acid saturation exceeds the PAS, sufficient lime is added to reduce it to the threshold value. In a pasture consisting of mixed species, the applicable PAS is that of the most sensitive species. Advisory services outside KwaZulu-Natal make wide use of pH as a determinant of lime requirement.

Factors involved in the efficiency of the liming reaction include the quality of the liming material and the depth of its incorporation into the soil. This may vary from plough depth (150 mm to 200 mm) at establishment to zero where perennial pastures are topdressed with lime.

In practice, the various factors influencing lime recommendations are taken into account in computerised advisory services. All that is needed is a soil sample and an indication of the intended or existing pasture species.

11.4.5.3   *When and how to apply lime*

When liming soils for establishment, two important factors should be borne in mind: (a) most limes take a month or more to react fully; and (b) moisture is necessary for the neutralisation reaction. Application of lime a week or two before planting, particularly where the soil moisture status is low will not, therefore, prevent acidity damage to acid-sensitive species. Furthermore, the lime needs to be uniformly spread and thoroughly mixed into the soil. Discing or harrowing is advocated in preference to ploughing to achieve adequate mixing (Farina & Johnston 1987).

Provided lime is thoroughly incorporated, there should be no danger of overliming unless recommended application rates are greatly exceeded. In experiments on heavy soils, up to 25 t/ha have been applied in a single dressing without detrimentally affecting plant growth.

While surface applied lime is unlikely to ameliorate soil below a depth of a few centimetres, the effectiveness of topdressing pastures with lime has been verified in both local and overseas research (Bromfield *et al.* 1987). Animal traffic and earthworm activity promote lime incorporation and thereby plant response to lime. However, topdressed lime takes longer to invoke a response than incorporated lime. It is advisable, therefore, to anticipate acidity build-up and so to apply lime as a 'preventative' measure.

11.4.5.4   *Calcitic or dolomitic lime?*

Although several Mg fertilizers are commercially available, dolomitic lime is often the most cost-effective means of increasing soil Mg levels. It should be used in preference to calcitic lime where soil Mg levels are less than 100 mg/L.

11.4.5.5   *Lime or gypsum?*

The calcium in gypsum is more readily available than that in lime, while the by-product phosphogypsum has the added advantage of containing about 0.5% P. The use of gypsum to combat subsoil acidity has reportedly 'revolutionised' lucerne production in the south-eastern USA (Hoveland pers. comm.), and it could conceivably play a similar role on inherently acid soils in this country. It needs to be appreciated, however, that gypsum and lime differ appreciably in their effects on soil properties. These differences are particularly important in the context of the nutrition of pastures. The data in Table 11.17 serve to illustrate the different modes of action of dolomitic lime and gypsum. These are:

(a)   although gypsum partially alleviates the toxic effects of aluminium by increasing soil Ca levels, it has little effect on soil pH. This is of particular significance in legume-based pastures, since the bacteria responsible for symbiotic N fixation operate effectively only within specific pH ranges;

(b)   topdressing with gypsum promotes a downward movement of Mg in the soil. This results in decreased Mg levels in herbage and a consequent increase in the likelihood of Mg deficiency in the forage.

Sulphur (contained in gypsum) is also known to depress selenium (Se) uptake, which may aggravate the widespread Se deficiency problem in animals in the higher rainfall areas of the country.

Table 11.17 Effects of dolomitic lime and gypsum on the chemical properties of a Griffin soil. Samples (150 mm depth) taken nine months after application of lime and gypsum.

Treatment	Ca (mg/L)	Mg (mg/L)	Exch. acidity (cmol$_c$/L)	Total cations (cmol$_c$/L)	Acid sat. (%)	pH (KCl)
Control	232	161	0.95	3.71	26	4.3
Dolomitic lime (5 t/ha)	424	295	0.38	5.17	7	4.8
Gypsum (4 t/ha)	487	97	1.04	4.49	23	4.3

The above considerations mitigate against the indiscriminate use of gypsum on pastures.

11.5   SULPHUR

11.5.1   Role and concentration in plants

Plants absorb sulphur (S) mainly in the form of the anion sulphate (SO_4^{2-}). Within the plant, sulphate is reduced and incorporated into organic molecules, principally amino acids, the building blocks of proteins. Particularly important S-containing amino acids are cysteine and methionine (Mengel & Kirkby 1987). Sulphur taken up in excess of the demand of the plant for the synthesis of organic S compounds is stored as sulphate-S.

Sulphur levels in plants are similar to those of P (0.15% to 0.4% in the dry matter) and decrease with the age of the plant material. The ratio of total plant N to total plant S is frequently used as an index of S deficiency. In general, for grasses and legumes, there is an adequate supply of S when this ratio falls between 14 and 18 (Cornforth & Sinclair 1984). An optimum N:S ratio of 10 to 11 has, however, been reported for hybrid Bermuda grasses (*Cynodon dactylon*) (Jones & Watson 1991).

11.5.2   Sulphur relations in soils

In most soils the bulk of the S is present in an organic form. The immediate source for plant uptake is inorganic S (sulphate-S), which occurs principally in the adsorbed form (chemically bonded to the surfaces of soil particles), with a small proportion as water-soluble S.

The S cycle in soils is in many respects similar to that of N, with mineralisation and immobilisation reactions governing the plant-available supply of S from the organic fraction. Because in many soils most of the S taken up by the plant during the growing season is derived from mineralisation reactions, routine soil tests cannot be used as a basis for determining fertilizer S requirements.

The strength of adsorption of sulphate to soil surfaces is intermediate between that of nitrate (very weakly held) and phosphate (strongly held). Substantial leaching of sulphate may occur in sandy soils in particular.

11.5.3   Sulphur nutrition of pastures

11.5.3.1   *Requirements*

Sulphur uptake by intensive pastures usually falls within the range of 15 kg to 40 kg S/ha/yr. An adequate supply is particularly important for promoting growth and N fixation by leguminous plants. Grasses and legumes often contain similar concentrations of S although grasses are frequently reported to be more vigorous accumulators of this element than legumes. Where S is not applied to grass/legume pastures, therefore, N fixation by the legumes may be negligible because of their inability to take in sufficient amounts of sulphur (Walker & Adams 1958).

In southern Africa the S requirements of plants have generally been met incidentally by using single superphosphate, ammonium sulphate, ammonium sulphate nitrate and S-containing fertilizer mixtures. However, as high grade fertilizers containing negligible quantities of S are increasingly used, incidences of S deficiency are likely to increase and planned use of S-containing fertilizers may become necessary.

11.5.3.2 *Factors influencing sulphur requirements*

Important factors influencing pasture S requirements are the reserves of S already present in soils, the extent of atmospheric deposition of S and the method of pasture utilisation.

As intimated earlier, organic matter levels and relations in soils have a major bearing on the amounts of sulphate made available for plant growth and hence on the need for fertilizer-S. Sandy, low organic matter soils are particularly poor S suppliers (Meyer 1985). However, of more importance than soil properties *per se* in determining fertilizer S requirements, is the extent of S deposition from the atmosphere. Coastal regions receive considerable amounts of S in sea spray, while in the vicinity of industrial and urban centres, the sulphur compounds (principally SO_2) released from burning fossil fuels may be deposited in rainfall, or may interact directly with the plants and the soil (Murphy 1980). These levels may reach as much as 50 kg S/ha/yr in the Mpumalanga highveld (Tyson *et al.* 1988) and in excess of 10 kg/ha/yr in adjacent areas in the Lowveld, Free State and KwaZulu-Natal. Cooke (1982) reports that S deficiencies are unlikely where rainfall supplies 12 kg S/ha/yr or more.

As with K and N, the method of pasture utilisation significantly influences the overall S economy of pastures. Under grazing, much of the ingested S is returned in dung and urine. As with N, the recycling of S through animals is relatively inefficient since concentrations of sulphate-S in excretal patches far exceed short-term plant requirements and so losses by leaching are generally substantial (Goh & Nguyen 1990).

11.5.3.3 *Sulphur fertilizers*

Appreciable amounts of sulphate-S are present in a number of the commercially available 'straight' (N, P or K) fertilizers. These are listed in Tables 11.7 and 11.11. In addition to these, calcium sulphate (gypsum), containing between 13% and 23% S, is a very cost-effective means of supplying S.

11.6 MOLYBDENUM

Molybdenum (Mo) warrants particular attention because of its crucial role in protein synthesis and in the symbiotic N_2 fixation process, even though it is taken up by plants in only trace amounts.

11.6.1   Function and concentration in pasture species

Molybdenum is a constituent of several essential enzymes in plants, including nitrate reductase, which is responsible for the reduction of nitrate in the cell cytoplasm (Römheld & Marschner 1991). Where it is deficient, plant protein concentration decreases, indicating the involvement of Mo in protein synthesis. The Mo-containing enzyme, nitrogenase, acts as a catalyst in symbiotic N_2 fixation. Therefore the Mo requirement of root nodules is particularly high, and under conditions of Mo deficiency, nodule weight and N_2 fixation decrease. Severely Mo deficient plants display symptoms characteristic of N deficiency, i.e. leaves are pale green to yellow and growth is stunted.

The required Mo concentration for optimum growth in the tops of most pasture species ranges between 0.1 mg/kg and 1.0 mg/kg dry matter. Under conditions of excessive Mo supply, plant concentrations may exceed 1.0 mg/kg with no harmful effects on growth (Martens & Westermann 1991). Such concentrations in herbage may, however, induce copper deficiencies in animals.

11.6.2   Relations in soils

Not all Mo in soils is available for plant uptake. That which is significant to plant growth exists as the anion MoO^{2-}_4. Soil Mo in this form is largely bound (adsorbed) to the surfaces of clays and iron and aluminium oxides, with water-soluble levels of MoO^{2-}_4 being very low. However, the strength of adsorption decreases sharply with increasing pH. Hence Mo deficiencies are restricted mainly to acid soils, with availability being increased by liming.

11.6.3   Correcting molybdenum deficiencies

Legumes generally have a substantially higher requirement for Mo than grasses. Lucerne, in particular, is extremely sensitive to a Mo deficiency. Specific methods of correcting this Mo deficiency include liming, soil application of Mo salts, foliar sprays and seed treatment.

Davies & Jones (1988) report that, on average, the uptake of Mo by a crop plant will increase two- to three-fold for each unit rise in soil pH between 5.0 to 7.0. Therefore liming can overcome a Mo deficiency but not where soils are inherently deficient in Mo.

Where Mo salts are employed, the choice is usually between sodium molybdate, ammonium molybdate and molybdenum trioxide (Table 11.18). These sources differ little in efficiency (Johansen *et al.* 1977). The amounts used at establishment usually range between 250 g and 500 g Mo/ha, with the higher rates in this range applying to acid clayey soils.

Foliar applications of Mo are effective in overcoming deficiencies, although residual effects may not be as long-lasting as soil applications. Rates of application in foliar sprays usually fall within the range 20 g to 100 g Mo/ha.

The incorporation of Mo into the lime coating applied to inoculated seed has proved to be an efficient means of satisfying the Mo requirements of legumes. The relatively insoluble molybdenum trioxide, applied at a rate of 100 g Mo/ha (152 g molybdenum trioxide/ha), is generally used for this purpose. It is believed that the soluble Mo compounds (sodium and ammonium molybdate) pose a hazard to seedlings and rhizobia,

Table 11.18 Compositions of commonly-used micronutrient sources.

Nutrient	Source	Nutrient content
Mo	Sodium molybdate	39% Mo
Mo	Ammonium molybdate	54% Mo
Mo	Molybdenum trioxide	66% Mo
B	Solubor	20% B
B	Boric acid	17% B
B	Borax	11% B
Cu	Copper chloride	47% Cu
Cu	Copper sulphate (pentahydrate)	25% Cu
Zn	Zinc chloride	48–50% Zn
Zn	Zinc sulphate (heptahydrate)	23% Zn
Mn	Manganese chloride	17% Mn
Mn	Manganese sulphate	23–28% Mn
Fe	Ferrous sulphate	19% Fe
Fe	Ferric sulphate	23% Fe

but in New Zealand sodium molybdate, at a lower rate than the above, is routinely applied to legume seed with no apparent deleterious effects (Lowther 1987).

Local opinion is that up to 100 g Mo/ha is required every three years for legume-based pastures on inherently acid, high clay content soils. In practice these small amounts of salt are applied as a spray solution or, alternatively, blends of NPK fertilizers can be used as carriers.

11.7 BORON, ZINC, COPPER, MANGANESE AND IRON

Deficiencies of B are rare in grass pastures but are not uncommon in high-yielding legume-based pastures, and in particular in lucerne stands used for hay production.

Boron, unlike most other nutrients, exists as an uncharged species in most soils, with an anionic form being found at a high pH where its availability to plants decreases. Deficiencies of B are more common on sandy soils than on clays and loams, since leaching losses in the former can be substantial.

A traditional problem with the use of B fertilizers is that different plant species differ in their B requirement. Legumes and root crops have large B requirements, while grasses have low requirements and are often highly sensitive to an excess of B (Finck 1982). Boron fertilizers must therefore be used with caution, especially in crop rotations which involve species with different requirements.

Some B fertilizer sources are listed in Table 11.18. Both soil applications and foliar sprays are used to correct B deficiencies, with the highly soluble solubor and boric acid being preferred for use in sprays. Boron seldom needs to be applied to pure grass swards but soil incorporation of B fertilizer at establishment is advisable for long-term legume-based pastures. The recommended rate is 2 kg B/ha. For high-yielding legume pastures such as lucerne, leaf analysis should be used to monitor plant B status. The optimum leaf B concentration in the tops of both lucerne and white clover ranges from 25 ppm to 30 ppm. Where deficiencies arise, foliar spraying at a rate of 0.2 kg to 0.4 kg B/ha is an effective means of supplying this element.

Zinc (Zn), copper (Cu), manganese (Mn) and iron (Fe) are classified chemically as heavy metals. They exist in cationic form in the soil and, because of their positive charge, they are retained against leaching on soil clays. Furthermore, Zn and Cu, in particular, are strongly bound to the soil organic matter. The availability of all four decreases sharply with liming.

Indications are that deficiencies of Zn, Cu, Mn and Fe seldom occur in pastures. Zinc applications are, however, recommended if soil levels are low. Copper deficiency is reportedly a problem in high organic matter soils in the southerly parts of the Eastern and Western Cape. Since excess Cu is toxic to both plants and animals, Cu containing fertilizers should be applied with caution.

Some inorganic fertilizer sources of Zn, Cu, Mn and Fe are listed in Table 11.18. In addition to these sources, chelates are often used to supply these nutrients. These are organometallic complexes in which the metal cation is bound on several sides (surrounded) by the organic molecule. The advantages of chelates are that they prevent fixation of the nutrient in the soil and, in the case of foliar sprays, facilitate absorption through the leaves.

11.8   SOME PRACTICAL GUIDELINES IN THE DIAGNOSIS OF NUTRIENT AND OTHER DISORDERS IN PASTURES

The diagnosis of nutrient disorders in pastures often requires a multifaceted approach requiring the assistance of an experienced adviser. What follows is merely a guide to the factors that should be considered in the diagnosis of problems.

11.8.1   General problems

The need for diagnosing nutrient disorders generally arises from a decreased production by either the animals or the pasture, and it is important to bear in mind that factors other than pasture nutrition may be implicated. These include, for example, the use of incorrect seeding rates, poor pasture utilisation practices, disease and pest problems, problems associated with herbicide residues or herbicide drift, poor water relations, poor soil physical conditions and soil salinity. These factors need to be eliminated as possible reasons for poor growth before an investigation is launched into possible nutrient disorders.

11.8.2   Indicators of nutrient disorders

(a) Soil fertility management. The history of fertilizer use and other practices can often provide helpful guidelines on nutrient sufficiency. This includes amounts of fertilizers, lime, manures and slurries applied in the past, and the use of pastures as holding areas while feed is brought in. Previous heavy applications of manure, slurry and fertilizer P, K and S may rule out deficiencies of these elements. A history of feeding hay or silage on a pasture makes P deficiency unlikely and points to a possible excessive soil K level. Liberal use of acidifying N fertilizers may have generated excessive soil acidity.

(b) Soil and climatic indicators. Knowledge of the rainfall and dominant soil types of the area can often point to likely problems. Soil acidity, low K reserves and strong P sorption are generally related to these factors. Sulphur and B deficiencies occur most commonly on sandy soils. Areas far from the coast and from industrial centres are most prone to S problems. Poorly drained soils are often subject to severe N losses by denitrification, whilst leaching of N may be common on sandy soils.

It is always worthwhile considering past soil tests along with subsequent fertilizer history. These may indicate inherent P, K or acidity problems e.g. high soil pH or overliming may indicate Zn, Mn, Cu and/or Fe deficiencies.

(c) Leaf and root symptoms. The age of the leaves exhibiting symptoms is an important indicator of the group of elements responsible. Stronger symptoms on old leaves indicate deficiencies of mobile elements (N, P, K, Mg and Mn), whereas symptoms mainly on young leaves suggest immobile elements (Ca, S, Fe, Cu, B and Mo). Light colour and necrosis of older leaf tips often indicates N or K deficiency, whereas dark green pasture with poor growth often reflects P deficiency. Purpling of the underside of ryegrass and fescue (not kikuyu) leaves is a sure sign of P deficiency.

Poor root development is often the most obvious symptom of soil acidity (Al toxicity) problems, but soil physical conditions (e.g. compaction) and root pathogens can also limit root development. In legumes, poor or ineffective nodulation (nodule interiors not pink), is often indicative of high soil acidity or Mo deficiency.

(d) Species-specific problems. Satisfactory establishment of lucerne and clovers, followed by poor persistence, is often a symptom of acidity. Poor growth of lucerne can also often be traced to B and Mo deficiencies. Poor persistence and productivity of hay crops is frequently due to a K deficiency.

(e) Invader weeds as indicator species. Invasion of pastures by *E. curvula* is often indicative of excessive soil acidity. Invasions of sorrel (an acid-tolerant herb of the genus *Rumex*) are usually associated with K deficiency.

(f) In-field variability. Growth variations in rows or strips usually indicate non-uniform past treatment while patchy distribution often suggests a variation in soil acidity or disease problems. Dramatically better growth in urine patches usually indicates a deficiency of N or K, whereas if growth is good in dung patches but not urine patches, P deficiency is more likely.

(g) Rapid herbage nitrate tests. Nitrogen test strips can be used to confirm N deficiency. The technique involves applying plant sap to the test strip (the most commonly used are the Merckoquant 10020 strips, made by the Merck Company). The intensity of colour developed on the strip is proportional to the concentration of nitrate in the sap. If the N status is high, apparent N deficiency can often be attributed to a K deficiency.

(h) Animal indicators. Reluctance of animals to graze pastures can be caused by excessive N and/or K fertilization. Magnesium tetany (grass staggers) and milk fever generally result from excessive K in the herbage: this may not affect grass production, but will depress animal production.

11.8.3 Confirming deficiencies and determining rates required

(a) Suspected P, K or Zn deficiencies or soil acidity problems. Soil sampling and testing is undoubtedly the most reliable way of determining requirements for P, K, Zn and lime. Plant samples may complement the soil samples.

The weakest link in the soil testing process is in taking the sample. It is important to use the correct sampler, and particular attention should be paid to such aspects as sampling depth, sampling pattern and the number of cores (sub-samples) per composite sample. More detail on these aspects is usually available from reputable soil testing laboratories.

(b) N, S and micronutrients other than Zn. Soil tests for these elements are seldom useful, and plant sampling is generally considered the most appropriate means of identifying or confirming deficiencies. However, soil analytical data such as soil texture and pH may be useful in making diagnoses (see section 11.8.2). On-farm experimentation, by spraying or fertilising test strips of pasture, can also be useful.

11.9 FERTILIZATION, LIMING AND ANIMAL NUTRITION AND HEALTH

The major components of forage carbohydrates, lipids, fibre and proteins (organic components composed mainly of C, H, O and N) are essential in the diet of grazing animals, in addition to a number of other elements. These include both macro- and micro-elements.

High quality intensive pastures improve animal production largely because their forage is high in metabolisable energy and protein. However, animal production on pastures producing high dry matter yields does not always meet expectations, as a number of factors may bring about imbalances in the organic and mineral components of the forage. Pasture species, stage of growth, soil type, climate, fertilizer and lime usage are all factors that may have a considerable influence on herbage composition.

Whereas southern African farmers using intensive pastures are largely aware of the importance of balance with respect to the major components of forage (carbohydrate, lipid, fibre and protein), there is little emphasis on herbage mineral composition. In the case of dairy herds this may be ascribed to the assumption that minerals in concentrate supplements compensate for mineral deficiencies or imbalances in consumed forages. However, as pointed out later in this discussion, the chemical composition of forages in southern Africa is often nutritionally far from ideal and there is a considerable potential for imbalances or deficiencies. Researchers in this field emphasise that a lack of clinical symptoms arising from toxicities or imbalances is not proof that problems do not exist. As noted by Spears (1991): 'Even marginal mineral deficiencies can reduce growth, reproduction or health in ruminants showing few if any clinical signs of deficiency.'

11.9.1 Proteins and nitrates

Increasing the supply of N to pastures usually results in marked increases in the protein and nitrate content of grasses. This is illustrated by the Italian ryegrass compositional data presented in Table 11.4. Accumulation of N in grasses increases sharply when N fertilization exceeds short-term plant N requirements for maximum growth. Applications above 50 kg/ha/application may, in cold- or moisture-stressed plants, result in crude protein levels of 25–34% and nitrate N levels as high as 1.0% in the herbage (Eckard & Dugmore 1993). Legumes, being independent of fertilizer N when effectively nodulated, tend to exhibit far less variation than grasses in their N contents.

Nitrogen levels greater than 4% of dry matter exceed both plant and animal requirements and are potentially toxic to ruminants due to the high levels of both protein and nitrate. Ammonia-induced bloat, ammonia toxicity and nitrate toxicity have all been responsible for ruminant fatalities and are often associated with moisture-stressed pasture and hungry or unadapted animals (Eckard & Dugmore 1993). These authors also point out that where conditions are less extreme, poor animal performance (which may include reduced fertility (Robinson 1990)) may result from energy stress in high-N pastures because rumen microbes require energy to metabolise the extra N. High N forage may also be temporarily rejected due to an ammonia appetite repression mechanism.

11.9.2 Non-structural and soluble carbohydrates

Fertilization practices may greatly influence the accumulation of non-structural and soluble carbohydrates in forage plants, and thereby their palatability and digestibility. Soluble carbohydrate levels have been widely reported to decrease with increasing N supply (Brown & Ashley 1974) and there is some evidence that soluble carbohydrate levels increase with an increasing P supply (Mays *et al.* 1980). The effects of increasing N and P supplies on total non-structural carbohydrate (TNC) levels in Italian ryegrass harvested in early winter are shown in Table 11.19. Particularly noteworthy in the context of the high rates of fertilizer N generally applied to irrigated Italian ryegrass, is the large decrease in TNC levels with increasing N rate. The fact that P increased TNC levels is consistent with on-farm observations that animals exhibit a preference for kikuyu grown with high levels of applied P. In a review article, Ozanne (1980) cited a number of studies in which animal preference, pasture digestibility, intake and body weight gains all increased with increasing P supply to the pasture.

Table 11.19 The influence of fertilizer N rate and soil P status on the total non-structural carbohydrate concentration (TNC, dry matter basis) in Italian ryegrass harvested in early May following establishment in March at the Cedara Research Station, KwaZulu-Natal.

N rate (kg N/ha)	Soil P test (mg/L)†		
	7	10	16
	TNC%		
0	16.0	19.8	22.1
35	12.9	13.8	15.9
70	11.8	11.7	12.7

† NH_4HCO_3/EDTA/NH_4F extraction

Potassium effects on TNC appear to be less dramatic than those of N and P. Indications are, however, that higher concentrations of soluble carbohydrates are found in forages grown under low K nutrition than in those grown at high K levels (Brown & Ashley 1974).

11.9.3 Phosphorus and calcium

Because of their important role in growth and bone formation, P and Ca must be available in adequate amounts to animals. Actual requirements vary appreciably between different classes of animals. Yearling beef animals with a mass of approximately 300 kg and gaining 0.5 kg per day require 0.22% Ca and 0.20% P in their diets (NRC 1976). Because of the high Ca and P contents in milk, lactating dairy cows have particularly high requirements for these elements. Required dietary concentrations depend largely on the amount of milk produced (NRC 1978). A 600 kg cow with a daily milk yield of 14 kg to 21 kg requires 0.48% Ca and 0.34% P in its ration, while the same size cow producing 21 kg to 29 kg of milk requires 0.54% Ca and 0.38% P in its diet.

Both Ca and P concentrations vary widely between pasture species. In addition, there are often marked seasonal variations in the concentrations of these elements. Fertilizer and liming practices also impact on the plant contents of these elements. Measured concentrations of Ca and P in some productive pastures are listed in Table 11.20. These data highlight important differences between species with regard to their ability to accumulate Ca and P. Legumes, such as white clover and lucerne, are rich in Ca relative to grasses, while amongst the grasses, significant differences in Ca and P contents are apparent. Italian ryegrass usually has a higher Ca concentration than fescue, *E. curvula* and kikuyu. In addition, Ca concentrations in Italian ryegrass increase sharply in response to increasing soil levels of this element, while the Ca content of kikuyu and *E. curvula* are much less responsive to changes in soil Ca (Table 11.21). The particularly low Ca and P concentrations in *E. curvula* are noteworthy. Clearly, therefore, mineral supplementation should be viewed as an essential practice where *E. curvula* hay is the mainstay of winter feeding programmes. In grazed pastures, marked seasonal variations in P supply to animals may result from the close relation between plant P uptake and soil temperature; thus forage concentrations of this element tend to be lower in winter in temperate species, and in spring in summer-growing tropical species.

Table 11.20   Concentrations of Ca and P (means of several samplings for each species) and Ca:P ratios in 4 to 6 week old regrowth of high yielding pastures.

Species	Ca	P	Ca:P
	%		
Italian ryegrass	0.49	0.29	1.7
Tall fescue	0.28	0.23	1.2
Eragrostis curvula	0.21	0.16	1.3
Kikuyu	0.25	0.35	0.7
White clover	1.76	0.32	5.5
Lucerne	0.98	0.26	3.8

Table 11.21   The influence of increasing soil exchangeable calcium levels on calcium concentrations in 4 to 6 week old regrowth of three pasture grasses.

Italian ryegrass		Kikuyu		Eragrostis curvula	
Soil Ca (mg/L)	Herbage Ca (%)	Soil Ca (mg/L)	Herbage Ca (%)	Soil Ca (mg/L)	Herbage Ca (%)
162	0.37	589	0.28	54	0.15
565	0.56	912	0.29	194	0.18
916	0.68	1 054	0.31	641	0.22
1 373	0.76				

A comparison of pasture mineral concentrations, such as those reported in Table 11.20, with dietary requirements is complicated by the fact that ruminants are often not able to effectively utilise all of a mineral contained in the herbage. This is because a portion of the minerals in the plant may exist in insoluble complexes. Holmes & Wilson (1987) report that the availability to animals of Ca and P in New Zealand pastures is approximately 68% and 58%, respectively. Various authors have suggested that high oxalate levels in kikuyu severely restrict the availability to animals of Ca in this grass (Reason *et al.* 1989; Marais pers. comm.). Viewed in the light of these findings, Ca and/ or P levels in a number of pasture species are likely to be inadequate for animals.

A widely held view is that the dietary Ca:P ratio is of great importance in the nutrition of ruminants, with the optimum ratio frequently reported to lie within the range 1:1 to 2:1 (McDowell *et al.* 1983). As indicated by the data presented in Table 11.20, this ratio typically varies widely between pasture species and is lowest in kikuyu and highest in the legumes. The particularly low ratio is thought by many to significantly affect the nutritive value of kikuyu forage. In this species the Ca:P ratio is less than one for much of the season and reaches a minimum as low as 0.65 in mid-summer. To add to this problem, high K levels typical of many kikuyu pastures will accentuate this problem, as will the high levels of oxalates found in kikuyu forage. Calcium combines with the acid to form very insoluble calcium oxalate, resulting in as much as a 50% reduction in the availability of the Ca in the ingested forage. It must be pointed out, however, that reports in the literature on the effects of widely varying Ca:P ratios in the diets of ruminants are highly conflicting (Little 1982), and indications are that this ratio has received undue emphasis.

11.9.4 Potassium and magnesium

Potassium and Mg are considered together since the two interact strongly, both in their uptake by plant roots and in their absorption in the gastro-intestinal tracts of ruminants.

Herbage K levels often exceed animal requirements, whereas herbage is often unable to supply sufficient Mg to animals. Potassium concentrations in pasture herbage, on a dry matter basis, generally fall within the range 1.5% to 6%, while Mg concentrations are typically between 0.1% and 0.4%. The required K and Mg concentrations in the diet of a dairy cow with a daily milk yield of 25 litres are 0.5% and 0.2%, respectively (Kemp & Geurink 1978).

Antagonistic effects of K on Mg uptake by plants and on its absorption by animals tend to increase the potential for Mg deficiency in animals grazing on pastures. Under conditions of favourable K supply in the soil many pasture species, and in particular temperate grasses, take up K in amounts far in excess of plant requirements for maximum growth. This so-called 'luxury-uptake' of K is associated with a marked decrease in plant concentrations of Mg, and also of Ca and Na. The Italian ryegrass yield and chemical composition data presented in Table 11.22 provide clear evidence of these effects. It is noteworthy that although a K concentration of approximately 3% is required for maximum yield of ryegrass, in this study its concentration increased to more than double this value at high soil K levels. From an animal health point of view, excessive K and low Mg levels in herbage are most undesirable. The problem is in fact more severe than is reflected by herbage elemental composition data such as those presented in Table 11.22, since K and Mg are differentially available for absorption within the gastro-intestinal tract of ruminants: the availability of K in pasture herbage reportedly varies from 80% to 100%, while that of Mg typically ranges from 17% to 20% (Kemp & Geurink 1978; Holmes & Wilson 1987). In addition, high K levels have been found to inhibit absorption of Mg. It is noteworthy that herbage cation imbalances, and in particular excess K and low Mg, have been widely implicated in the incidences of hypomagnesaemic tetany (Reid & Jung 1974; Wilkinson & Stuedeman 1979), infertility (Dugmore *et al.* 1987; Beringer 1988), bloat (Turner 1981; Reason *et al.* 1989) and a reduction in milk yields (Holmes & Wilson 1987). Recent studies in the United States indicate that low physiological availability of Ca can occur in cows fed high K diets in the last three to four weeks of pregnancy, and that these cows are at a much greater risk of suffering from retained fetal membranes, sub-clinical hypocalcaemia and milk fever (Beede 1992).

Table 11.22 The influence of increasing soil K status on the chemical composition of irrigated Italian ryegrass harvested in autumn on a highly leached soil (32% clay) in the KwaZulu-Natal Highland Sourveld.

Soil K	DM yield	Herbage composition				
(mg/L)	(kg/ha)	K	Mg	Ca	Na	K/Ca + Mg*
		%				
66	1 323	2.83	0.35	0.73	0.18	1.15
125	2 043	4.40	0.27	0.66	0.11	2.08
176	1 838	5.38	0.24	0.60	0.08	2.83
426	1 939	6.19	0.20	0.48	0.05	3.89
AU (0.05)	375	0.55	0.04	0.10	0.03	0.58

* Ratio calculated on equivalents basis

Local research points to luxury-uptake of K and its attendant effects on herbage levels of other cations being particularly marked in the cases of kikuyu and Italian ryegrass. Unfavourable Ca:P ratios in kikuyu have been shown to be aggravated by an excess of K in the soil. With regard to Italian ryegrass, the very high K uptake reported in Table 11.22 is by no means unique: K levels of more than 6% in this species have been reported from other parts of the world (Hylton *et al.* 1967; Robbins & Faulkner 1983). A salient feature of the data contained in Table 11.22 is the marked increase in the K/Ca + Mg ratio (ion concentrations expressed on an equivalents basis) with increasing K supply. This ratio is widely used as an indicator of the tetany potential of herbage, with values in excess of 2.2 being linked to an increased incidence of tetany in grazing animals (Kemp & t'Hart 1957).

In the light of the evidence discussed before, it is obviously extremely important to prevent the build-up of excessive K in the soil and to maintain a favourable Mg supply. Judicious K fertilization, the rotation of winter feeding camps (in order to evenly distribute recycled nutrients), and regular soil testing are necessary to ensure that soil K levels do not become excessive in grazed pastures. Adequate soil Mg levels may, in general, be ensured through the use of dolomitic lime in the control of soil acidity.

11.9.5 Sodium

There is considerable variation in the Na content of pasture plants. Some plants readily translocate Na from their roots to their leaves while others preferentially accumulate Na in their roots and translocate very little to their leaves. The former are referred to as natrophiles, and the latter as natrophobes (Smith *et al.* 1978). A classification of important pasture and fodder crops based on these terms is presented in Table 11.23. Tall fescue is considered to be intermediate between the natrophiles and the natrophobes, since the concentration of Na in the initial leaf growth of this grass is low, but increases markedly with successive cuts.

Table 11.23 A classification of pasture and fodder species into natrophilic and natrophobic type plants.

Natrophiles	Natrophobes
Rhodes grass	*Eragrostis curvula*
Perennial ryegrass	Kikuyu
Italian ryegrass	*Paspalum* spp.
Cocksfoot	Lucerne
White clover	Red clover
Barley	Millet
Oats	Maize
	Soyabean

The recommended Na requirement for grazing ruminants is between 0.04% and 0.18% in the dry matter, with the higher level recommended for lactating dairy cows (McDowell *et al.* 1983). As with K, most of the Na in herbage is available for assimilation by animals. Sodium levels in natrophobic plants are inadequate to satisfy the dietary needs of grazing animals and so direct supplementation of animals with Na is necessary where natrophobes form a substantial part of the diet. Sodium levels in natrophiles may be adequate for grazing animals provided Na availability in the soil is not limiting. Highly-weathered inherently acid soils tend to have very low levels of available Na and on such soils the plants, even if they are natrophilic, may contain inadequate levels of Na.

In recent years there has been increasing interest in the use of Na salts as fertilizers on soils low in Na (Mundy 1983; Chiy & Phillips 1991; Manson 1995). In the case of natrophilic species, it has been found that part of the K requirements for growth can be replaced by Na, thereby improving the quality of herbage available to animals and reducing the K fertilizer requirement. Typical rates of application are 100 kg to 200 kg Na/ha. Interestingly, increased Na levels in herbage have been reported to improve the palatability of temperate grasses (Horn 1988) and significantly increase milk fat content (Chiy & Phillips 1991).

11.9.6 Other essential elements

The availabilities in pasture herbage of S, chlorine (Cl) and trace elements do not usually pose constraints on animal production. Possible exceptions are Cu, Mo and Se, which may require particular attention in animal systems on pastures.

Sheep, especially wool sheep, and dairy cows generally have a higher S requirement (0.14–0.26% S in their diet) than beef cattle (0.10% S in their diet) (Goodrich & Garrett 1986). Sulphur deficiencies commonly occur when non-protein N sources, or those with low-quality protein, are used to meet a large percentage of an animal's N requirements, or when diets low in protein are offered. Under these conditions, supplementation with S has been shown to be necessary for optimal functioning of the rumen microbial population.

Goodrich & Garrett (1986) also state that supplementation with S is beneficial if forages with low S availabilities (such as spear grass (*Heteropogon contortis*), forage sorghum and fescue) are fed. Sulphur may also aid in the detoxification of the cyanide that may occur in some pastures. On the other hand, Goodrich (1978) found that dietary sulphur levels in excess of 0.30% may reduce feed intake and production in sheep and dairy cattle.

A close inter-relation exists between Cu and Mo in ruminant nutrition. Whereas Cu supplies may sometimes be insufficient for animals, the problem with regard to Mo is usually one of toxicosis (McDowell *et al.* 1983; Holmes & Wilson 1987). In forage with a high Mo content, Cu availability to ruminants is depressed, and animals suffer from a condition known as molybdenosis. Since forage plants, particularly clovers (Mengel & Kirkby 1987), tend to accumulate high concentrations of Mo where the supply of this nutrient is excessive, great caution must be taken with Mo fertilization. Holmes & Wilson (1987) report that the following situations are likely to give rise to Cu deficiencies in dairy animals:

(a)   Cu below 5 mg/kg on a dry matter basis;
(b)   Cu:Mo ratio less than 2:1;
(c)   Cu marginal (5–10 mg/kg), Mo normal (1–3 mg/kg) and a high S content associated with 20–30% protein in fresh forage.

Application of copper sulphate (5 kg/ha) is effective in increasing the Cu content of pastures, especially on acid soils. Herbage should be analysed to establish the need for and required frequency of this treatment. However, as indicated by the diagnostic criteria outlined above, herbage analysis must also include Mo and possibly S in order to be useful in establishing the sufficiency of Cu supply to animals. In fertilising with Cu and supplying licks containing Cu to animals, the well-known susceptibility of sheep to Cu toxicity should be taken into account.

In recent years widespread confirmation of Se deficiencies in livestock, particularly on the eastern seaboard of South Africa, has led to increased interest in this element. Signs of pronounced Se deficiency include reduced growth and nutritional muscular dystrophy (often referred to as white muscle disease in lambs and calves), poor reproductive performance in older animals and high incidences of retained placentas. A recent survey of pastures and veld in KwaZulu-Natal indicated that herbage Se levels were generally well below the critical level (0.1 mg Se/kg DM) for animal nutrition (Higgins 1992). Selenium supplementation of livestock by the addition of Se to licks, oral dosage or injections is complicated by the fact that a very narrow range exists between sufficiency and toxicity. Recently, topdressing pastures with Se has been found to be an effective and convenient alternative to direct administration to animals (Millar & Meads 1987; Rimmer *et al.* 1990). The generally recommended application rate is 10 g Se/ha, with the source being either sodium selenate or commercially available 'slow release' Se pills. Herbage treated in this way has been found to contain satisfactory levels of Se for grazing animals for periods of 9 to 12 months, after which there is a requirement to repeat topdressings. Interestingly, Se fertilization of pastures grazed by dairy animals in KwaZulu-Natal has been reported to significantly reduce the incidence of mastitis.

11.10 STRATEGIES FOR IMPROVING NUTRIENT-USE EFFICIENCY IN PASTURES

The principles underlying aspects dealt with in the following discussion have, in most cases, been considered in detail earlier in this chapter. The purpose here is to provide a brief summary of practical measures for improving nutrient-use efficiency on pastures.

11.10.1 Utilisation of soil fertility built up under perennial pastures

Nutrients accumulate in soils of long-term perennial pastures such as kikuyu, especially when they are used as holding camps in winter when conserved feed (hay or silage) is being fed. Excessive K accumulation is a recognised problem, but organic matter also accumulates, storing N, P and other nutrients. These nutrients could be far better utilised if farming systems were developed with nutrient cycling in mind. The ideal option would be one of rotating such pastures with pastures from which nutrients are removed. For example, an ideal system would be to convert, after several years, kikuyu pastures to annual ryegrass or maize silage lands. Alternatively, conserved feeds should be fed on lands depleted of K, and slurries and manure should not be used where levels of P and K are already high. Furthermore, it should be borne in mind that high K levels in pasture topsoils may be rapidly depleted by taking a number of hay or silage cuts while not applying any K fertilizer.

11.10.2 Use of organic fertilizer sources

Manure produced on-farm is generally an under-utilised, low-cost source of nutrients. Off-farm sources of organic fertilizers should also be considered. In many situations it may be more economic to utilise an organic source than exclusively using inorganic fertilizers.

Several factors should be considered when assessing the viability of using organic fertilizers:

(a) the cost of transporting and spreading the material;

(b) the cost of the source per kilogram of nutrient required. If soil P and K are already high, and only N is needed, the nutritional benefit of the organic fertilizer will be limited to its N content;

(c) possible benefits may accrue from improved soil physical properties such as higher infiltration rates. Some irrigated annual pastures may be particularly responsive to organic fertilizers; and

(d) beware of 'magic' mixtures which cost far more than the elemental nutrient content would if bought as bagged inorganic fertilizer. Very few of these have been shown to be truly cost-effective.

11.10.3 The importance of liming

Because liming involves moving and applying large quantities of material per unit area, it is costly and, as a result, lime is not always applied when recommended. The economic consequences of the under-application of lime can, however, be serious and it is usually advisable to cut back on fertilizer inputs rather than on lime if capital is limited.

Frequently, benefit can be derived from lime applications in excess of recommended amounts, for the following reasons:

(a) re-acidification of soils often leads to a recurring requirement for lime;
(b) applications in excess of the recommended amounts can reduce the required frequency of application; and
(c) liming in excess of recommended amounts can enhance soil conditions by promoting biotic activity. The mineralisation of organic N, P and S is enhanced and soil physical properties often improved through the increased activity of earthworms and other organisms.

11.10.4 Emphasis on legume nutrition

Most pasture legumes have higher fertility requirements than associated grasses. The benefits in terms of N fixation and improved animal production that can be gained are, however, well worth the investment required to establish legumes. The correct liming of perennial legumes is essential for persistence, particularly of lucerne and clovers. Optimal P and Mo nutrition is also essential for satisfactory nodulation and N fixation. Potassium nutrition is especially important in grass-legume mixtures and in cut-and-remove systems such as lucerne grown for hay. Sulphur and B deficiencies are also potential problems. Where supplementary N is used, special attention should be given to the timing of applications. Little benefit is gained from N applications when N fixation is peaking, but in cooler periods when legume growth is slow, N applications can boost production.

11.10.5 Flexible nitrogen management

Nitrogen applications should be adjusted according to the production capabilities of the pasture through the season and the demand for forage. High rates of N in winter, when grass production is limited by low temperatures, can reduce the acceptability of pasture to grazing animals. High N applications to maintain high production of pasture in summer is also pointless where the herbage cannot be utilised effectively or conserved. This is especially pertinent in the light of indications that excessive herbage N concentrations may be detrimental to animal production in certain situations.

11.10.6 Choosing most cost-effective sources

Prices and availability of fertilizers, both inorganic and organic, and of ameloirants such as lime and gypsum, often vary considerably from one season to the next, as do the costs of transporting and spreading them. Farmers should, therefore, never make the mistake of assuming that what was cheapest last year is necessarily the cheapest this year. Hidden costs of using particular forms of fertilizers or ameloirants, too, should be kept in mind. Examples are the increased soil acidification when ammonium sulphate is used rather than limestone ammonium nitrate or urea, and the higher transport and application costs (per kilogram of nutrient) of organic fertilizers than of inorganic fertilizers, and of low-grade limestones than highly effective limes.

12

The management of planted pastures

CONTENTS

The success or failure of a pasture programme is largely dependent on the defoliation management applied. The varied and pronounced effects of defoliation determine the productivity of the pasture and the performance of the animals utilising the forage it produces. However, there is no universal system of grazing (defoliation) management to suit all pasture species or all animal production systems. Pasture species differ in many respects and these differences need to be taken into account when planning grazing programmes. The pasture species in common use in South Africa can be conveniently grouped into six categories (section 12.1.1.3) on the basis of their season of growth and growth habit. Within each of these categories plant responses to defoliation management are essentially similar. This section is concerned primarily with the objectives of pasture management, the principles underlying management, grazing management practice, management for seed production and the regeneration of degraded pastures in the major rainfall regions of South Africa.

12.1 HUMID REGIONS

12.1.1 Summer rainfall region

P.E. Bartholomew

12.1.1.1 *Management objectives*

In order to get the most out of pastures the grazier must have specific management objectives which need to be both pasture and animal orientated.

233

Pasture orientated objectives include:
(a) maximum dry matter production per unit area, at minimum cost;
(b) the provision of herbage of appropriate quality to different categories of livestock;
(c) efficient utilisation of the pasture (i.e. minimum wastage);
(d) the extension of the growing season of the pasture to the maximum length possible; and
(e) the maintenance of the density and vigour of the pasture.

These objectives need to be met without the need to resort to complex management procedures.

Included among the animal orientated objectives are the achievement of:
(a) appropriate levels of animal production for the production system envisaged; and
(b) appropriate levels of animal production per unit area of land.

Animal management programmes, like pasture programmes, should be as simple as possible.

Having defined the objectives of a production system, it should be posssible to identify the most appropriate pasture species and the most appropriate pasture utilisation practices and grazing system to adopt. Here a certain amount of compromise is usually necessary. The grazier will generally see the pasture objectives as being met by management strategies designed to:
(a) harvest the pasture in such a way that its growth rate remains at an acceptable level;
(b) utilise the pasture at a stage of growth at which it will provide the quantity and quality of herbage required by the type and class of animal utilising the pasture; and
(c) promote the longevity and density of the pasture.

To do this effectively, the grazier needs to understand the basic principles of pasture growth, the response of plants to defoliation and the implications of different defoliation regimes when applied to different pasture species.

12.1.1.2 *Some basic principles*

12.1.1.2.1 *Growth rate*

Assuming adequate temperature, soil moisture and nutrients for the species in question, the growth rate of pasture plants is largely a function of the amount of potentially active leaf tissue exposed to light. This is so because plants require the energy produced by the leaves for growth. Within limits, therefore, the greater the leaf area the higher will be the growth rate of the plant and so any treatment which reduces leaf area will reduce pasture growth rate. Heavy grazing is therefore invariably followed by a period of slow growth until sufficient leaf area has again accumulated. However, an excessive accumulation of leaf material will also reduce growth rate. When the pasture canopy has become closed, the basal leaves become shaded and senesce. These senesced leaves either accumulate or they may decay, depending on the environmental conditions at the time. If they decay, then new leaves which continue to be produced at the top of the closed pasture canopy may no more than balance those which decay at the base of the canopy. Where they do not decay but accumulate at the base of the sward, total yields will continue to increase but forage quality will decline. From the point of view of the yield of good quality forage, there is, therefore, no advantage in resting the pasture once the canopy has closed (Tainton 1973).

12.1.1.2.2 Quantity and quality of dry matter

The frequency and severity at which pasture plants are defoliated have pronounced effects on both the quantity and quality of available forage. To maximise production, the pasture should retain sufficient leaf to allow for rapid growth for as much of the growing season as possible. In a rotational grazing system animals should be removed from a pasture before they have grazed all the leaf material. Where the pasture is cut, the cutting height should be such that a reasonable amount of residual leaf is left on the pasture (usually 8 cm to 10 cm of stubble).

Because of the wide ranging growth characteristics of different pasture species, there are, however, exceptions to the above recommendations. For example, the leaf canopy of *Trifolium repens* rises in a horizontal plane as the pasture grows. Most of the leaf material is therefore removed whether the pasture is defoliated leniently or severely. Also, where the basal leaf of any pasture has senesced there may be little advantage to lenient defoliation.

From this discussions it follows that, as a general rule, the more severely a plant is defoliated, the more slowly it will recover and the less severely it is defoliated, the more rapid will be its regrowth.

Defoliation frequency is also important in determining regrowth rate. The more frequently a pasture is severely defoliated, the lower its dry matter yield will be. Clearly, however, there are innumerable permutations with respect to height and frequency of defoliation. Table 12.1 gives a generalised indication of the reaction of plants to the broad categories of defoliation.

Table 12.1 The effects of frequency and intensity of defoliation of pasture plants on herbage dry matter yield and quality.

Defoliation frequency	Defoliation severity	Dry matter yield	Herbage quality
Frequent	Severe	Low	High
Frequent	Lenient	High	Medium
Infrequent	Severe	High	Low
Infrequent	Lenient	Medium	Low

While the principles presented in Table 12.1 apply to all pasture species, the magnitude of these effects will differ among different species. For example, the decline in herbage quality with the age of the regrowth is far more dramatic with a grass like *Eragrostis curvula* than it is with a clover like *T. repens*. Similarly*, Pennisetum clandestinum* is far more 'tolerant' of severe defoliation than *Dactylis glomerata*. It is important for the grazier to understand the reaction to management of the species being dealt with and to apply harvesting principles within the limitations of that species.

12.1.1.2.3 Pasture utilisation

Much emphasis has been directed by grassland scientists in South Africa at the production of large quantities of dry matter, but relatively little attention has been paid to the efficiency with which this dry matter is harvested by animals. As a general rule, a large proportion of dry matter produced is not consumed during grazing and so does not contribute to the pastoral enterprise.

Animals will select herbage of a higher quality and palatability than the 'average' of that on offer. They will first crop the youngest and most nutritious material at the top of the sward before returning to graze the older and less nutritious material lower down in the sward (Frame 1992). As the pasture is grazed down, so the animals are left with herbage that is increasingly less palatable, less acceptable and of lower nutritional value. Dry matter intake will also decrease as the amount of herbage on offer decreases. Thus, as the pasture is grazed down, not only will the quality of intake decline but also the quantity of the material consumed.

In practice, the objectives of maximising dry matter yield from a pasture (which requires a long regrowth period) and of offering the animals high quality herbage (which requires a short regrowth period) are in conflict. Also, should a pasture be leniently utilised, a large amount of herbage will be left following grazing. This accumulated material will reduce the quality of the herbage available to animals at the following grazing by 'contaminating/diluting' the new growth. Since animals select for new regrowth, their intake may well be severely restricted. This 'dilution' of new growth can be particularly important in high producing dairy animals requiring a high dry matter intake. In addition, such accumulation can result in considerable loss of material through trampling and result in the accumulation of large amounts of organic matter on the soil surface. This organic matter can promote the development of fungi and mould, particularly under moist conditions, thus further reducing the acceptability of the herbage on offer.

The loss of available herbage through wastage has been estimated by Du Plessis (1978) to be as high as 60% in some pasture situations in the Highland Sourveld of KwaZulu-Natal (Table 12.2), whereas the general estimate used in feed budgeting by Jones & Arnott (1977) and Klug & Webster (1993) is 20%. In tropical pastures, inefficient use of the dry matter may be unavoidable under grazing. However, in the higher quality, more palatable temperate species such losses are unnecessary.

Table 12.2   Estimated wastage of dry matter on a variety of pasture types grazed under practical farming conditions (Du Plessis 1978).

Pasture type	Type of animal grazed	Wastage
Italian ryegrass – irrigated	Beef steers	28.6
Italian ryegrass – irrigated	Dairy cows	29.0
Italian ryegrass – dryland	Beef steers	22.3
Italian ryegrass & clover – irrigated	Dairy cows	46.0
Italian ryegrass & clover – dryland	Beef cows	19.0
Italian ryegrass & oats – dryland	Dairy cows	50.0
Ariki ryegrass – irrigated	Dairy cows	39.0
Ariki ryegrass & clover – irrigated	Beef heifers	43.0
Cocksfoot – dryland	Beef steers	26.8
Cocksfoot – dryland	Dairy cows	60.0
Tall fescue – dryland	Dairy heifers	26.5
Tall fescue – dryland	Beef cows	58.0
Tall fescue – dryland	Dairy cows	34.0
Tall fescue & clover – irrigated	Dairy cows	27.0
Kikuyu – irrigated	Dairy heifers	51.4
Kikuyu – dryland	Dairy heifers	26.8
Kikuyu – dryland	Dairy cows	17.0

The main contributory factor to wastage in intensive pasture systems arises from trampling and fouling of herbage, particularly on well grown-out pastures where the objective is to maximise yield. While such losses are difficult to eliminate, they can be substantially reduced by strip grazing, by the use of a two herd/flock follower system and by forward creep grazing systems. These are discussed in section 12.1.1.4.

12.1.1.2.4 Pasture condition

The regrowth of pastures that are made up largely of young tillers and in which there is a dense population of young leaves at the base of the sward will, normally, be rapid where a reasonable amount of leaf remains after utilisation. These remaining leaves should be photosynthetically active and will contribute to rapid regrowth. If, however, most of the tillers are reasonably old, the basal leaves will be too old to contribute substantially to photosynthesis if they remain after grazing. Similarly, if the pasture has been at full canopy for any length of time the basal leaves (even of pastures composed of relatively young tillers) will have degenerated and will not recover if exposed to light. There is little point, therefore, in attempting to leave effective leaf on such pastures after grazing, although the retention of large amounts of stubble which contain large amounts of carbohydrates may make some contribution to regrowth rate. The amount by which yields are increased as a result of such an increased regrowth is unlikely to match the amount of potential livestock feed lost in the ungrazed stubble, which will soon decay.

In practice, it is possible to restrict the period of slow growth which normally follows the utilisation of a pasture having a closed canopy, by grazing it as soon as possible after full canopy has developed. This condition can easily be recognised by viewing the pasture from above. If the leaves almost completely cover the soil or if few sun-flecks reach the soil surface at midday, or if there is some yellowing of the lower leaves, full canopy will have been reached. The pasture will then need to be grazed so that a new set of young active leaves can develop to provide the energy needed for a new flush of regrowth.

12.1.1.2.5 Retention of pasture density

Pasture density is developed in newly sown stands and retained in older stands by the process of tillering (refer to Chapter 2). For the development of a dense sward and the retention of tiller density, management must be designed to stimulate tiller development. Furthermore, management must ensure that young tillers are given the opportunity to develop to maturity. This entails ensuring that the parent tiller has sufficient leaf to provide its daughter tillers with their energy needs while they are still young and wholly dependent on it. In addition, conditions should be such that sufficient light penetrates to the level at which the young tillers' leaves are situated. This will ensure that, as the young tillers produce leaves, these leaves can photosynthesise and take over the function of supplying carbohydrates for further leaf and root growth. Generally, a single heavy utilisation treatment, whether by grazing or mowing, will activate a large number of previously dormant buds. The tillers which develop from these buds will grow slowly at first. If the plant has not stored sufficient reserves or cannot accumulate sufficient leaf to 'feed' these new tillers within a short period following treatment, they may die. In most circumstances sufficient energy for the growth of the young tillers can be supplied by the old tillers, provided the latter are not regrazed at frequent intervals. A single hard grazing treatment will, therefore, normally increase the density of a sward. However, swards that are repeatedly heavily grazed tend to become sparse; new tillers develop but are starved of carbohydrates and therefore die.

Young tillers may also fail to survive in swards that retain a dense canopy for any length of time. This is because young shaded tillers are not able to produce their own photosynthate. Such tillers become spindly, their leaves yellow and die and their root systems fail to develop. If light is not allowed to penetrate to the level of the plant crown for any length of time, new tillers may begin to develop high up on the parent stem and well above the soil surface, where they will be unable to develop an effective root system. Therefore, while lenient grazing may be practised at times to encourage rapid recovery growth, it should not be continued indefinitely. Occasional hard grazing should be incorporated in the defoliation programme to stimulate the development of a new population of tillers low down on parent stems, particularly in autumn and spring when conditions normally favour such development.

12.1.1.2.6   Growth rate and pasture utilisation

The higher the growth rate of a pasture, the more frequently it should be defoliated. Frequent light defoliation allows for unrestricted intake, a rapid regrowth, a high quality of herbage for livestock and ensures that the pasture does not remain at full canopy for any length of time. In addition, frequent light defoliation guarantees a large residual leaf area. High growth rates are associated with the high rates of photosynthesis, providing the potential for 'surplus' carbohydrates for the development of lateral tillers. Of particular significance during such periods of abundant pasture is wastage (section 12.1.1.2.3). In practice, high growth rates imply the need to introduce additional animals or to reduce the number of camps in the grazing system. Camps that are excluded can be used by other animals or for the production of hay or silage.

During periods of slow growth, the grazing cycle should be lengthened by slowing down the rate at which animals are rotated through the camps. Because growth is slow, defoliation during these periods is usually severe. Lengthening the grazing cycle provides the plants with an opportunity to accumulate a reasonable amount of leaf before the next defoliation; allows the plants to replenish the reserves used in initial regrowth; and provides a greater bulk of material for the next grazing. Such a slowing down of the grazing cycle can normally be achieved only by increasing the period of stay and may lead to an increased severity of defoliation. Animal numbers may therefore need to be reduced or supplementary feed provided. Should a leader-follower system be in operation, the follower group should be removed from the pasture at such times and returned only when growth rates have increased sufficiently to again warrant their use.

12.1.1.3   Pasture types and their characteristics

Pastures in common use in the humid regions of South Africa can conveniently be classified according to their growth habits and defoliation management requirements (refer to Chapter 9).

12.1.1.3.1   Grass pastures

(a)   Tropical/subtropical tufted grass species
      The warm season tufted grasses such as weeping lovegrass (*E. curvula*), blue buffalo grass (*Cenchrus ciliaris*), Smuts' fingergrass (*Digitaria eriantha*), kleingras

(*Panicum coloratum*), Guinea grass (*Panicum maximum*), Dallis grass (*Paspalum dilatatum*) and others have two major disadvantages when compared with the cool season tufted species like the ryegrasses (*Lolium multiflorum* and *Lolium perenne*), tall fescue (*Festuca arundinacea*) and cocksfoot (*D. glomerata*). With few exceptions the warm season grasses have an inherently low digestibility (Fig. 12.1). In addition, their tufted growth habit is associated with a lack of resistance to and slow recovery from heavy grazing.

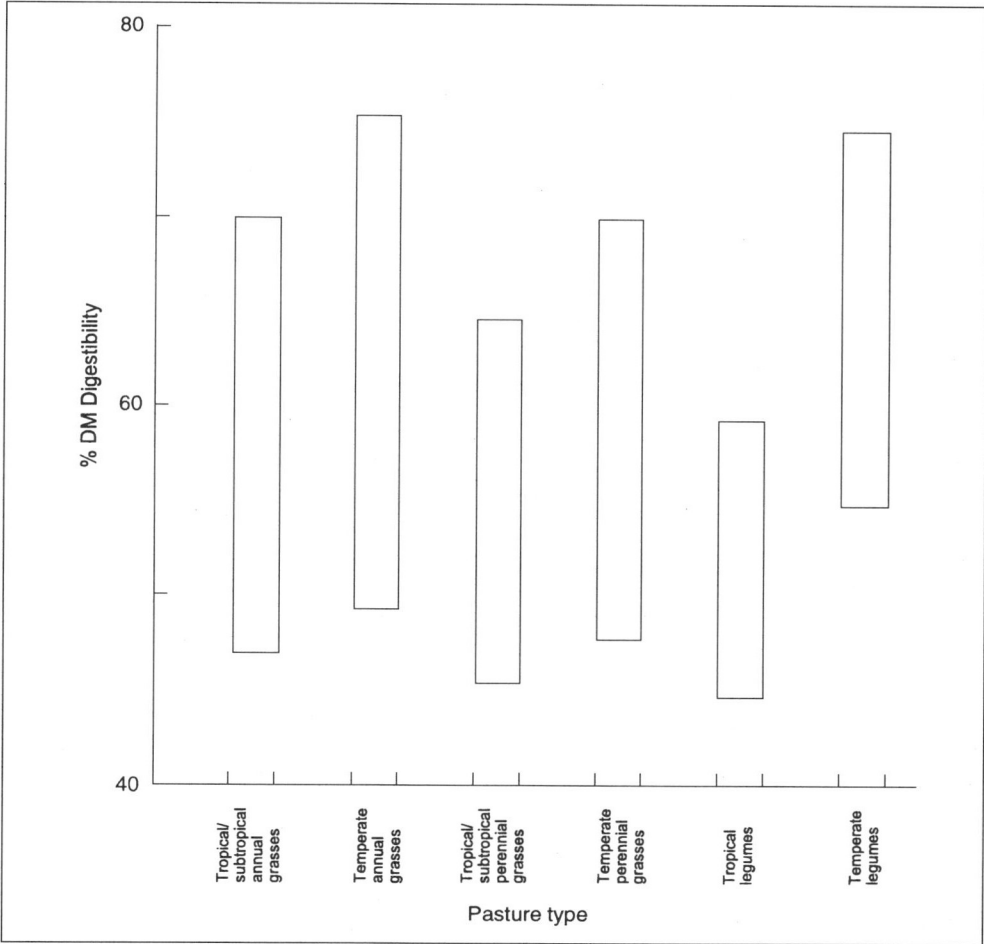

Figure 12.1 Ranges in dry matter digestibility of different categories of pasture species (data for tropical legumes adapted from Minson (1977) for *Desmodium intortum, Desmodium uncinatum, Neonotonia (Glycine) wightii* and *Stylosanthes humilis;* remaining data from Humphrey 1978).

Because of the low digestibility of the tropical/subtropical species, animals do not perform well if they are required to utilise a large proportion of the standing herbage, particularly since digestibility declines markedly down the canopy. If, however, grazing is lenient so that the animals can select only the more nutritious material, their performance can be satisfactory. Such lenient grazing implies an inefficient use of the forage so that wastage is high if animals are to perform satisfactorily. At first glance this would seem to be wasteful, but lenient use has a major secondary advantage. Dry matter production from pastures that are leniently grazed is consistently higher than from intensively grazed pastures. Therefore, lenient grazing, even with a lower efficiency of utilisation, may lead to greater overall intake of dry matter and so greater animal production per unit area of pasture. However, where grazing is lenient, a large quantity of low quality stem material may accumulate as the season progresses. It may often be advisable to remove this material in mid-summer by mowing.

Bearing these aspects in mind it appears that continuous grazing at lenient to moderate stocking rates may best suit this type of pasture (Tainton 1976). If grazing is rotational, then animal movement should be relatively rapid so that a 10 cm to 15 cm stubble remains when the animals are moved to new grazing.

(b) Tropical/subtropical creeping grass species

Pastures composed of prostrate sod-forming (rhizomatous and/or stoloniferous) species, which include kikuyu (*P. clandestinum*), couch, coastcross II and Star grasses (*Cynodon* spp.), Nile grass (*Acroceras macrum*) and others, are generally well adapted to grazing with the exception of the Star grasses, which should be more leniently grazed than the other species. Sod-forming grasses generally recover well from heavy use. Not only do they retain a dense sod even under repeated heavy grazing (particularly kikuyu and *Cynodon dactylon*), but sward density may decrease if they are not periodically grazed hard. Like the warm season tufted species, they have a low digestibility relative to the cool season grasses (Fig. 12.1). To ensure satisfactory animal performance they should be leniently grazed to allow the animals to select the more nutritious herbage. However, such treatment will not be compensated for in these species by as great an increase in dry matter production as in the tufted species. Hard grazing will in any event be required periodically to retain sward density. Therefore the pasture should be grazed hard and the animals supplemented if high individual animal performance is required. Alternatively, high producing animals could be permitted to select the high quality components of the forage before being followed by low producers which should graze the remaining stubble (refer to section 12.1.1.4.4). Generally, therefore, while these pastures may be continuously grazed, they can best be managed using rotational grazing which allows for better control of the frequency and the height to which they are grazed.

(c) Cool season grass species

The cool season grasses commonly grown in South Africa include oats (*Avena sativa*), stooling rye (*Secale cereale*), tall fescue (*F. arundinacea*), cocksfoot (*D. glomerata*), Italian and Westerwolds ryegrass (*L. multiflorum*), perennial ryegrass (*L. perenne*), and the crosses between the annual (*L. multiflorum*) and perennial (*L. perenne*) ryegrasses (*L. x boucheanum*) such as Augusta, Bison and Geyser. These species are moderately resistant to hard grazing although they do vary in this respect. The perennial ryegrasses would generally be considered the most resistant of these to heavy grazing during spring and autumn, but not during summer in the warmer areas. Tall fescue, oats, stooling rye, Italian and Westerwolds ryegrass and the ryegrass crosses are less resistant to severe defoliation, with cocksfoot the least resistant. Like perennial ryegrass, tall fescue should not be defoliated severely during summer, particularly in the warmer areas.

Qualitatively these cool season grasses are much better than the warm season species (Fig. 12.1). The ryegrasses, for example, have crude protein levels ranging from 20% to 25% compared with the 10% to 12% for lovegrass and 10% to 17% for kikuyu. They are also much less fibrous (20% crude fibre as against the 30% for the warm season kikuyu). It is therefore possible to achieve a much higher utilisation of the herbage produced by the pasture and still maintain better animal performance than is possible with the warm season species.

Management of the cool season pastures involves the following:
1) management to achieve maximum dry matter production;
2) methods of ensuring an adequate supply of forage during periods of slow growth (e.g. in mid-winter);
3) management to achieve maximum utilisation of the dry matter produced; and
4) the retention of pasture density, particularly in the perennial species.

12.1.1.3.2 Legume pastures

(a) Tropical legume species

The tropical legumes, which include such twining species as greenleaf desmodium (*Desmodium intortum*), silverleaf desmodium (*Desmodium uncinatum*), *Glycine wightii* and others have not proved successful in South Africa outside the frost-free coastal areas and the Mpumalanga lowveld (Grunow 1979). In view of their apparent success in the tropical and subtropical regions of Australia, this lack of success is somewhat surprising. No doubt the abundance of vigorous pioneer grass species (*E. curvula, Eragrostis plana, Sporobolus pyramidalis, Sporobolus africanus, P. maximum* and others) which readily invade any stand that is not highly competitive and their general lack of tolerance of hard grazing are partly responsible for this lack of success.

Unlike the temperate legumes, these tropical species are not particularly acceptable to animals and when mixed in with grass they are often avoided, particularly during spring. They are generally more readily grazed during autumn when they can provide moderate quality grazing when that of most summer grass species is declining rapidly. This lack of acceptability is fortunate for they are generally not tolerant of heavy grazing and recover very slowly from heavy use (*Stylosanthes humilis* is a notable exception). Pastures that comprise mixtures of the twining legumes and grasses are therefore best adapted to lenient grazing systems. The animals will, in any event, tend to select a grass diet in the spring and largely ignore the legume, providing the stocking rate is not too high. This will allow the legume to grow largely unhindered and will reduce the competition afforded by the grass.

If rotation systems are used, periods of absence should be adequate to allow the legumes to grow out sufficiently to suppress the growth of the associated grasses. Without such suppression, grasses are likely to dominate the pasture and legume survival may be poor.

(b) Temperate legume species

These legumes, which include such well-known species as white clover (*T. repens*) and red clover (*Trifolium pratense*), have a distinct quality advantage over the grasses. They are also normally less costly to maintain because of their low fertilizer N requirements. Because of their very high quality (Fig. 12.1), intense utilisation will not materially reduce the performance of individual animals, at least not during the spring period (Fig. 12.2). Because of its stoloniferous habit, white clover can tolerate intense utilisation provided it is not subject to excessive competition from associated unpalatable species that may remain largely ungrazed.

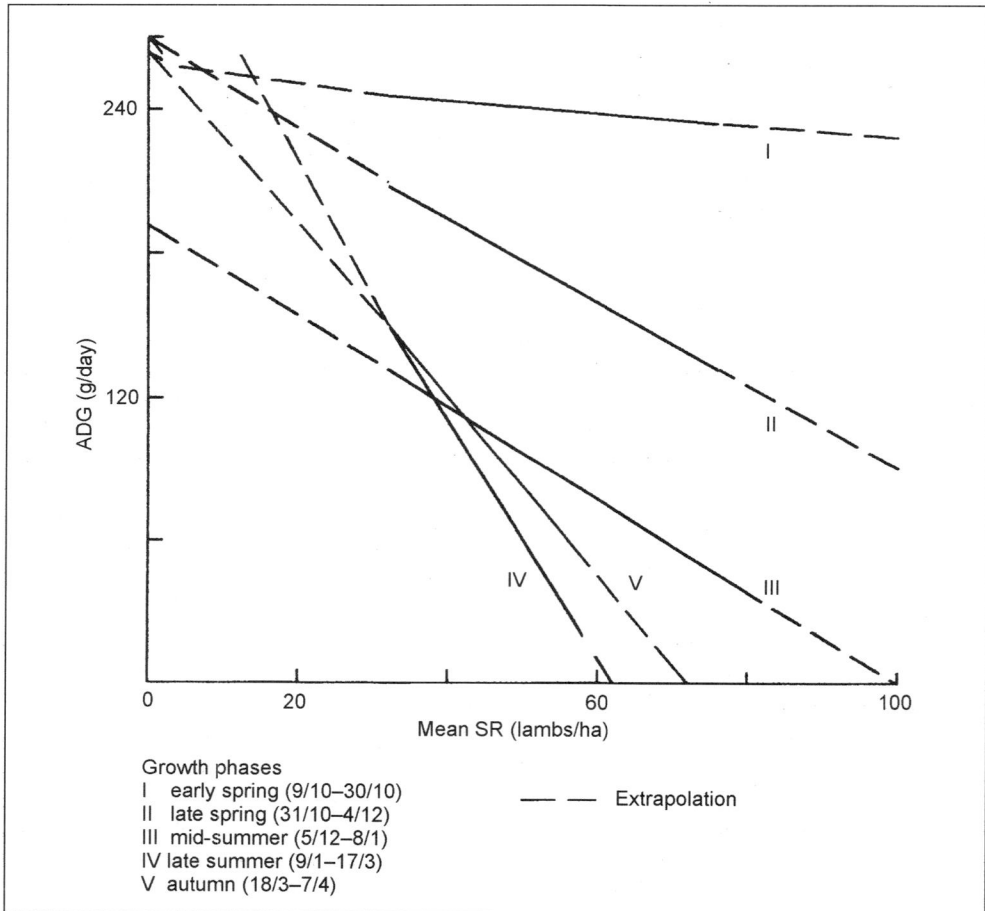

Figure 12.2 Relation between stocking rate and average daily gain of lambs on *Trifolium repens* pasture during different periods of the grazing season (Brockett 1978).

The major disadvantage of clover pastures is undoubtedly their tendency to produce unpredictable bloating in cattle. This disadvantage is a very real one but can be controlled provided adequate precautions are taken. The following safeguards may be incorporated in the management programme on bloat-prone pastures:
1) animals should be kept off young over-succulent clover-dominant pastures, particularly during wet weather;
2) drenching twice daily with anti-foaming agents (either with synthetic products marketed for this purpose, or with natural oils such as peanut or sunflower oil)

provides an effective method of control. Pastures may be sprayed with these agents before they are grazed, but this is usually wasteful. Other less reliable methods include treating the drinking water or the application of an anti-foaming agent to the coats of the animals, from which it is licked;

3) supplying a coarse dry roughage, such as hay;
4) the required area of clover pasture can be cut and left to wilt for a day before being grazed; and
5) hungry or unadapted animals should not be given access to clover pastures.

The advantage which clover has of fixing N both for its own use and for use by the companion grasses is perhaps of doubtful value in intensive irrigated pastures. To take full advantage of the ability of clover to fix N, little fertilizer N should be used. This immediately limits the yields that can be expected from the pasture and may make for inefficient use of the high-cost irrigation facilities. This disadvantage is very real on farms that have limited acreage of irrigable land. Even in these situations, the inclusion of clover in a sward that is fertilized with N may be warranted because of its high quality, even though little N may be fixed.

In semi-intensive situations and particularly in the cooler areas at high altitude, clover will produce well during the spring and autumn. In these areas it may be more economical to rely largely on clover to supply N to the pasture and to accept a lower-yielding pasture than to rely entirely on fertilizer N. Fertilizer N may still be used strategically on such clover-dominant swards to boost production when forage is in short supply.

Clovers, when mixed with grass, are generally best grazed rotationally. They are invariably more palatable than the companion grass species and so are likely to be selectively and intensively grazed in continuous systems. Where clover is grown alone, animal performance may be significantly better from continuous than from rotational systems (Brockett *et al.* 1979).

12.1.1.3.3   Grass/legume pastures

The characteristics of the grass/legume pasture will depend on the particular species of grass and legume used. Hence, the behaviour of a white clover/ryegrass pasture will differ considerably from that of a *Glycine/Panicum* pasture. Management must therefore depend on the particular species included in the mixture, and should be based on the requirements of each species. Typically in these pastures great difficulty is experienced in maintaining the balance between grasses and legumes. Appropriate grazing management procedures will contribute to maintaining such a balance. The uniform defoliation of both components of the sward is particularly significant. Where, for example, the pasture comprises a mixture of tufted grasses (such as one of the ryegrasses or *E. curvula*) and a creeping legume (such as white clover), heavy continuous grazing will promote the dominance of the legume and lenient grazing the dominance of the grasses. In contrast to this, heavy continuous grazing will promote the grass component at the expense of the legume component in mixed pastures comprising grazing sensitive legumes (such as lucerne and many of the rambling tropical legumes) and grazing tolerant prostrate grasses (such as kikuyu or Bermuda grass). It may be necessary, from time to time, to mob graze with a follower herd or to use a mower to uniformly defoliate the pasture. In addition, a balance between grass and clover can be achieved by strategically applying N (to boost the grass component) or P (to boost the legume component). The management of these pastures therefore revolves around both grazing management and the manipulation of soil fertility.

12.1.1.4 Planning grazing programmes

12.1.1.4.1 Continuous grazing

Animals can be introduced to newly established pastures when the plants are sufficiently well established and a reasonable leaf area has developed. Pasture species differ with respect to the timing of this initial grazing. *Lolium multiflorum* can be grazed lightly at a relatively 'immature' stage, whereas *F. arundinacea* and *D. eriantha*, for example, must be allowed to develop relatively good canopies before being utilised for the first time. Grazing can commence on fully established pastures once sufficient leaf area has developed to sustain high growth rates.

Aspects of continuous grazing systems include:

(a) a low labour requirement;

(b) a much lower requirement for fences and watering points than in other grazing systems;

(c) the ability to keep grazing records is easier;

(d) there is less disturbance of the animals but animals may tend to become 'wild' when infrequently handled; and

(e) most importantly, good production per animal can be expected provided stocking rates are low.

Continuous grazing systems require the careful adjustment of animal numbers so that the pasture does not become over-utilised during the growing season. As growth slows down towards the end of the season more intensive grazing can be allowed, but animals should be removed in time to allow for 10 cm to 15 cm of recovery growth before the pastures become dormant. This will reduce the tendency for weeds to invade and will promote more rapid growth of the pasture when conditions again become conducive to active growth.

Ideally, where grazing is continuous, the stocking rate should be adjusted regularly throughout the season so that the pasture can be kept in as ideal a growing condition as possible. This is seldom possible in practice, although it may be possible to occasionally introduce large groups of animals for short periods of time to prevent the pasture from becoming excessively rank. On the other hand, all or at least some of the animals should be removed when there is insufficient pasture.

Unfortunately, there are some serious drawbacks to continuous grazing. These include:

(a) the difficulty of evaluating the amount of available forage on hand at any time. The need to remove or introduce additional animals is often only evident when it is too late;

(b) area selection or patch grazing can result in poor pasture utilisation. Some areas of the pasture may be regularly overgrazed while other areas become moribund (particularly with sheep and especially if animals are introduced to a pasture which carries a large amount of forage). This may seriously affect overall pasture productivity;

(c) the system does not lend itself to forage conservation and so this must be done on separate pastures;

(d) forage cannot be rationed to the animals, so that it is difficult to balance forage demand and supply;

(e) applying fertilizer, and particularly N, can cause poisoning (both from excess nitrate N in the herbage and from animals eating spilt fertilizer);

(f) parasite problems may become accentuated under continuous grazing in highly intensive irrigated pastures;

(g) since it is not possible to exclude animals from recently irrigated areas, puddling and compaction can adversely affect water infiltration and the productivity of irrigated pastures; and

(h) because the animals are handled infrequently, they tend to become increasingly less tolerant of being handled.

The disadvantages of continuous grazing may, therefore, often be such as to warrant the use of rotational systems.

On cool season pastures the aim is often to maximise production per animal. Where a steady growth rate is maintained by the pasture for long periods of time during the growing season, stocking rates can readily be equated to pasture growth rate. Under these conditions, continuous grazing has often provided good results. Selective grazing may become a problem if there are wide differences in the acceptability of herbage. Topping (generally with a mower), even though it involves a certain amount of wastage, may be advisable in order to maintain pasture condition and encourage uniform grazing. Stocking rates must be strictly controlled during periods of slow growth to prevent overgrazing and the resultant decline in animal and pasture production. During periods of rapid growth, additional animals should be introduced or temporary fences erected to enclose a portion of the area for fodder conservation. There seems to be general agreement, however, that higher stocking rates and greater animal production per hectare can be achieved with rotational systems than with continuous systems, but that this often entails a reduced performance of the individual animal.

12.1.1.4.2 Rotational grazing

There is much to be gained from the rotational use of pastures. However, these advantages do not arise automatically once a rotation system is instituted. In the words of Professor M.McG. Cooper: '. . . one has to use one's head in top-level grassland farming for there is no one recipe that fits the variety of circumstances that are encountered, especially when there are very high stocking intensities. There is really no satisfactory substitute for brains and a preparedness to use them to best advantage.' (Cooper & Morris 1973). This is particularly true of intensive rotation systems if they are to be used to full advantage. Rotational grazing should aim to achieve as many of the following potential benefits of such systems as possible:

(a) the grazing intensity should be so controlled that the leaf area index, on average, is sufficiently high to sustain a rapid growth rate;

(b) losses through leaf senescence should be reduced to a minimum by grazing before the pasture becomes too mature;

(c) plants should be allowed an adequate opportunity to recover from grazing;

(d) species composition, particularly in mixed grass/clover pastures, should be controlled by varying the intensity and frequency of grazing;

(e) efficient utilisation of the standing forage should be achieved by preventing excessive trampling and fouling losses;

(f) the quality of the forage presented to the animals should be closely controlled by introducing the animals to the pasture when it is in an appropriate condition. Better quality control can therefore be achieved with rotational than with continuous grazing;

(g) the available forage should be rationed by adjusting the size of the daily allocation of pasture relative to the period of stay, or vice versa;

(h) a proportion of the pasture should be set aside for fodder conservation, for foggage or for seed production during peak growth periods;

(i) the amount of forage on hand should be assessed regularly so that animal numbers can be adjusted in time; and

(j) fertilizers, particularly N, should be applied immediately after grazing when they will not constitute a danger to the animals.

12.1.1.4.3 Strip grazing

Strip grazing is a 'refinement' of rotational grazing. Here the animals are allowed access to only a narrow strip (break) of ungrazed pasture at any particular time. New breaks may be offered to animals from half-hourly to daily. The animals are also allowed access to recently grazed breaks. Fouling and trampling is thus largely confined to recently grazed areas. No break should be included within the grazed area for more than three days so as to limit the period that any area of pasture is exposed to grazing, trampling and fouling. Electric fencing is well suited to this type of system.

Strip grazing requires more labour than fixed camp systems. The operator needs to be trained to judge when additional pasture should be allocated and the size of each new break. There is also a need for strict supervision when fertilizer, particularly N, is applied.

If at all possible, grazing should be delayed until the pasture has become well established and has produced a reasonable amount of leaf. If this is not possible because there is a shortage of other feeds, only lenient grazing should be permitted. This can be achieved with a rapid rotation. Alternatively, the whole pasture area can be opened to grazing and rotational grazing delayed until forage begins to accumulate on the pasture. This will typically be accompanied by uneven utilisation of the pasture as the animals begin to graze selectively. When rotation is implemented, it is important to balance forage demand with the amount of forage allocated to the animals in each 'break'. This balance should ensure that, while the animals are able to 'fill', the pasture is evenly and efficiently utilised. In a strip grazing system, a back fence should follow behind the forward fence and should include only that area grazed one day, or at the most two days, previously. This protects those strips which are more than one or two days into their recovery period and provides a loafing area so that the animals spend less time on the newly allocated pasture. Fouling is therefore reduced to a minimum.

If it is not possible to achieve uniform grazing with high producing animals, either a section of the pasture could be closed off and its forage conserved as hay or silage; the high-producers could be followed in the rotation by low-producers; or the pasture could be topped with a mower.

As the growth rate of the pasture continues to increase as the plants become well established, the size of each new allocation ('break') must be reduced or the period of stay increased. The rotation cycle is therefore lengthened. When the first camp grazed in the rotation has regrown to a stage at which it is again ready to be grazed, it should be opened up for grazing and any ungrazed camps used to produce conserved feed or managed for a seed crop. This process continues throughout the season, with intake and the extent to which the pasture is selectively grazed being controlled by frequent adjustment of each new allocation of pasture.

As growth rate slows down at the end of the growing season, the size of each new allocation must again be increased. The cycle therefore shortens until the pasture has insufficient time to recover before it is due for its next grazing. Conserved feed may then be fed back on the pasture. The older swards, or those which are due to be ploughed out for cropping or re-establishment, should preferably be used for this purpose. Movable racks are useful for this process because they can be moved regularly to prevent animals from concentrating in specific areas for long periods of time.

12.1.1.4.4   Leader-follower grazing system

The leader-follower grazing system is an adaptation of rotational grazing. Efficient utilisation of herbage on offer can be achieved by using a two herd/flock system. A leader group, which should comprise the most productive animals, is allowed to graze the pasture first. The animals will have access to the high quality palatable material and so should perform well. They are followed by a second (or even a third) group of lower producing animals with lower nutritional requirements, to 'clean up the pasture' to the desired level. This system is useful only where a lower level of production is acceptable for the follower group. A third group could be used as pasture conditioners and would be expected to perform poorly. An alternative would be to use a mower to clean up the pasture but this would not be conducive to efficient pasture utilisation. Where a mower is used, care should be taken to prevent large amounts of surface mulch from accumulating on the pasture. Such mulch may reduce the regrowth potential of the pasture, it can promote fungi and mould and so reduce acceptability, and may reduce the efficiency of applied N.

The use of a leader-follower system is particularly effective on temperate pastures because of the relatively high quality of the material at lower levels within the sward (i.e. as the sward is grazed down) and efficient utilisation is relatively easily achieved. In addition, reasonable animal performance can be expected from the follower herd on such pastures.

A flexible system of pasture utilisation, allowing for efficient use of available herbage and well suited to a two herd/flock grazing system, is illustrated in Fig. 12.3 (Bartholomew *et al.* 1994). The system allows for leader and follower animals, provides for the allocation of different sized camps, and allows for the easy fertilization of the camps, once grazed. It is nonetheless convenient to have camps of the same size, as the amount of fertilizer applied to each would then remain constant, so reducing errors. A further advantage of the system is that it is easy to assess pasture availability within the separate panels, thus facilitating the timing of animal withdrawal from camps or of changes to the grazing cycle. Entire panels can be removed from the cycle to allow, for example, for the production of seed, foggage, hay or silage. The design is also compatible with irrigation systems, allowing for efficient pasture production. Animals are watered and receive licks in the passages, thus reducing trampling of the pasture area. Passages allow the regrowth of previously grazed areas to be protected during their rest periods.

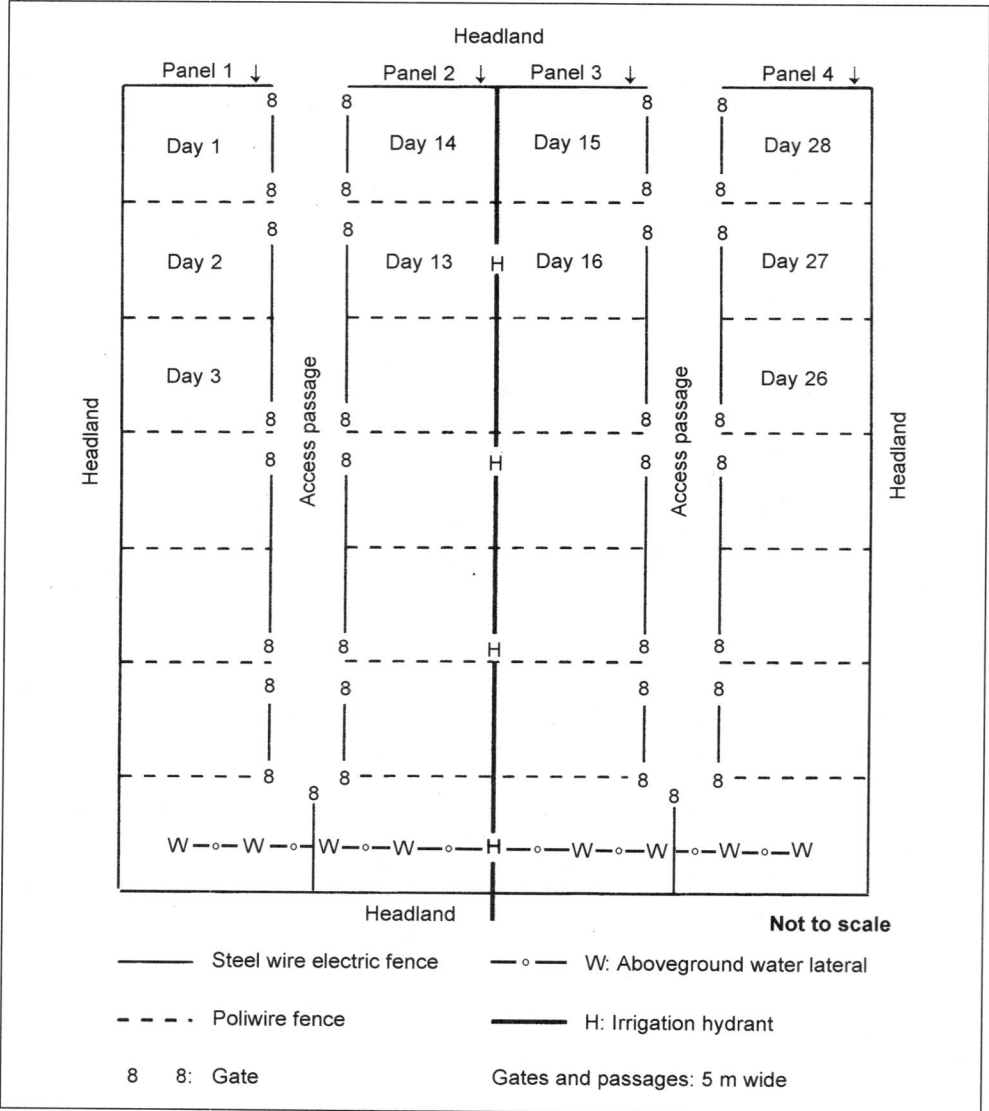

Figure 12.3 Fence network design providing for efficient and effective pasture management.

12.1.1.4.5 Forward creep

Forward creep grazing is particularly suited to ewe/lamb systems, although it can also be used for cow/calf systems. The lamb/calf is allowed to move forward, through gates that prevent the dam from following, into the next camp to be grazed in rotational grazing systems. In so doing, the young animal is given first choice of the herbage on offer. The dams then follow and clean up the pasture, with the young animals being allowed to creep into the next camp in the rotation. Many variations of pasture creep feeding systems are possible. For example, if the potential irrigable area available for cool season pastures is limited and is confined to flat land adjacent to streams, as it often is, then low-cost pastures can be established on adjacent non-irrigable sloping land. The dams can graze these lower quality pastures while the young are allowed to creep laterally into adjacent high quality pastures.

Where the dams follow their young, careful attention should be paid to their condition. This is particularly important as the offspring approach weaning age. The dry matter intake of the offspring increases considerably towards weaning so that the dams may be subjected to considerable nutritional stress.

12.1.1.4.6   Zero grazing

Zero grazing is the practice of feeding freshly cut forage to animals in sheds or yards. This method was made practicable with the advent of the forage harvester which replaced more laborious methods of harvesting the material. The pasture needs to be harvested at least once, but preferably twice, daily, unless some adequate storage facility is available. The labour requirement of the system, even when a forage harvester is available, is high but may be offset by a number of important advantages:

(a) the dry matter can be utilised very efficiently, so there is little wastage;
(b) the pasture can be defoliated at a height conducive to rapid regrowth of the species in question;
(c) trampling, which in wet weather in particular may suppress regrowth, is eliminated;
(d) forage from bloating pastures can be rendered safer if the material is mixed with less bloat-inducing forage or if wilted before feeding;
(e) it will normally reduce the amount of walking required of the cows;
(f) daily intake can be more accurately controlled; and
(g) animal performance is enhanced since the animals will expend less energy walking and grazing.

There are, however, a number of important disadvantages to zero grazing. These will vary in magnitude according to the conditions prevailing on any farm. Costly machinery and high labour input is required in the harvesting and feeding operations. Animal management will demand greater attention under such intensive feeding conditions than where animals are allowed out on pasture. The labour requirement for maintaining the feeding yards in a suitable condition and the requirement for large supplies of clean bedding would in most instances be costly. In addition, the fertilizer requirement of the pastures would be considerably higher than for grazed pastures unless the dung and urine are returned to the pastures.

While zero grazing may be economically justifiable in certain instances, it is unlikely to warrant serious consideration on most South African farms. In many situations it is likely that greatly improved utilisation of pasture dry matter can be achieved at low cost by improved methods of grazing management. It may be an appropriate procedure to use where there is a limited area of land suited to the production of winter pasture or where this land is situated some distance from the cow-byre (for dairying) or where it is important to maximise dry matter production per unit area of land.

12.1.1.5   Management for seed production

South Africa has in the past relied heavily on imports for its supply of forage-grass seed. Locally produced seed has largely been a by-product of the livestock industry rather than a product of the planned production of a seed crop. Seed yields and seed quality have therefore generally been poor. Reasonably high yields of good quality seed can, however, be produced provided the species is suited to the environment and appropriate management techniques are employed. Although this can best be done by growing specialist seed crops, good seed yields can be obtained from dual purpose (forage and seed producing) pastures. To do this, certain basic principles concerned both with the management of the pasture and with techniques of harvesting and treating the seed prior to storage need to be implemented.

12.1.1.5.1 *Management of the pasture*

(a) Dense stands of forage grasses, with individual plants small and closely spaced, do not produce good seed crops. The best crops are produced from relatively sparse stands, with large individual plants with vigorous tillers. Such tillers produce large inflorescences which bear not only many seeds, but individual seeds are well filled, resulting in a high thousand seed weight (TSW). Specialist seed producing crops should therefore be planted at a much lower seeding rate than those established for their forage. They should also preferably be planted in rows to facilitate mechanical operations that may be required to keep the stand weed-free and vigorous.

(b) For good seed production a large number of tillers need to develop early in the crop's seasonal growth cycle. Autumn and winter initiated tillers produce more seed in the following season than do spring initiated tillers. Tiller development should therefore be stimulated in autumn by cutting or grazing and by providing sufficient N and water.

(c) Tillers should be allowed to develop into vigorous units before flowering commences by not grazing frequently or severely.

(d) Warm season species should not be harvested for forage late in autumn but should be allowed to carry a canopy of aftermath (9 cm to 15 cm) into the winter. This canopy should not be removed until spring rains have fallen. Local results have shown that burning *E. curvula* in spring is more beneficial to seed production than is cutting in spring (Field-Dodgson 1973; Rethman & Beukes 1988). The effect of burning on seed production in other warm season species has not yet been investigated.

(e) Early spring growth should be harvested for forage before the stems begin to elongate. This will reduce the tendency for the pasture to lodge during the flowering stage and so will facilitate efficient pollination and seed set. It will also reduce tiller mortality associated with shading, as well as facilitate harvesting operations.

(f) When the stems begin to elongate, the pasture should be closed until the seed is ready for harvesting. Any harvest at this stage will destroy the developing flowers.

(g) Conditions for growth should be as ideal as possible while the inflorescences are developing and irrigation may be necessary at this time. Once pollination commences, care should be taken not to wet the inflorescence as this will adversely affect the distribution of pollen. Once pollination is complete, irrigation may continue.

(h) A dressing of N should be applied at the commencement of seed fill (Phillips 1994).

12.1.1.5.2 Harvesting

Harvesting time is crucial if a reasonable proportion of the seed is to be collected. Local trials have shown that only 40% to 50% of the seed of commercial seed crops is actually harvested. The losses can be attributed to uneven seed ripening, incorrect timing of the harvest operation and the use of inefficient harvesting techniques (Field-Dodgson 1973, 1974).

The best time to harvest the seed crop will vary according to the harvesting techniques used. Where the crop is cut and allowed to dry in the swathe or windrow, or where it is stooked prior to later threshing, it is advisable to cut the crop relatively early (medium-dough stage), i.e. before shedding commences. Provided adverse weather conditions do not interfere with subsequent operations, such techniques will normally allow for a good recovery of the seed crop. However, the crop remains vulnerable while it is in the swathe or windrow and heavy losses may be experienced if wet weather or heavy thunderstorms follow cutting. By stooking the material, the danger of crop damage caused by unfavourable weather is reduced. The labour requirements of this system are high and so this method is warranted only under exceptional circumstances.

The correct timing of the seed combine operation is particularly difficult, especially in those crops whose seed ripens unevenly. Uneven ripening is a characteristic of many forage grasses, including such important species as *F. arundinacea*, *E. curvula*, *D. eriantha*, *P. maximum* and *P. dilatatum*. A general guideline is to delay harvesting until one quarter to one third of the seed has shattered from the majority of the seed heads. At this stage a wheat combine or a conventional mower can be used. The latter should have an attachment (e.g. hessian threaded to a steel rod and attached to a beam lying behind the mower blades) to collect the flowering stems after they have been cut. Here the stems will need to be stored under cover to dry before the seed is threshed. An alternative method is to mount a beam, onto which a series of half-drums have been attached, to the front of a tractor. The height of the beam should be such that, as the tractor moves through the pasture, it will agitate the seed-heads as it hits them, whereupon the ripe seed should shatter and drop into the drums. This process can be repeated at intervals as the crop ripens. This method has proved effective in harvesting seed of *P. maximum* (Drewes pers. comm.). There is no doubt that direct combining results in low recovery of the seed crop from the species listed here. Drying in the swathe, windrowing or stooking deserves careful consideration for maximising seed harvest from these species.

12.1.1.5.3 Seed drying

Freshly-harvested seed, and in particular seed which has been combined directly, will normally be reasonably moist. Such seed will deteriorate rapidly if it is not dried sufficiently before it is stored. Within as little as one hour, freshly-harvested seed stored in sacks or bins may heat sufficiently to cause it permanent damage. Seed should therefore be dried as soon as it has been harvested.

For relatively short periods of storage (up to 12 months) the moisture content of the seed needs to be reduced to about 13% to 14%. For longer periods of storage, further drying to a moisture content of about 10% is advisable. Such seed should be stored at low humidity and temperature or sealed in moisture-proof bags if its quality is to be retained for long periods.

12.1.1.6 *Regeneration of degraded pastures*

At best, the regeneration of degraded perennial pastures must necessarily involve the correction of the factor, or factors, which led to their degradation in the first place. Degradation by way of a replacement of the planted species by invading species will have resulted from a change in the competitive balance between the pasture species and the invaders, to the advantage of the latter. In many pasture types recovery will be slow should the competitive balance be swung in favour of the planted species unless these plants have some mechanism of spreading rapidly throughout the pasture. In these species the pasture will need to be completely re-established.

12.1.1.6.1 *Grass pastures*

(a) Tropical/subtropical tufted (bunch-type) species

Warm season grasses, such as *E. curvula, C. ciliaris, D. eriantha* and *P. maximum*, are not as tolerant of grazing mismanagement as are the creeping sod-forming types, nor are they capable of spreading rapidly by vegetative means. They are therefore less competitive than the sod-formers and are more difficult to maintain in the pasture. Degradation of pastures frequently results from a run-down of soil fertility. However, it may also result from bad grazing practice or from soil compaction (Grunow pers. comm.).

Most pastures comprised of tufted species cannot spread vegetatively or produce new seedling crops (*D. eriantha* being an exception) unless the seed is buried, the soil is compacted around the seed and adequate nutrients are supplied to the seedlings. Hence such pastures should normally be ploughed out and re-established by conventional means, or fresh seed sod-seeded into the old pasture. If complete re-establishment is envisaged, the land should be adequately fertilized, re-established to the new pasture and managed in the manner appropriate to the bunch-type warm season species. Every effort should be made to ensure that a full stand develops since gaps will provide sites for the invasion of unwanted species. To do this, it may be advisable to crop the field for one or two years to provide sufficient opportunity for the preparation of a suitable seedbed and to control weeds.

As an alternative to complete re-establishment, the pasture may be renovated by sod-seeding, provided unwanted weeds are absent or can be economically eliminated in some other way. The success of this method will depend on a number of factors and, in general, will be less assured than where re-establishment is complete. For sod-seeding to succeed the competitive effect of all existing vegetation must be substantially reduced and the seed suitably placed in the soil. If these conditions can be met, sod-seeding may be advisable since it is less costly than conventional establishment. If not, the more costly but more assured method of complete re-establishment should be contemplated.

(b) Tropical/subtropical creeping species

Sod-forming pasture species such as *P. clandestinum*, the *Cynodon* species and others, are inherently highly competitive and are able to spread rapidly when conditions for growth are favourable. These species are also relatively tolerant of a wide range of grazing practices. To perform well, however, they require an adequate supply of plant nutrients and given fertile conditions they are generally able to retain or regain their density in a pasture. Renovation of these pastures often requires little more than the provision of sufficient plant nutrients. Tillage operations have proved unnecessary and while mowing may aid in their renovation, it is not essential. Mucking in winter will generally be advantageous in that it provides an often much needed supply of plant nutrients. Subsequent applications of fertilizer, particularly N, together with occasional heavy grazing which these species can tolerate (with the possible exception of Star grass), will stimulate these grasses and allow them to exercise their high competitive ability.

(c) Temperate species

Temperate species range from those which can be successfully re-established from the seed produced by surviving plants (such as *L. multiflorum* cultivars) to those which are more difficult to establish (such as *F. arundinacea*). Sod-seeding may prove successful provided suitable conditions can be provided for the seed and weeds do not create a problem. If suitable conditions cannot be provided, then conventional re-establishment procedures will need to be adopted. In KwaZulu-Natal, the productivity of perennial ryegrass pastures has been improved from the second year by annually applying 10 kg of seed per hectare. The fertility and moisture conditions must be good and the pasture should be overseeded in autumn following relatively severe defoliation.

Irrespective of the success of germination in any renovation programme involving temperate species, seedling survival is often a problem because of the highly competitive nature of the many summer-growing weed species in most areas. Management should, therefore, be designed to favour the cool season species relative to the warm season species. Since these cool season species are more active than the warm season species in autumn and early spring, applying fertilizer and irrigation water during these periods, coupled with moderately lenient grazing, may contribute much to the survival of the pasture plants.

12.1.1.6.2 Legume pastures

(a) Twining tropical legumes

As a result of their ability to spread vegetatively, twining tropical leguminous pastures can often be rejuvenated by following generous late winter dressings of P with long rest periods during spring. This should permit the legumes to smother, at least partially, any pioneer grasses that may have invaded. Subsequent grazing should be lenient so that legume vigour will increase sufficiently to provide effective competition to encroaching plants.

If the density of the legume population has declined substantially and perennial grass weeds have invaded, it may be necessary to re-establish the pasture.

(b) Temperate legumes

In upright 'tufted' red clover and lucerne pastures, regeneration by methods other than partial (sod-seeding) or complete re-establishment are unlikely to succeed since these species have a limited capacity to spread vegetatively or by natural reseeding. In pastures of the prostrate legumes, such as white clover, regeneration is theoretically possible because of their ability to spread vegetatively. It is also possible that at least some of these species can regenerate from seed produced by the pasture plants themselves. Unfortunately no definite information is available on seedling regeneration in these pastures under South African conditions.

The treatment which should be applied to recently sown white clover pastures that are in poor condition is similar to that recommended for the twining tropical legumes. Phosphate should be applied in late winter, and grazing should be lenient to promote the development of vigorous clover plants. In older white clover pastures, regeneration from new seedling populations would appear to be essential since individual white clover plants are relatively short-lived. The death of the seedling taproot, usually within 18 months to 2 years of seed germination, appears to signal a reduction in the vigour of the plant and its eventual death. This problem is aggravated by an inherent lack of available P in most South African soils. Consequently, the life of these pastures is much reduced in the absence of generous dressings of P.

Sod-seeding white clover in autumn has proved successful in maintaining stands of this species, provided suitable soil fertility and soil moisture conditions prevail.

12.1.1.6.3 Grass/legume pastures

Poor fertility, poor timing of N applications, inadequate soil moisture and uneven defoliation of the different components of the sward are generally the major causes of imbalances between grass and clover. To reconstitute these pastures it may be necessary to sod-seed grass into legume-dominant pastures, or legume into grass-dominant pastures, but before this is done the factor or factors that led to the imbalance in the first place will need to be removed. Unless this is done, the pasture will rapidly revert to single species dominance, and the effect of the treatment will therefore be short-lived.

12.1.1.7 Irrigation

Irrigation serves two functions in the humid summer rainfall areas. It is used in winter for temperate pastures such as ryegrass and strategically in summer as a supplement to the natural rainfall.

Winter irrigation

In the summer rainfall areas, temperate pastures are almost completely dependent on irrigation water for their survival and so these pastures can be justified only for highly intensive production systems such as dairy and the production of fat lambs. Irrigated winter ryegrass has, in the past, also been used to finish weaners for the market, but the anticipated increase in the cost of irrigation water, the high costs of fertilizer N and the low net return for beef is likely to make weaner finishing on irrigated pasture increasingly unattractive in the future.

While the amounts of water needed by the pasture can be reasonably accurately determined using, for example, the A-pan to determine daily evapotranspiration, as a general rule a sufficient amount of supplementary irrigation water is applied to ensure that the pasture is provided with a total (from natural rainfall and irrigation) of 25 mm/week. This amount is based on a 3 mm to 4 mm evaporative loss of water per day. Where evaporative losses are likely to exceed this amount, the amounts of water applied should be increased accordingly and where shallow rooted pasture crops, such as white clover or perennial ryegrass are irrigated, the frequency of irrigation should be increased to about five days. Producers should, however, guard against applying excessive amounts of water too frequently since not only does this lead to a wastage of water, but it leads also to a loss of N through denitrification and leaching and to a general decline in the vigour of the pasture as air is excluded from the rooting zone for extended periods of time. Also, the rate at which water is applied should not exceed the infiltration capacity of the soil, since this is likely to result in an unnecessary wastage of water through surface runoff.

Neither the quality of the soils nor water quality should cause problems in the humid areas of South Africa. Soils used for intensive pastures are, with some notable exceptions, usually sufficiently deep and freely drained and the natural waters sufficiently free of salts to allow them to be effectively used for irrigation.

Summer supplementary irrigation

Strategic supplementary irrigation is often provided for summer growing dairy pastures, and for kikuyu pastures in particular but also for perennial ryegrass which may be irrigated year-round (Brockett pers. comm.). The requirement for irrigation in these pastures is often sporadic and infrequent so that a permanent irrigation system can seldom be justified for these pastures. More commonly, irrigation pipes are 'borrowed' from winter irrigated ryegrass pastures where winter irrigation is practised.

Supplementary irrigation is particularly useful in these summer pastures where spring rains are late, although the irrigation pipes may still be required on the ryegrass pastures at this time. It can also be particularly useful during mid-season droughts where, because temperatures will normally be high, an extremely good response may be expected from the summer pasture. The response of these pastures to autumn irrigation is likely to be poor and it would seldom be cost-effective to irrigate at this time.

12.1.2 Winter rainfall areas *J.M. Van Heerden & M.B. Hardy*

Most of the pastures grown in the winter rainfall region are grown in association with cereal crops in the semi-arid rather than the humid areas. However, some pastures are grown in the humid areas and these tend to have a strong legume component. Because of this, the management guidelines presented here are based largely on the needs of the legume rather than the grass component of the sward since it is this component that is the more vulnerable of the two.

Broadly speaking, the main aims of effective management systems in the winter rainfall region are:

(a) to retain an acceptable botanical composition, and in particular an acceptable amount of legume in the sward;

(b) to produce the highest possible yield of high quality material as uniformly throughout the year as possible; and

(c) to utilise the material produced as efficiently as possible.

Many of the procedures described previously for the summer rainfall areas are equally appropriate to the winter rainfall region and will not be repeated here. In this discussion emphasis will be given only to those aspects of management which directly affect the utilisation of the pasture by grazing animals in the winter rainfall areas, namely stocking rate and grazing procedure.

Both rotational and continuous grazing are recommended for pastures in the region, depending on individual circumstances. Rotational grazing is widely recommended in intensive dairy operations and is used almost exclusively on pastures grown under irrigation and for species such as Italian ryegrass when used to bridge gaps in the forage flow. Here rotation is preferred to continuous grazing largely because it allows the producer to (a) ration forage more efficiently; (b) make more effective use of the available forage; and (c) provide a greater level of control over the intensity and frequency of defoliation of the pasture plants. Lucerne pastures also need to be grazed rotationally if they are to persist (McKinney 1974; Van Heerden & Tainton 1988).

Pastures comprising white clover and perennial ryegrass mixtures are, on the other hand, often grazed continuously because of the beneficial effect of frequent grazing on the development of clover stolons. This improves the stability of the composition, and particularly the relative proportions of clover and grass, of such pastures. This in turn provides a more constant quality of diet to animals through the season than is provided by rotationally grazed pastures. Extensive dryland annual medic and subterranean clover pastures are generally also grazed continuously in both South Africa and Australia (Morley *et al.* 1969; Fitzgerald 1976).

There are innumerable ways in which rotational grazing can be applied in practice and the performance of any particular system will, in any event, be dictated by the stocking rate used. Therefore, before any firm recommendations can be made, the influence of stocking rate and grazing system on different pasture types needs to be examined.

12.1.2.1 The influence of stocking rate and method of grazing on pastures

12.1.2.1.1 Pasture quality

The legume pastures grown in this region provide forage of excellent quality. Medic and medic/clover pastures, for example, provide material having crude protein (CP) levels between 15.4% and 30.8% (Brand *et al.* 1991). In addition to providing high quality winter grazing, dry matter residues and mature pods are well used by sheep during the dry summer months (Wasserman 1980). Up to 500 kg DM/ha of these pods may be available on medic pastures at the start of summer.

Forage quality in the temperate and Mediterranean type legumes and grasses used in the winter rainfall region declines only slowly as the material ages during extended rest periods (Van Heerden 1986). The grass component of irrigated grass/clover pastures tends to become unpalatable in pastures which are only lightly utilised over long periods (Van Heerden & Tainton 1988). Nevertheless, changing quality associated with the ageing of the material should not play a major role in determining the grazing procedures used in these pastures.

12.1.2.1.2 Botanical composition

As discussed previously, the major legumes used in pastures in the winter rainfall region, viz. white clover, lucerne, the medics and subterranean clover, respond very differently to management. The legume content of annual legume pastures, such as subterranean clover and the medics (Morley *et al.* 1969; Fitzgerald 1976; Van Keuren & Hoveland 1985), and of perennial white clover based pastures (Van Keuren & Hoveland 1985; Van Heerden & Tainton 1988) is most stable under continuous grazing. In contrast, rotational grazing is necessary if lucerne is to be retained in the pasture (McKinney 1974; Van Heerden & Tainton 1988).

In rotational systems the contribution of both white clover and lucerne increases as the length of the rotation, i.e. the length of the recovery growth period between each successive period of occupation increases, at least up to a period of absence of six weeks (Van Heerden *et al.* 1989). Here the influence of a particular grazing strategy is accentuated by increasing the stocking rate (Tainton 1974). In irrigated grass/legume pastures, the grass content has been shown to increase, and the content of white clover to decrease, with increased grazing pressure. The amount of lucerne seems to be unaffected by grazing pressure (Van Heerden & Tainton 1988, 1989). A reduction in the amount of white clover is apparent even at low stocking rates when the stocking rate is increased under rotational grazing. Under continuous grazing, white clover content initially increases as the stocking rate increases from low to moderate levels, but then declines as stocking rate is increased further.

Under dryland conditions also, the persistence of lucerne is promoted by rotational grazing (Langenhoven pers. comm.). Different lucerne varieties differ, however, in their tolerance of grazing. The local cultivar, SA Standard, is extremely grazing-tolerant and is damaged much less by continuous grazing than the newer aphid resistant cultivars such as Cuf 101, Granada and Baronet, which cannot tolerate a high grazing pressure.

12.1.2.1.3 Pasture production and grazing capacity

Because the grazing system in use and the stocking rate applied influence the length of the rest period and the intensity to which the pasture is defoliated, it is logical that they will also influence pasture production (Van Heerden 1986; Van Heerden & Tainton 1988, 1989). The influence of these factors on production is brought about largely by their influence on the leaf area index of the pasture. The length of the rest period required to allow the pasture to accumulate sufficient leaf to enable it to effectively intercept all the radiant energy incident on it, varies among different grasses and legumes (Van Heerden & Tainton 1989). For example, the production of white clover/tall fescue, white clover/cocksfoot and perennial ryegrass pastures in the winter rainfall region is considerably higher when these pastures are allowed four-week periods of recovery between grazings than when this period is extended to six weeks. In contrast, pure fescue pastures and those containing a mixture of lucerne and either fescue or cocksfoot, benefit from an extension of the recovery period to six weeks (Van Heerden & Tainton 1989).

The recovery period which will give rise to maximum production varies seasonally. Brougham (1970) showed that maximum production of white clover/perennial ryegrass pastures is achieved with a 6–8 week recovery period in winter, reducing to 3–4 weeks in spring, before again extending to 4–5 weeks in summer and 4–6 weeks in autumn. He also recommended that such pastures be grazed much shorter in autumn and winter (25–50 mm) and spring (25–75 mm) than in summer (75–100 mm). Although it is unlikely that managers will be able to apply such guidelines with great precision, the closer they are adhered to, the higher the production of the pasture is likely to be.

Under dryland conditions with lucerne pastures (Langenhoven pers. comm.) and under irrigation with mixtures of white clover, lucerne and temperate grasses (Van Heerden & Tainton 1988), more forage is produced at near optimum stocking rates in rotational systems applying a period of absence of 5 weeks than is produced under continuous grazing. However, the stocking rate used greatly affects production from either system. At low stocking rates under irrigation, production has been found to be higher in white clover, lucerne and temperate grass pastures under continuous than under rotational grazing. As stocking rate is increased, production declines more rapidly under continuous than under rotational grazing so that, at moderate to high stocking rates, rotational grazing is superior to continuous grazing.

In pastures comprising annual temperate legumes, seeds start to germinate following the first adequate autumn rains. Subsequent pasture productivity will then depend heavily on moisture availability, on temperature and on the size of the soil's seed bank. If the autumn rains are late, the seedlings will emerge during the cooler winter months. Dry matter accumulation will then be slow and annual dry matter production may be as low as 3 tons/ha. Carrying capacity will therefore be low. However, with sufficient moisture in the early growing season (early to mid-April) and with good follow-up rains, the pastures can produce up to 7 tons DM/ha.

Generally, these annual pastures are able to maintain a high productive potential as long as seed reserves exceed about 200 kg/ha and in the region of 400 to 500 plants/m² emerge at the start of their growing season (Carter & Lake 1985; Carter 1987; Jones & Carter 1989; Kotzé 1999).

The annual leguminous pastures are entirely dependent on seed for their survival (refer to Chapter 2). Defoliation management must therefore be designed to allow for seed production. Flowering and seed set is maximised when these pastures are continuously grazed and a defoliation height of 6 cm to 8 cm is maintained (Carter 1987). In medic pastures such management ensures that the canopy remains prostrate and open. This enhances flowering and seed production. Heavy grazing in the early season opens up the canopy and in so doing benefits seed production by increasing inflorescence number and, in subterranean clover, by also promoting the burial of seed burrs (Frame *et al.* 1998). Here rotational grazing is not advocated since it tends to promote a more upright growth form and there is therefore a greater potential to remove a large proportion of the season's seed production if, as is common in rotational grazing systems, the pasture is grazed at a high stocking density in an attempt to minimise the wastage of dry matter.

12.1.2.1.4   Irrigation

The area referred to here is that situated along the foothills of the Outeniqua mountain range from Riversdale in the south-west to the Tsitsikama in the north-east. Here there is a high probability of rain falling in all months of the year and the monthly average A-pan evaporation is only in the region of 180 mm even during the hottest month of the year (January). Permanent pastures in this region therefore have a relatively low additional moisture requirement. In theory, about 75 mm of irrigation water is required per month during December and January and only between about 30 and 40 mm during the cool winter months. In spite of this apparent low requirement for irrigation, the region experiences frequent dry spells and occasional berg-wind conditions when irrigation will impact considerably on pasture performance. The high price of land and of the irrigation infrastructure and the fact that the intensive dairy industry relies almost exclusively on pasture for its forage requirements, means that irrigation systems need to be well planned so as to make maximum use of the available resources, and particularly of the available water. To do this a permanent sprinkler system is generally recommended. It is also recommended that land-owners install tensiometers in each pasture to enable them to regularly monitor soil moisture availability. Water should be applied when soil moisture tensions exceed about 20 kPa.

The main pastures irrigated in this region are those comprising mixtures of the temperate grasses (*L. perenne, D. glomerata* and *F. arundinacea*) and temperate legumes (mainly *T. repens* and *T. pratense*). To a lesser extent, pure stands of perennial ryegrass, lucerne and kikuyu are also irrigated. In recent years the practice of oversowing irrigated kikuyu pastures with annual ryegrass and clovers during the cool winter months, in an effort to improve the amount and quality of fodder produced by the pastures, has become more widely adopted.

Irrigated pastures are used almost exclusively by the dairy industry although there is a potential for both fat-lamb (Van Heerden *et al.* 1989) and beef production, given that suitable markets are available.

12.1.2.1.5 Animal production

Because management will have some influence on the quality of the material on offer to the animal (Van Heerden 1986) and because it generally has a major influence on the legume content of the pasture (Van Heerden & Tainton 1988, 1989), it will also affect the performance of the individual animal. Management also influences dry matter production and therefore grazing capacity, so it will also influence total production per unit area of pasture.

Because the legume content of a pasture has such an overriding influence on the quality of the material it produces (Gibb & Treacher 1983, 1984; Van Heerden & Tainton 1987, 1988), the influence of management on the legume component of the sward often has a major influence on animal production. In lucerne pastures, where both production and persistence are promoted by rotational grazing, rotation will invariably allow for better animal performance than will continuous grazing (McKinney 1974; Langenhoven pers. comm.). With the annual legumes (e.g. subterranean clover) (Morley *et al.* 1969) and with white clover (Van Heerden & Tainton 1988), however, the legume content and the level of animal production are normally higher under continuous than under rotational systems of management.

12.1.2.2 Practical grazing systems for the winter rainfall region

On the basis of the previous discussion, some firm grazing management recommendations can be made for pastures in the winter rainfall region. Due to the considerable influence of stocking rate on pasture and animal performance, and the variable seasonal animal requirements and production of pastures, it is imperative that appropriate stocking rates be applied at all times.

A number of practical camping systems for different types of pastures typically grown in the winter rainfall region is presented in Table 12.3. Note that the number of camps is the number required for each group of animals. The total number of camps required on any property will therefore also depend on the number of animal groups being run.

Table 12.3 Practical paddocking systems for legume pastures.

Pasture type (legume)	Paddock number	Rest period (days)	Grazing period (days)
Mediterranean clover	1	–	–
Balansa clover	1	–	–
Persian clover	1	–	–
	6	35	7
	11	35	3 or 4
Medics	1	–	–
White clover	1	–	–
	6	35	7
	11	35	3 or 4
Red clover	6	35	7
	11	35	3 or 4
Lucerne	6	35	7
	11	35	3 or 4

12.2 SEMI-ARID REGIONS

R.H. Drewes, N.F.G. Rethman &
L.G. Du Pisani

12.2.1 Summer rainfall areas

In the summer rainfall semi-arid cropping areas, cultivated pastures comprise palatable perennial tropical grasses such as *P. maximum* (Guinea grass*), Anthephora pubescens* (Anthephora or Wool grass), *D. eriantha* (Smuts' fingergrass), *C. ciliaris* (Blue buffalo grass), species of *Sorghum* and *Cynodon* (Kweek or couch), *Chloris gayana* (Rhodes grass), Bana grass (a cross between *Pennisetum purpureum* and *Pennisetum glaucum*) and lucerne. In the more easterly areas, *E. curvula* (weeping lovegrass) is incorporated into the forage flow programme to a greater or lesser degree and may make an important contribution to the forage programme. To a large extent the management of these pastures revolves around their integration into the existing pattern of forage production by veld in the area. More than 50% of the area is still under veld which is sweet and can therefore supply year-round grazing. In much of the semi-arid area, maize stalks make an extremely important contribution to the forage requirements of the animals during winter.

The inherent palatability of the summer pastures used in this region allows them to be grazed during summer. This provides an opportunity to rest out the often degraded veld for use in winter, so providing it with an opportunity to recover its composition and vigour. Alternatively, the pastures can be used for the production of either hay or haylage for winter feeding. In this way they can play an important role in providing a fodder bank in an area of extremely unreliable rainfall. When used in these ways they are generally able to increase the carrying capacity of properties in the area, many of which have been repeatedly sub-divided over the years and are no longer economic units as purely veld properties, particularly in view of the often poor condition of the veld.

Because of the expansion of pasture farming in the area, there is a strong demand for seed of adapted grasses so that seed production is an additional financially rewarding enterprise, and is likely to remain so for the foreseeable future. Although fertilizer applications will invariably increase their yields, these grasses provide useful forage whether or not they are fertilized. Costs of production may therefore be kept relatively low.

Lucerne has been widely used in the past under irrigation for the production of hay but, during the past few years, large areas have been planted to dryland lucerne which has been successfully used for grazing. It has the advantage over grasses of being a low-cost crop since it normally requires the addition only of P, although lime is sometimes needed and, very occasionally, K.

Because of the potential of the grasses to fulfil both the summer and winter forage needs of animals, it is generally advisable to regard the growing season (summer) management and dormant season (winter) management as separate operations. Grass forage has the potential to support a relatively high level of production during the summer, but during the winter these pastures are generally capable of providing little more than maintenance. Seen in this light, summer grazing should be preferentially given to the animals with the greatest production potential and to those which can realise the highest prices. By the same token it would be logical to stimulate the production of the pastures by providing adequate levels of fertilizer, a sufficient number of camps to facilitate management and to mow the pastures, when appropriate, in order to effectively utilise their production potential. By way of contrast, it is unlikely that high level inputs into those pastures used during the dormant period, when they are likely to do no more than provide for animal maintenance, can be justified.

12.2.1.1 *Summer utilisation*

12.2.1.1.1 *Grazing*

Green grazing can be expected from about the middle of October onwards, although the characteristically variable climatic conditions during the spring can severely delay spring production should, for example, a late frost or a dry spell be experienced. Because of the unpredictability of early growth it is always advisable to allow a green fodder bank to accumulate on the pasture before summer grazing commences. This approach has the advantage of helping to bridge the inevitable periods of food shortage which arise during periods of low rainfall in summer.

Under normal circumstances, summer grazing can safely begin on grass pastures when the plants reach the early flowering stage and, on lucerne pastures when the plants reach the late flowering stage. This provides ample opportunity for the plant to recover its carbohydrate reserves and for the development of an extensive root system so necessary for persistence and continued production in this relatively harsh environment. In lucerne, such delayed grazing also almost totally eliminates the danger of bloat in animals. When grazed, the principle which should be adopted is to graze relatively leniently (take half and leave half). This serves to provide a 'fodder bank' to cover summer drought periods and it encourages a more rapid recovery following droughts.

The summer grazing capacity of pastures in this area has, to date, been established only for Smuts' fingergrass and weeping lovegrass. The mass of Simmentalers at Potchefstroom (Drewes pers. comm.) increased by 326 kg/ha at a grazing capacity of 2.5 weaners/ha for the summer period. The steers gained 0.9 kg/day. Gains of 0.7 kg/day have been recorded in young Afrikaner oxen at a stocking rate of approximately 3 animals/ha (Beckerling pers. comm.). In the vicinity of Bapsfontein, Smuts' fingergrass grazed at 2.9 weaners/ha has produced an average livemass gain over a two year period of 297 kg/ha at an ADG (average daily gain) of 0.73 kg, while weeping lovegrass stocked at 3.8 weaners/ha has produced an average of 178 kg/ha at an ADG of 0.41 kg (Grunow & Pieterse 1984).

Lucerne is predominantly a grazing pasture for sheep, and particularly for Merino sheep (which are not predisposed to bloat) in this area. With the necessary attention, and particularly if grazing is always delayed until the late flowering stage, lucerne pastures can also be used as grazing for other breeds of sheep and for dairy cows. A mean grazing capacity of 6 ewes with lambs/ha for the 8 month summer period can be anticipated in the cropping areas of this region (Anon. 1986).

There is generally no need for more than 3 to 4 camps for cattle grazing systems, but better management can usually be applied in sheep systems where 5 to 6 camps are used. Rotational grazing provides the needed flexibility in this area of unpredictable forage availability since it allows for feed rationing and is therefore preferred to continuous grazing. In both situations it is necessary to introduce a mower from time to time, either to remove accumulated residual material from grazed camps, or to produce a hay crop from camps which are rested during periods of rapid growth.

12.2.1.1.2 Hay production

Palatable hay can be made from all of the grasses listed in the introductory paragraphs of this chapter. At least two rain-free days are needed to allow for adequate curing of the succulent grass stems. For this reason, the use of rotary cutters is advocated as this permits faster drying.

Where hay is made to overwinter beef animals, its quality is not as important as it would be, for example, if made to feed sheep or young reproductive stock. In the former case, consideration can be given to delaying the hay-making operation until March when there is little likelihood of rain spoilage. In a normal rainfall year the flowering stems of the grasses will have produced seed and will have died by this time, but the leaves are still likely to be green.

Because the spring forage produced by pastures in this area is normally in such great demand for grazing, the pastures are normally not closed for hay until late spring or early summer. In good seasons, when the veld is able to supply the needs of the animals through the summer, all available pastures should be closed off for hay following the spring grazing period. The late season growth, following the hay cut, should normally be reserved for foggage.

It seems that animals will readily consume hay whether or not the pasture from which it is produced has been fertilized. If pastures are used regularly for hay production, a considerable quantity of nutrients will be removed from the area so that frequent monitoring of the fertility status of the soil is advisable. Also, compaction by the machinery used in hay operations may detrimentally affect the production of the pasture so that occasional ripping may be advisable.

12.2.1.1.3 Seed production

For general information on seed production, refer to section 12.1.1.5.

12.2.1.2 Winter utilisation

12.2.1.2.1 Foggage

The most difficult periods of the year for livestock in the semi-arid summer rainfall areas are late winter, spring and early summer. In the cropping areas, crop residues will generally already have been fed to the livestock and land preparation for spring planting will have commenced, whereas in the non-cropping areas forage accumulated during the previous season will often have been fully utilised during the winter. In all likelihood the hay reserves will also have dwindled and because of low temperatures and/or unpredictable rainfall, no reliance can be placed on production from the veld at this time. The symptoms of the poor feeding conditions are commonly seen in a decline in the condition of animals, particularly if they calve during this period, as well as in poor reconception of the cows, and in the appearance of bare patches in the veld and the increase in the amount of unwanted species of *Aristida, Eragrostis, Elionurus* and others.

The provision of foggage from sown pastures would seem to be the most cost-effective means of bridging the difficult August to November feeding period on most farms. Not only does it eliminate the need for mechanical harvesting, but it also reduces the labour demand compared with traditional methods of winter feeding since the animals harvest the material themselves. Generally, however, foggage is not particularly suited to feeding sheep, although with appropriate management they can make effective use of this material.

The strong recommendation for the use of Smuts' fingergrass and species such as Guinea grass, Blue buffalo grass and Rhodes grass for foggage in this area is not because these species have a particularly high nutritive value, nor because they are better able than other grasses to retain their nutritive value after they have matured. Data from fistula samples from animals suggest that both crude protein and *in vitro* digestibility of foggage from these species is no better than that from good veld. Their advantage lies in the fact that they remain palatable to livestock after maturity, even after having been heavily frosted. These grasses are therefore able to provide the bulk roughage needs of the animals through the winter. Good winter supplements need to be fed with such roughage as an integral part of any overwintering strategy.

Another aspect which goes hand in hand with the provision of foggage is the provision of hay. Because of the extremely variable summer-growing conditions, pasture growth rates vary greatly over the season so that it is essential that hay be available at all times to buffer the forage system. The hay should normally be stored close to the fields in which it will later be fed. The round bale is particularly suited to this. It has been found that the animals make good use of both hay and foggage when these are fed together in such a system. Alternatively, the hay may be fed back on the foggaged pastures once the foggage has been fully utilised.

Questions are often asked about the bare patches which develop where hay is stored in the grazing camps and in the vicinity of self-feeding hay racks. Careful observation has shown that so much seed accumulates in these areas, particularly where Smuts' fingergrass is present, that the bare patches rapidly recover during the following summer. Even in patches which develop under hay stacks, recovery is generally so rapid that by the end of the next season such patches can be recognised only because the grass growing on them is taller than the rest of the pasture. Implicit in such rapid recovery, however, is that heavy accumulations of residual material should be thinned out. Excess material can best be spread over any bare patches which have developed.

The need to concentrate animals on foggage, with or without hay fed back on the pasture, until such time as the veld has grown out sufficiently in the spring (as late as mid-November in some years), raises problems of possible compaction damage to the entire pasture. Once again, observation shows that this causes no physical damage to the pasture, but it does cause some physiological damage to the grasses. This results in slower spring growth and a reduction in the amount of seed produced in the following summer. Even this effect, however, is normally apparent only on the last camp used in the winter feeding cycle and where it occurs it is to the advantage of the quality of the next winter's foggage. This material is typically much less stalky than on those camps which have grown out rapidly in the spring.

The height to which the pasture grows out in summer, together with its density, largely determines the extent to which animals will use accumulated material efficiently in the following winter. There are indications that this material will be very efficiently used if the leaf canopy is no taller than 20 cm and the stems no taller than about 1 m. There is therefore no advantage to be gained by fertilizing heavily, or indeed of experiencing a particularly good rainfall season, since the material then becomes too dense and stemmy and is poorly used as foggage. Animals perform poorly on such material. As part of the strategy to ensure that good quality material is carried into the winter as foggage, one or both of the following corrective management manipulations must be adopted:

(a) either no fertilizer, or at best only small quantities of fertilizer, must be applied; and
(b) the pasture must be either grazed, or a hay cut taken, in December or January.

The sub-division of an area of foggage into three to four camps per group of animals is sufficient to allow for efficient utilisation of the material. In the Potchefstroom area, with an average rainfall of 600 mm, such pastures can carry approximately 1 cow/ha from August to mid-November. The foggage itself can usually carry the animals until the end of September, whereafter the animals can revert to the hay produced from the same pasture during the previous season and to any regrowth which accumulates on the pastures.

An overwintering strategy using maize stalks, foggage and, in the late winter, hay mixed with regrowth, has been tested at Potchefstroom with both large and small stock for a number of years. Over a four-year period, Afrikaner cows gained an average of 14 kg each during the three months leading up to calving (mid-June to mid-September). Their Condition Index declined from 3.7 to 3.6 (out of a maximum of 5) over this period and at the end of winter (after calving) averaged 3.4. Simmentaler cows each gained 15 kg over the same period prior to calving and achieved a Condition Index of 3.2 at the start of the winter, 3.1 immediately before calving and 2.8 at the end of the winter after having calved. The conception rates and calving percentages of these animals, and their calves' weaning masses, were equivalent to animals overwintered in stalls on a ration of silage and hay.

When the foggage material produced by sown pastures is reserved for feeding until the late winter and early spring, the veld can be given an adequate opportunity to grow out before it needs to be grazed. The role that sown pastures can play in permitting the veld to recover, by reducing spring grazing pressure, is probably one of its most important roles on mixed farms.

12.2.1.2.2 Short spring foggage

Winter stall-feeding should not continue beyond the start of the rainy season in about October. The veld will normally not have recovered sufficiently to be grazed by this time, and many farmers would dearly love to have pastures which can supply a green bite. No such pasture is available in the summer cropping areas, although in good rainfall years the lands grasses (annual grasses which invade fallowed land) will provide some green grazing at this time. A better alternative would seem to be to use Smuts' fingergrass for this purpose. This is done by cutting hay early enough in the previous season to allow for regrowth of approximately 15 cm of leaf in the autumn. If such pastures are 'put-up' too early, the resultant foggage will contain a large proportion of stem and its quality and acceptability will be poor at the time of feeding.

Early spring rains will stimulate early spring growth and by the middle of October a mixture of senescent leaf and young green leaf material will be available to the animals. Even if frost or drought dries out the material, it remains a perfectly acceptable forage for animals. Hay also needs to be made available during this period to ensure against possible forage shortages from the pastures. Animals which have been fed on silage in stalls over the winter and on pastures treated in this way until the veld is ready for grazing in early summer, have shown no visible signs of stress.

12.2.1.3 Fertilizing

The use which is made of the pasture dictates the amount that should be spent on fertilizers in this region. If they are used in summer to produce hay or for summer grazing by potentially productive animals, then fertilizer levels need to be relatively high. However, if they are used largely as winter foggage, then low-cost methods of production, particularly with respect to fertilizer usage, can be adopted.

There is reasonable agreement on fertilizer practices that should be adopted for temperate winter pastures, lucerne and for tropical grasses such as weeping lovegrass and kikuyu. It is also generally agreed that soil tests should form the basis of determining the amounts of P, K, Ca and Mg which should be applied, and there is agreement on the threshold soil test levels for K, Ca and Mg. However, whether soil test levels for P should be as high in this region as the generally recommended 12 mg/kg for indigenous tropical grasses has yet to be firmly established. The most recent recommendations for grain sorghum on black turf soils are, for example, as low as 8 mg/kg (Beukes & Kotze 1988). This raises the question of whether the levels currently recommended for the tropical indigenous species are realistic in this region.

An even bigger question surrounds the levels of N which should be applied. Recommendations are often too readily based on traditional attitudes and standardised procedures which, in the current difficult economic climate and in the light of other uses to which pastures are put, may no longer be applicable.

In the heavy clay soils of the summer cropping areas it is often claimed that rain is the best form of fertilizer. The wide annual fluctuations in the yields not only of agronomic crops, but also of veld, support this view. Many farmers have produced grain sorghum for many years without applying any fertilizer while the veld continues to produce good yields year after year without nutrient shortages seeming to limit production. According to trials undertaken by Beukes & Kotze (1988), vertic soils have the capacity to provide as much as 40 kg N/ha/year. One also needs to take into account the fact that past experience, supported by the work of Du Pisani *et al.* (1987), shows that breeding cows cannot be maintained economically on highly fertilized pastures.

A possible approach to fertilizing cultivated pastures in these semi-arid areas is as follows:
(a) attempt to use summer pastures to their maximum potential by using relatively high fertilizer levels and use them only for the most productive types of animals (dairy cows, weaners, slaughter lambs, cull animals and ewes with lambs); and

(b) attempt to keep the costs associated with management and fertilizer practice as low as possible on pastures being used for foggage, since this material can be expected to do no more than maintain animals through the winter period.

According to available data it would appear that summer pastures should receive 60 kg N/ha in the spring, with a further 40 kg N/ha in January, provided weather conditions are such that late summer growth is likely.

Where pastures are to be used for foggage, fertilizer practice should be based on an evaluation of the expected demand for forage and the available hay supplies on the farm. There may, for example, be sufficient hay in storage which, together with the expected yields of foggage, should adequately cater for the needs of the animals through winter. In such circumstances no fertilizer need be applied in that particular season. Should additional hay be needed, further applications of N would be advisable. Here it is recommended that 20 kg N/ha be applied for each ton of hay that is required, assuming the hay will have an average protein content of 12.5%.

The above recommendations apply particularly to the heavy clay soils. On sandy soils relatively high levels of N are required, at least until reasonable levels of organic material have been built up in the soil. Thereafter, lower levels can be used.

Fertilizer practice on cultivated pastures can therefore be seen as an aid to management. In any season fertilizer may or may not be applied, depending on the past history of the pasture and on the forage needs of the farm. Fertilizers should not be applied as a matter of course, but levels should be adjusted according to the circumstances which apply at any time.

12.2.1.4 *Regeneration of degraded pasture*

Some of the Smuts' fingergrass pastures on the Potchefstroom research farm have been grazed every year since 1944. What is remarkable about these pastures is that they have survived for this period under relatively unfavourable soil conditions (level mudstone derived soil which tends to waterlog).

Considering the fact that most of the grasses used for pastures in these areas (Smuts' fingergrass, Guinea grass, Anthephora and Rhodes grass) are all indigenous (or naturalised in the case of Rhodes grass) and have survived for centuries in the veld, it is not surprising that pastures comprising these grasses appear to have an indefinite life-span. In any event, they produce so much seed each year that regeneration through seedling development is no doubt an ongoing process. They are also not dependent on fertilizer for their survival.

Loosening the soil with a tine implement during periods of low rainfall improves yields from these pastures. The effect of this practice is to reduce the plant population and promote infiltration of moisture. Such treatment has, however, not been found to be essential for the long-term survival of the pastures.

12.2.1.5 *Irrigation*

Because of the large amounts of irrigation water required to guarantee reasonable pasture growth in these areas, irrigation at the farm scale in the summer rainfall areas is normally restricted to small patches of special purpose pastures such as lucerne. These small irrigated pastures are often located immediately downstream of farm dams from which water can be led by gravity. Costs are therefore low and crops are irrigated opportunistically, when water is available.

12.2.1.6 *The economics of dryland pastures in semi-arid regions*

Planted pastures are unlikely to be able to support viable livestock production systems on their own in the semi-arid regions because of their relatively high cost and their unreliable seasonal production. Their main value generally lies in their ability to provide forage during strategic periods of the year, when a bottleneck in forage availability may seriously limit the number of animals which can be carried. They may also play a useful role in stabilising forage flow by providing a reasonably reliable source of forage, by increasing the carrying capacity of a property or by providing high quality forage for particular categories of animals, such as those being finished off for the market. The economics of pasture use must therefore be judged not only on the basis of the production which is achieved by the animals during their period on pasture, but also on the role they play in the overall feeding strategy of the property.

Seen in this light, the profitability of pastures needs to be judged by comparing the cost of the feed they produce with the cost of other potential sources of feed which would meet the needs of the livestock. Therefore, while it is generally true that efforts to maximise dry matter production (by, for example, applying extremely high rates of N) cannot be justified economically (Archibald 1990), this may not be true of pastures in those areas where forage may carry an extremely high value at certain times of the year. Each system needs to be judged on its own merits. The contribution that the forage which is produced can make to the production system will greatly be influenced by the physical properties of any property and the livestock system which is being employed.

A word of warning with respect to the stocking rates which should be used on pastures in these semi-arid regions: If over-optimistic stocking rates are used, pastures may contribute little to the stability of the livestock system (Du Pisani 1992) and may in fact worsen conditions during the inevitable dry periods.

12.2.2 Winter rainfall areas

M.B. Hardy

The mean rainfall in this region ranges between 250 mm and 500 mm. Approximately 90% falls in late autumn, winter and early spring along the western seaboard. Here spring and autumn droughts are common. Along the south-western seaboard winter rains predominate, but between 20% and 50% of the annual rainfall occurs in summer (Anon. 1999).

Most pastures grown in the predominantly winter rainfall semi-arid region are legume based. Lucerne is grown under dryland conditions only in areas which experience some summer rain. Annual temperate legumes are grown in rotation with cereal crops in both the western and the south-western seaboard regions. This system provides benefits by way of improvements to soil fertility, soil structure, crop hygiene, opportunities for weed control and in the diversification of farming risk (Wheeler 1987; Mcewen *et al.* 1989), as well as reducing the amounts of N fertilizer required by subsequent grain crops (Herridge 1982; Mcewen *et al.* 1989; Petersen & Varvel 1989). Annual pasture legumes such as the medics (*Medicago* spp.) and clovers (*Trifolium subterraneum* – subterranean clover and *Trifolium balansae* – Balanse clover) are reported to provide between 40 kg N/ha/annum and 100 kg N/ha/annum to the soil profile, up to 40% of which is available to the subsequent crop (Clarke 1980; Ladd *et al.* 1981).

The annual medic pastures which form part of the cereal rotation in the semi-arid winter rainfall region are used mainly by sheep. These pastures provide a good quality green grazing through the winter months and foggaged leaf and stem material and, in the case on the medics, often large quantities of pods during the summer months. For further details on their role in sheep production systems of the region, the reader is referred to Chapter 14 section 14.2.3.

Irrigation requirements in this region are far greater than in the humid regions. The monthly A-pan evaporation during the hot, extremely dry summers ranges from about 280 mm (Riviersonderend district) to 360 mm (Moorreesburg district) during January and from 60 mm to 70 mm, in these two districts, during June. The soils mostly tend to be sandy loams so that irrigation needs to be frequent and irrigation scheduling carefully planned. Because of the scarce water resources of this region and the fact that the water that is available is used preferentially by the fruit industry, relatively little pasture is irrigated except some which is grown exclusively for dairying. The most important pasture types used here are kikuyu, lucerne and perennial ryegrass/clover (Hardy pers. comm.).

In the more westerly regions which receive between about 20% and 50% of their rain in summer, pasture irrigation is more commonly practised. Here irrigation practice closely resembles that of the humid winter rainfall region (section 12.1.2.1.4) except that two to three times as much water needs to be applied if a lack of water is not to limit pasture yield.

12.3 ARID REGIONS   *C.H. Donaldson & N.M. Tainton*

12.3.1 Summer rainfall areas

This discussion will be confined to the management of the three main drought fodder crops grown in the arid regions and the use of strategic irrigation as a means of producing high quality fodder. Acknowledgement is made to Jordaan (undated) for the information presented in this section.

12.3.1.1 *Drought fodder crops*

12.3.1.1.1 *Spineless cactus* (Opuntia *spp.*)

Spineless cactus provides only a sub-maintenance diet to sheep (see Table 12.4), requiring that the animals be supplemented with both energy and protein. The plant is also vulnerable to overgrazing and to damage by insects and is best fed as a chopped feed. Plantations of spineless cactus, with a production capacity of 3.3 t dry matter/ha/annum in a 480 mm rainfall season, are reported to be able to maintain up to 20 sheep per hectare when the animals are supplemented daily with 500 g to 600 g lucerne per head (Steynberg & De Kock 1987).

Table 12.4 The nutritive value and dry matter intake of fresh American aloe, Spineless cactus and Old Man Saltbush (Jacobs 1977).

	American aloe	Spineless cactus	Old Man Saltbush
Moisture (%)	83.2	90.0	75.0
Crude protein (%)	4.8	4.3	10.9
TDN (%)	65.0	64.7	50.0
Fat (%)	1.5	1.6	2.0
Minerals (%)	10.3	19.3	10.3
Fibre (%)	16.2	9.7	37.9
DM intake (g/sheep/day)	300	346	1 466

Spineless cactus can be used in one of four main ways (Juritz 1920; Aucamp & De Kock 1970). It can be grazed directly, pre-sliced before feeding, turned into a concentrate meal or turned into silage.

Direct grazing

When grazed directly, the cactus should not be too tall, nor the material too old. Grazing at intervals of three years has proved satisfactory. Because the material is relatively palatable, steps need to be taken to protect plants from being overgrazed (De Kock 1980a). Spineless cactus can best be grazed by dividing the cactus plantations into 0.4 ha to 0.9 ha camps, which can then each be grazed intensively for short periods of time (Aucamp & De Kock 1970). Sheep take about 2 weeks to become accustomed to utilising the plant.

Pre-sliced material

Here the material should be harvested and sliced into approximately 30 cm by 30 cm pieces, which should then be allowed to wilt for one to two days before feeding (Aucamp & De Kock 1970; De Kock 1980a). The material can be fed in the inter-row spaces within the cactus plantations or in kraals, where it should preferably be fed in troughs to reduce wastage. This procedure is less wasteful than direct grazing.

Spineless cactus meal

The high moisture content of spineless cactus limits dry matter intake by animals and therefore affects its usefulness as a feed (De Kock 1980a). Sliced material can easily be sun-dried and hammer-milled to produce a meal. Such a meal is easily stored for times of drought and can in any event be used more efficiently by animals than can undried material.

Spineless cactus silage

Good quality silage can be made by chaffing spineless cactus material together with oat straw, lucerne hay or other forms of roughage. These should be mixed in a ratio of five parts (by weight) of spineless cactus to one part of roughage. Molasses (2% by weight) should be added unless the cactus material contains a reasonable amount of fruit.

Spineless cactus is not in itself a complete feed. When fed alone it provides a diet which contains about 80% of the energy, 36% of the protein, 32% of the P and 2% of the salt required by sheep. To balance the animals' diet, lucerne hay (85 g per sheep per day in summer and between 198 g and 227 g per day in winter) or maize grain (113.5 g per sheep per day in summer and between 227 g and 284 g per day in winter) should also be fed and the animals should be fed a lick containing 60% bonemeal and 40% salt (Bonsma & Maré 1942; Aucamp & De Kock 1970).

12.3.1.1.2 *Old Man Saltbush* (Atriplex nummularia)

Old Man Saltbush has a high yield potential (2.0 t to 2.6 t/ha/annum in an area receiving 240 mm to 750 mm/annum), is disease resistant, makes efficient use of available moisture and nutrients, is drought tolerant and is adapted to a wide climatic range and a range of soil types. It has been reported to have a crude protein content as high as 21% (Jacobs & Smit 1977), although Jacobs (1977) has also reported much lower values

(10.9%) (Table 12.4). However, much of this protein is apparently degraded in the rumen to ammonia so that the real protein value is apparently considerably lower than the laboratory values would suggest (Weston *et al.* 1970). The plant is reported to be low in available carbohydrates (Hassan *et al.* 1979), necessitating supplementation of the diets of producing animals with some energy source. The plant has proved useful for both sheep (Barnard 1985) and ostriches (Anon. 1990). The use of halimus saltbush (*Atriplex halimus*) is reported to reduce the ostrich feed costs by as much as 50%. It can meet the maintenance needs of mature merino wethers, but not of merino ewes with lambs. Supplementation with 120 g maize per day to the ewes will, however, not only provide for the maintenance needs of the ewes, but also allow for a daily liveweight gain of 133 g per day (Jacobs 1977). The data of Table 12.4 suggest that intakes of sheep grazing Old Man Saltbush plantations are not unreasonable. Recent research suggests that intake can be increased further by restricting the crop to low salt-content soils, by providing the animals with salt-free drinking water and by utilising the crop in conjunction with veld so that the animals can gradually become accustomed to utilising the saltbush.

Saltbush is also able to meet the minimum maintenance requirements of pregnant Angora ewes, although Hobson *et al.* (1986) nonetheless recommended the supplementation of 300 g alkali-ionophore treated whole maize per head per day as a preventative measure against abortion.

Old Man Saltbush should not be grazed during its establishment year, but soon thereafter to induce a shrubby growth form. Plants should not be allowed to grow out beyond the reach of the animals because of the wastage which then results. If the plants grow too tall they should be trimmed back to approximately 1.5 m (De Kock 1980b). Plantations are best grazed only once a year, when they can be grazed reasonably severely (Aucamp & Cloete 1970). The plant is susceptible to frequent defoliation and is best incorporated into the forage flow programme of a property by reserving it specifically for periods immediately following rain. This would permit the veld to be rested at this time to facilitate recovery.

Steynberg & De Kock (1987) have reported that a realistic stocking rate for Old Man Saltbush is 4 to 7 sheep/ha/annum in areas with a minimum rainfall of 240 mm per annum. A carrying capacity of 1 sheep per hectare can be expected where the rainfall is about 150 mm per annum.

12.3.1.1.3 *American aloe* (Agave americana)

The nutritional value of American aloe is similar to that of spineless cactus (Table 12.4). It provides a sub-maintenance diet and requires supplementation with both energy and protein. It is fed as a chopped feed. When used as the only feed, however, it causes paralysis in sheep but this can be prevented by the alkalinisation of the chopped material (Marais pers. comm.). It is capable of producing 1.2 t edible dry matter/ha/annum where the rainfall is 315 mm/annum and of meeting the maintenance requirements of up to 9 sheep/hectare/annum, provided the animals are each supplemented with approximately 800 g lucerne per day (Steynberg & De Kock 1987).

The leaves of this plant should be cut off close to the stem so as to harvest as much of the fleshy and most palatable part of the leaf as possible. The fibrous tip and edges of the leaves should then be removed and the remainder sliced into approximately 50 mm to 100 mm pieces for feeding (Lamprecht 1957; De Kock 1980a).

American aloe should be reserved specifically for feeding during periods of severe drought. Extended periods of undisturbed growth should be provided to allow as much drought forage as possible to accumulate. Defoliation should, therefore, be infrequent.

12.3.1.2 Irrigation

Irrigation on a farm scale in the arid summer rainfall region is, as in the semi-arid regions (section 12.2.1.5), generally confined to small patches of special purpose pastures such as lucerne. On a larger scale, however, lucerne in particular may be produced under irrigation in large scale irrigation schemes associated with some of the major storage dams (such as the Vaal-Hartz irrigation scheme) or on the banks of some of the large rivers. This is a specialised form of production and is outside the scope of the discussion in this text.

12.3.2 Winter rainfall areas

Pastures are not used extensively in the arid parts of the winter rainfall region although medics may be used in a rotation with cereals in areas receiving a relatively low rainfall. The reader is referred to section 12.2.2 for a discussion of the use of medic pastures in rotation with cereals in these areas.

13 Fodder conservation

13.1 HAY

R.H. Drewes[1]

13.1.1 Introductory comments

A large proportion of the hay which is made on South African farms is of relatively low quality. The reasons for this are many and varied but generally seem to result from one or a combination of the following:

(a) a lack of appreciation of the important role that good quality hay can play in a production system, and the extent to which this role is negated when poor quality hay is used;

(b) a lack of appreciation of the management procedures needed to produce good quality hay. This is aggravated by the fact that losses during hay-making are invisible;

(c) inclement weather; and

(d) the allocation of insufficient management time and equipment to the hay-making operation.

Hay-making is often given low priority relative to other farming operations and is often merely fitted into the farming programme when and if time is available. Consequently, crops may often be too mature to produce good quality hay. The resultant poor performance of animals fed this hay then reinforces the general lack of appreciation of the important role that hay can play in any feeding programme.

[1] Acknowledgment is made to the staff of the Pasture Section at the Nooitgedacht Research Station, Ermelo, for much of the information presented in this section.

Plate 5   Response of lucerne (*Medicago sativa*) to reduced soil acidity through liming. The area between and to the left of the two white pegs was unlimed. The soils of the area surrounding this plot had previously been limed.

Plate 6   Response of Italian ryegrass (*Lolium multiflorum*) to reduced soil fertility. The level of acid saturation of the soil of the plot to the right was 67%. That of the plot on the left had been reduced by liming to 42%.

Plate 7 Response of Italian ryegrass (*Lolium multiflorum*) to potassium (K). No K fertilizer had been applied to the plot shown on the left. The plot on the right had received 100 kg K/ha.

Plate 8 Response of Italian ryegrass (*Lolium multiflorum*) to nitrogen (N). No N had been applied to the plot shown on the right. That on the left had received 70 kg N/ha.

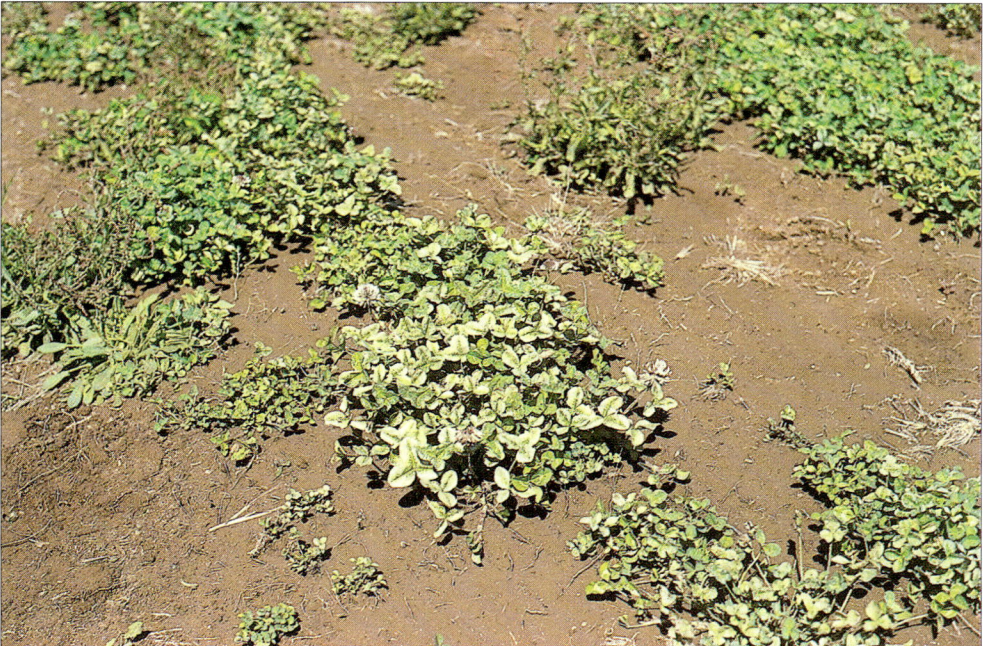

Plate 9   White clover (*Trifolium repens*) showing severe potassium (K) deficiency.

Plate 10   White clover (*Trifolium repens*) response to phosphorus (P). The plot shown on the left had been provided with 20 kg P/ha. That on the right had received 80 kg P/ha.

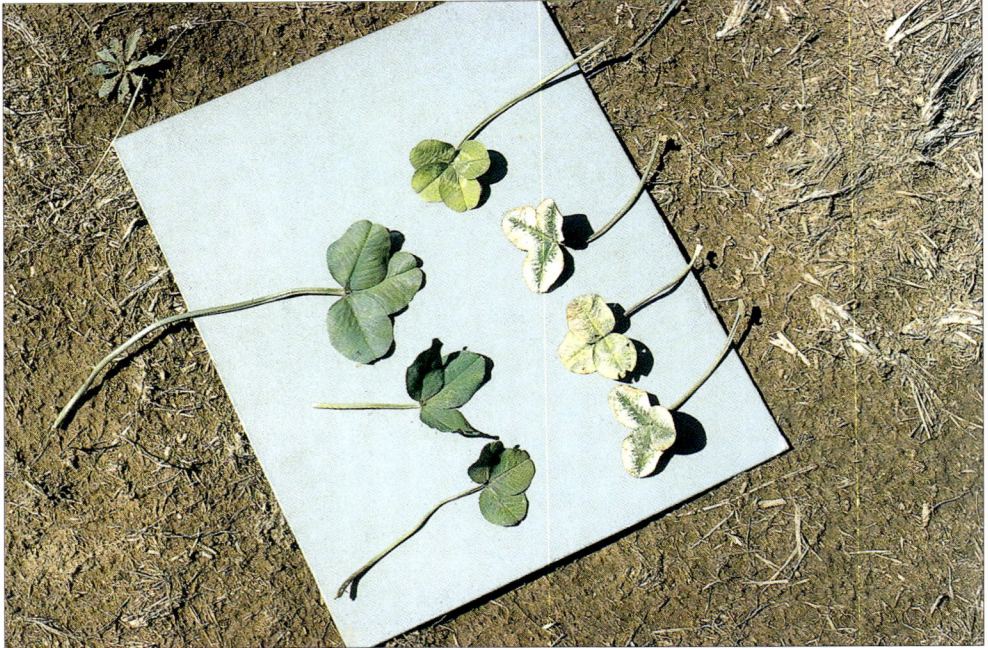

Plate 11 White clover (*Trifolium repens*) leaves of potassium (K) sufficient (to the left) and K deficient plants (to the right).

Plate 12 Frost burn on potassium (K) deficient plants of kikuyu (*Pennisetum clandestinum*) shown on the left. K fertilizer had previously been applied to the area shown on the right.

13.1.2 Requirements for the production of good quality hay

(a) The quality of the crop from which hay is made must be good (leafy, digestible etc.). The crop from which it is produced should be adequately fertilized and the hay harvested before the crop becomes too mature and stemmy.

(b) The crop must be dried as quickly as possible to reduce respiration losses, exposure to bleaching and to the possibility of rain damage.

(c) The hay should be dried to a moisture content of between 12% and 15%, to prevent the development of mould during storage.

13.1.3 Factors affecting the decision on whether or not to make hay

(a) Will the role of hay in the feeding programme justify its cost of production?

(b) Is the necessary equipment available to make good quality hay?

(c) Is there a crop which will justify the expense of making hay?

(d) Are there soils on the property which will support forage yields sufficiently high to justify the expense of making hay?

(e) Is the climate of the region conducive to producing good quality hay?

(f) To what extent will hay-making conflict with other operations on the farm?

(g) Can hay be purchased at a lower price than the cost of on-farm production?

13.1.4 Potential hay-making crops

Because of the need for rapid curing, hay crops need to be fine stemmed and relatively non-succulent. For this reason, the premier hay crop in the humid regions of South Africa is weeping lovegrass (*Eragrostis curvula*). Even though it produces a relatively poor quality hay, it can be cut, dried, baled and stored on a normal summers day in these regions. Lucerne (*Medicago sativa*), however, remains the premier hay crop because of its high quality, but due to its high moisture content it can be strongly recommended only in the more arid regions under irrigation. In the more humid regions the risk of spoilage during curing is relatively high. Other potentially useful hay grasses are Smuts' fingergrass (*Digitaria eriantha*), although it is relatively wet and may be difficult to cure, Rhodes grass (*Chloris gayana*), Teff (*Eragrostis tef*), Blue buffalo grass (*Cenchrus ciliaris*) and species of *Cynodon* (e.g. Star grass) or the *Cynodon* crosses (e.g. coastcross II).

13.1.5 How to make good quality hay

In producing good quality hay it is essential to harvest the crop when it is relatively young and leafy. Forage quality declines rapidly during flowering, and although this is compensated for by increased yields and easier curing, delays in harvesting beyond the 10% flowering stage cannot generally be recommended unless the specific intention is to provide a large bulk of relatively poor quality material at low cost. As a general rule farmers are advised not to delay harvesting beyond 10% flowering.

If good quality hay is to be produced, the material must be dried as rapidly as possible. Hence after cutting, usually in the early morning, the material needs to be fluffed up with a tedder once or twice before it is raked into windrows for later baling. Excessive handling of dry legume forages should be guarded against since this is likely to lead to a loss of rapidly drying and high quality leaf material. During drying the material should not be exposed to long periods of direct sunlight since this will cause bleaching and a loss of carotine.

Various additives (desiccants) are available which will speed up drying and, where the material is stemmy, as for example in most lucerne crops, it is advisable to run the material through a crimper immediately after it has been cut. This crushes the stems so that they dry at much the same rate as the leaves and this reduces leaf loss. An alternative procedure which, because of the costs involved, is suited only to high quality material, is to dry the material artificially using diesel or electrically driven heaters. This procedure is normally reserved for such crops as lucerne, where a very high quality product can be produced when young forage is dried in this way.

Mower blades should always be kept sharp. Blunt blades tend to tear rather than provide a clean cut. This reduces the rate of crop regrowth.

13.1.6   Storage

The amount of material cut on any one day should never exceed the amount that can be baled and, where appropriate, moved into storage. The material should be well dried before it is stored in bulk since there is always a danger of spontaneous combustion of material stored at high moisture contents. Storing hay under cover is expensive and so a large proportion of hay which is now made is stored as large round bales which are stored in the open.

Irrespective of how hay is stored, its quality will decline as it ages. The oldest material should therefore always be fed first.

Cognisance should always be taken of the possible danger of the carcases of dead rodents contaminating hay during storage. This raises the danger of botulism in animals fed the material.

13.1.7   Soil fertilizer requirements

The hay-making operation depletes soil fertility and this may reduce the productive life of a pasture. A 4 ton crop of hay, for example, contains approximately 80 kg N, 8 kg P, 60 kg K and 16 kg Ca, or the equivalent of 267 kg LAN (30% N), 96 kg superphosphate (8.3% P), 120 kg muriate of potash (KCl) (50% K) and 46 kg lime (35% Ca). Pastures from which hay is cut must, therefore, be fertilized more heavily than grazed pastures if they are to remain productive. Feeding this hay back on the pasture will serve to reduce these nutrient losses but may lead to severe trampling damage in the vicinity of the feed troughs, and is therefore practicable only on pastures which can tolerate such treatment. Such feeding should therefore be reserved for sacrifice camps or fed out on summer grazed sod-forming pastures.

13.2   SILAGE   *M. De Figueiredo*

13.2.1   Silage and principles of ensilage

Silage is formed when fodder containing less than 50% dry matter is stored anaerobically (in the absence of oxygen). Storage takes place in a silo. The main objective of ensiling any crop is to preserve it, with a minimum loss of nutrients, for use when other feeds are not available. This preservation takes place by means of fermentation.

In natural fermentation, the most important objective is to achieve anaerobic (oxygen-free) conditions. Under aerobic conditions losses will result from the respiratory activity of both plant cells and oxygen-loving microflora. Respiration depletes the sugars contained in the material and usually results in the loss of energy in the form of heat. Since losses are generally from the digestible fraction of the forage, a reduction in nutritive value can be considerable.

During preservation, the fodder will normally undergo acid fermentation when the bacteria produce lactic, acetic and butyric acids from sugars present in the raw material. The result is a reduction in pH (the material becomes more acid) and a total or partial elimination of the predominantly acid intolerant micro-organisms which cause spoilage.

While the creation of anaerobic conditions eliminates losses due to the activity of aerobic micro-organisms, clostridia, which are anaerobic bacteria usually present as spores on harvested forage, start multiplying as soon as conditions become anaerobic. These organisms produce butyric acid from sugars and lactic acid and also degrade amino acids into a variety of products of poor nutritional value. Consequently the efficiency of preservation is commonly judged by the relative proportion of the different fermentation acids produced. The greater the ratio of lactic acid to butyric acid, the higher is the efficiency of the ensiling process and the better the quality of the silage (McDonald 1981).

Clostridial growth is, however, inhibited by acid conditions. The lactic acid bacteria, also normally present on harvested crops, are facultative anaerobes, i.e. they are able to grow both in the presence and absence of oxygen. These bacteria ferment the naturally occurring sugars (mainly glucose and fructose) in the crop into a mixture of acids of which lactic acid is the major one. Lactic acid is a stronger acid than butyric acid and so the material becomes more acidic. Since clostridia require very wet conditions for active growth, an alternative method of inhibiting their growth is to wilt the crop before ensiling it.

13.2.2   Crops used for silage

Silage can be made from a variety of materials but forage grasses, cereals and legumes are used most commonly. Some of these crops, such as maize, are purposely grown for silage, but silage is often also made from multi-purpose pastures. The removal of surplus grass (for silage) will ensure a succession of high quality regrowths for grazing and can be used to prepare an area for the production of foggage. Whatever system is adopted, the crop used should be selected according to the class of animal which will be fed the silage, as well as the equipment available and the costs involved.

13.2.3   Types of silos

There are many types of silos in use. They range from tower silos to walled bunker (clamp), pit (which may or may not be lined) or stack silos. It may also be made in bales.

13.2.3.1   Tower silos

Tower silos can be made of concrete, steel plate or fibre glass with a conical or domed roof. They can extend to a height of 20 m, with a diameter of approximately 5 m.

13.2.3.2 Bunkers

Bunkers usually comprise rectangular structures with three walls which can be fabricated from concrete, brick, plywood sheets or other similar materials. They often have a concrete base with drainage channels to facilitate the collection of effluent. Pit or stack silos are modifications of this, the former comprised of no more than an unlined pit, and the latter a stack constructed above the ground.

13.2.3.3 Big bale silage

Big bale silage is made in polythene bags. Round bales of 1.2 m width by 1.2 m diameter, each weighing 300 kg to 500 kg, are most commonly used. Once the bags have been filled, they are sealed and stacked on top of one another. This technique has a number of useful features. The silage is produced in small packages so that there is little wastage when feeding. Little capital is required for storage. However, special care needs to be taken to protect the bags from damage by wind, hail or animals during storage. Should the plastic bags be punctured, the silage will be exposed to aerobic conditions and will rot and cure as compost rather than as silage. Preventative measures should also be taken against rodent damage.

13.2.4 The mechanical aspect of ensilage

With the exception of big bale silage, where special machinery is used to harvest and ensile the crop, crops are normally harvested for silage in one of two ways. They may be cut, lifted and transported to the silo in one operation. Alternatively the crop may be wilted after having been cut and then needs to be lifted in a separate operation. Wilting can be speeded up by frequent turning (tedding). In both methods, a forage harvester whose cutting action may vary from a flail-type to a double or metered-chop, is commonly used. The selection of equipment depends on the procedures used to process and transport the crop.

The procedure used in filling a silo will depend on the type of silo and on the available equipment. In all cases the material must be well compacted to exclude air. In bunker, pit or stack silos this can be done with a tractor, and in a tower silo by means of tramping or other more modern devices. Large bunkers which cannot be filled in one day should be sealed between successive fills using the wedge method (Fig. 13.1). When a silo is full it should be sealed with plastic sheeting kept in place with evenly distributed weights.

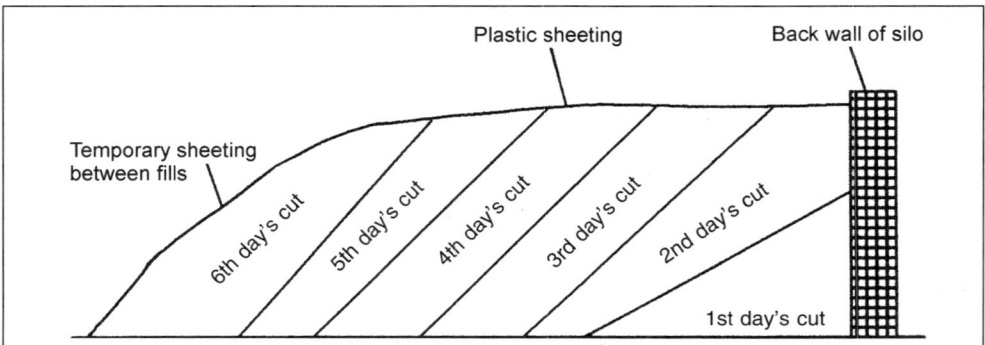

Figure 13.1 Cross-section of a silo filled by the wedge method (Woolford 1984).

13.2.5 Silage quality

The quality of silage essentially depends on the quality of the crop being ensiled and on the changes which take place during preservation. The higher the nutritive value at ensiling, the higher will be the quality of a well made and well preserved silage. In addition, the carbohydrate content of the ensiled material influences preservation during storage. Its content varies with plant species, plant maturity, time of day that the material is cut and with the amount of N applied to the crop. Cereals have a higher carbohydrate content than do temperate grasses, which in turn have a higher carbohydrate content than tropical grasses or legumes. Non-structural carbohydrate content increases with maturity as well as during the day until about 2 pm, when it reaches a plateau. It decreases during the night (Butler & Bailey 1973; De Figueiredo 1987).

The carbohydrate content of ensiled material is inversely related to the amount of N applied to the crop. This is particularly true for temperate grasses. This correlation is, however, not well defined for the subtropical kikuyu (De Figueiredo 1987).

The amide and nitrate content are also influenced by the level of N fertilizer applied (McDonald 1981). The total amount of N in fresh herbage is largely determined by the amount of fertilizer N applied to the crop and the stage of maturity of the herbage at harvest. Some 70% to 90% of this N is normally in the form of protein. The remainder consists of non-protein compounds. These include free amino acids, peptides, amines, chlorophyll, ureides, nucleotides, nitrates and the amides glutamine and asparagine.

The digestibility of all grasses decreases as they mature, but the rate at which this happens varies among different species. It also varies with time of year. In the northern hemisphere the initial fall in digestibility of temperate grasses is quite slow (0.5 units/day) but increases rapidly from mid-May (late spring) (Minson 1971). In subtropical grasses, which inherently have a lower digestibility than do temperate grasses, the decrease in digestibility with increasing maturity is appreciably slower than this (0.1 to 0.2 units/day) (Minson *et al.* 1960). This is important because the yield and dry matter content of any crop increases with maturity and it is often necessary to reach a compromise between quantity and quality. In the whole maize plant, digestibility increases as the cobs begin to form and remains high as the crop matures. Yield and dry matter content also increase with crop maturity. These factors, combined with the high carbohydrate content of maize, makes it an ideal silage crop.

13.2.6 Changes during preservation

The complete preservation of a crop as ensilage is impossible. This is because many kinds of micro-organisms which cannot be excluded completely from the plant material will attack it to a greater or lesser extent, depending on the success achieved in its preservation.

13.2.6.1 *Important micro-organisms in silage*

Four types of micro-organisms are commonly found in silage:
(a) Lactic acid bacteria. These bacteria have been described as micro-aerophilic, gram positive, non-spore forming organisms which ferment sugars to lactic acid (Rogosa 1974). They are generally divided into two major categories, the homofermentative and heterofermentative types.

(b) Clostridia. These have been described as gram positive, spore-forming, usually mobile rod-shaped bacteria that grow under anaerobic conditions, fermenting sugars, organic acids and proteins (Smith & Hobbs 1974; Beck 1978). They comprise two main physiological groups: saccharolytic clostridia, which ferment mainly lactic acid and sugars and have limited proteolytic activity, and the proteolytic clostridia, which ferment mainly proteins and possess only a limited activity for carbohydrates. Some clostridia have both a high saccharolytic and a high proteolytic activity.

(c) Enterobacteriaceae. These are gram-negative, non-spore forming, facultatively anaerobic, usually motile, non-pathogenic, rod-shaped bacteria which ferment carbohydrates (Beck 1978). The optimal pH for those commonly associated with silage is around 7. They are usually active only in the early stages of fermentation, when the pH favours their growth (Brierem & Ulvesli 1960; Gouet & Chevalier 1966). A rapid development of lactic acid bacteria will suppress their growth (Gibson *et al.* 1961), but where their numbers are insufficient to produce sufficient lactic acid to lower the pH, enterobacteriaceae might dominate the fermentation process. The silage will then develop high levels of acetic acid, the main fermentation product of enterobacteriaceae (McDonald 1981).

(d) Fungi. These are eukaryotic micro-organisms which can grow either as single cells – the yeasts, or as multi-cellular colonies – the moulds. They obtain nutrients by secreting extra-cellular enzymes such as amylases, lipases, cellulases and proteases. The enzymes break down complex organic molecules to simple monomers which they absorb through their cell membranes (McDonald 1981). Their presence is linked to the aerobic deterioration of silage. Yeasts can, however, grow under anaerobic conditions and compete with lactic acid bacteria for sugars, which they transform into ethanol (of little, if any, preservative value in silage), propanol, iso-butanol, iso-pentanol, acetic, propionic, butyric and iso-butyric acids, as well as small amounts of lactic acid (Woolford 1976; Kibe *et al.* 1977).

Success in the natural preservation of a crop as silage is essentially dependent on the amount of readily fermentable carbohydrates, the buffering capacity and the dry matter content of the fresh crop. Crops with a high content of fermentable carbohydrates, high dry matter content and low buffering capacity, preserve well. So, for example, whole-crop cereals will produce high quality silage relatively easily because they contain high levels of highly fermentable carbohydrates and, if cut at the appropriate time, they have a high dry matter content. Their silages are, however, low in crude protein (generally 7% to 9% on a dry matter basis) and, for most livestock fed these silages, protein supplementation is necessary. Crops which do not fulfil these requirements need to be wilted before ensiling, or additives will be needed.

13.2.6.2 *The chemistry of ensilage*

Carbohydrates

High levels of readily available fermentable carbohydrates (fructose, glucose, sucrose and the fructosans) are a prerequisite to successful crop preservation because they are important sources of energy for lactic acid bacteria. Structural carbohydrates (cellulose and hemi-cellulose) may provide some water-soluble carbohydrates, but their role is minor (Dewar *et al.* 1963). While sucrose is the principal fermentable substrate in tropical and subtropical grass crops (including maize) (McAllan & Phipps 1977), fructosan predominates in ryegrass (MacKenzie & Wylam 1957).

The pathways by which sugars are fermented depend on the type of lactic acid bacteria present and the nature of the sugars. While bacteria of the homofermentative type ferment hexoses entirely to lactic acid (1 mole of glucose or fructose being converted into 2 moles of lactate) (Fig. 13.2), heterofermentative bacteria ferment the two hexoses, glucose and fructose, differently. If glucose is fermented, 1 mole will yield 1 mole each of lactate, ethanol and carbon dioxide. If fructose is fermented, then every 3 moles will yield 1 mole each of lactate, acetate and carbon dioxide and 2 moles of mannitol (Fig 13.3). Therefore the heterofermentation of hexoses produces less acid than does homofermentation. Also, heterofermentation from fructose is less efficient than from glucose. Therefore, since some grasses have higher fructose than glucose levels, the dominance by the heterofermentative type of lactic acid bacteria can only be detrimental to silage preservation.

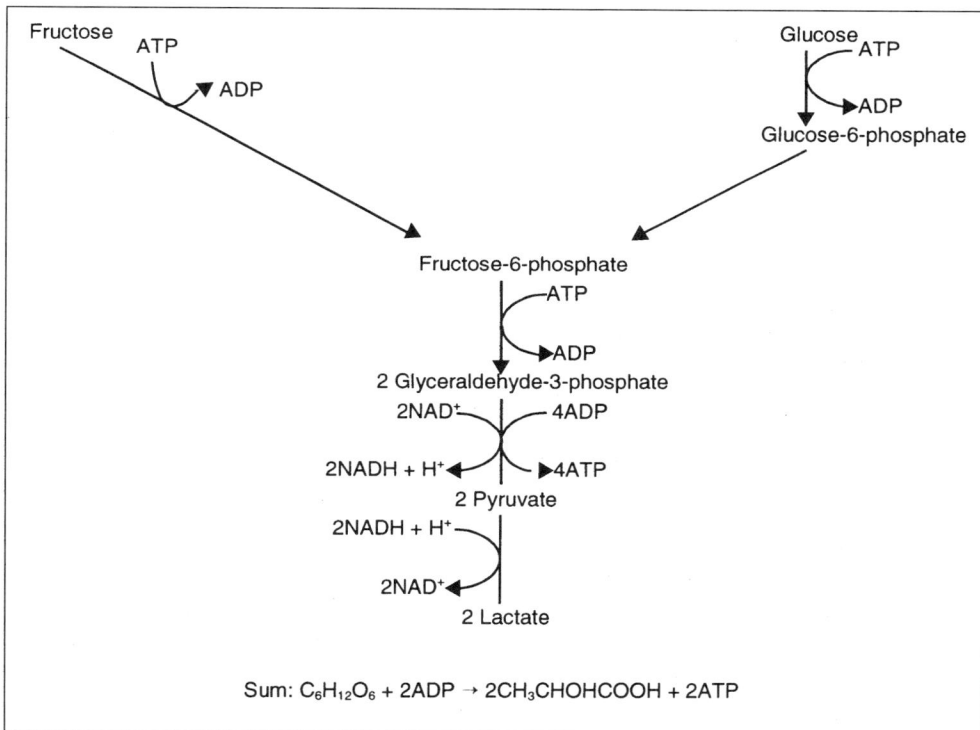

Figure 13.2 Fermentation of glucose and fructose by homofermentative lactic acid bacteria (McDonald 1981).

Pentoses are found in small amounts in forage crops as a result of hemicellulose hydrolysis. It is unfortunate that these sugars are not more abundant, as they are fermented only to acid by both homofermentative and heterofermentative lactic acid bacteria. Here 1 mole of a pentose is converted to 1 mole each of lactate and acetate without any loss of dry matter as carbon dioxide (Woolford 1984).

$$NADPH + H^+ \rightarrow NADP^+$$

Glucose   Fructose ———→ Mannitol

ATP   ATP

ADP   ADP

Glucose-6-phosphate ◄——————— Fuctose-phosphate

NADP$^+$

NADPH + H$^+$

6-phosphogluconate

NADP

CO_2 ◄   ►NADPH + H$^+$

Ribulose-5-phosphate

Xylulose-5-phosphate

ADP   ATP

Glyceraldehyde-3-phosphate   Acetyl-phosphate ———→ Acetate

NAD$^+$ 2ADP   NADPH + H$^+$

NADH + H$^+$ ◄   ►2ATP   NADP$^+$

Pyruvate   Acetaldehyde

NADH + H$^+$ NADPH + H$^+$

NAD$^+$ ◄   NADP$^+$

Lactate   Ethanol

$$C_6H_{12}O_6 \text{ (glucose)} + ADP \rightarrow CH_3CHOHCOOH + C_2H_5OH + CO_2 + ATP$$

$$3C_6H_{12}O_6 \text{ (fructose)} + H_2O + 2ADP \rightarrow CH_3CHOHCOOH + 2C_6H_{14}O_6 + CH_3COOH + CO_2 + 2ATP$$

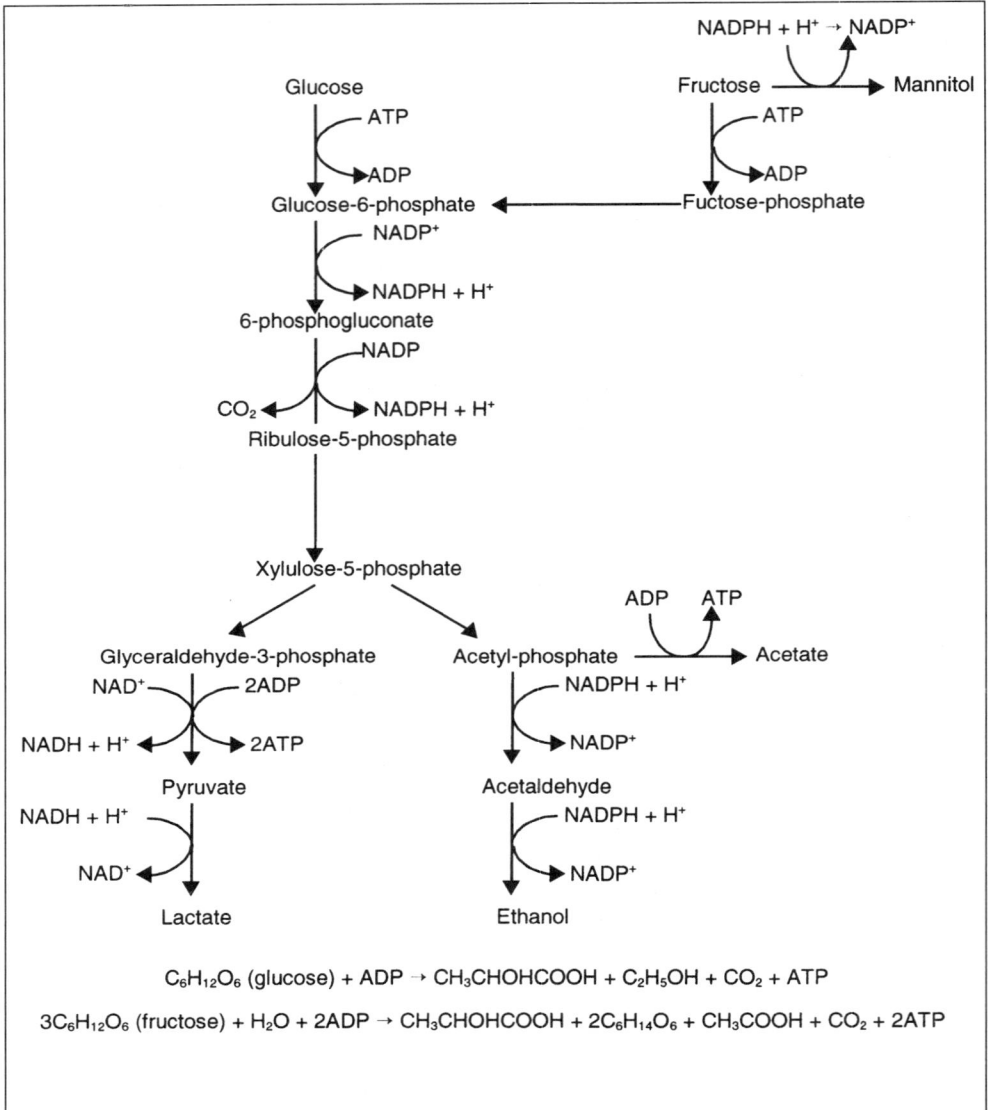

Figure 13.3 Fermentation of glucose and fructose by heterofermentative lactic acid bacteria (McDonald 1981).

Sugars and lactic acid are utilised by clostridia as a source of energy to produce butyric acid, carbon dioxide and hydrogen. In silages in which lactate fermenting clostridia are dominant, little or no lactic acid or sugars remain in the silage. Here the main fermentation acid is normally butyric, although large amounts of acetic acid are also frequently present.

Apart from acetic acid, enterobacteriaceae ferment sugars to formic acid, ethanol and 2,3-butanediol (McDonald 1981).

Buffering capacity

Another prerequisite for the good preservation of a crop is a low buffering capacity, i.e. an inability to resist pH change (Playne & McDonald 1966). The more quickly the pH drops, the sooner conditions are created for the proliferation of desirable micro-organisms and for the elimination of the undesirable ones. Different herbage species have different buffering capacities. In general, legumes are more highly buffered than are forage grasses and these grasses more so than maize (Nash 1959; Dewar *et al.* 1963).

The most important buffering compounds of plants are anions, organic acid salts, sulphates, nitrates, chlorides and orthophosphates (Smith 1962). Plant proteins account for only 10% to 20% of the buffering capacity. While a number of organic acids have been identified in grasses, their concentration is usually low. Where they occur in high concentrations they are likely to contribute greatly to the buffering capacity.

Season and maturity also influence the buffering capacity of forage (Greenhill 1964). During ensiling, organic acids are rapidly and completely dissimilated into fermentation acids and carbon dioxide. Because organic acids buffer most strongly at the normal pH of grasses (pH = 6), the pH in silage does not drop rapidly enough. Thus it might seem that the rapid dissimilation of organic acids is beneficial. However, these acids are dissimilated into fermentation acids of even stronger buffering capacity as well as into carbon dioxide, with a subsequent loss of dry matter. This results in an eventual two to four times increase in buffering capacity during ensiling (McDonald & Henderson 1962; Greenhill 1964; Playne & McDonald 1966; De Figueiredo 1987).

Nitrogen compounds

Once herbage is cut, protein breakdown (proteolysis) begins. It continues inside the silo with its extent being influenced by the rapidity with which oxygen depletion and acidic conditions are established. The significance of proteolysis is that amino acids (e.g. lysine, arginine and histidine) and protein can undergo considerable changes. Appreciable losses of aspartic acid, threonine, tyrosine, proline, glutamine, methionine, serine, glutamic acid and cystine have also been reported (Woolford 1984). Other amino acids, such as amino-butyric acid, tryptophan and ornithine, together with propionic acid, iso-valeric acid, histamine, tyramine, ammonia and carbon dioxide are also products of amino acid breakdown by plant enzymes and/or the silage microflora. However, even if the amino acids do not themselves change during preservation, proteolysis needs to be minimised because, from the point of view of animal production, it is preferable that protein *per se* instead of free amino acids enter the duodenum.

In addition to sugars, lactic acid bacteria have complex nutritive requirements for amino acids, peptides, nucleic acid derivatives, vitamins, salts and fatty acids. It is not surprising, therefore, that these bacteria have been found to de-aminate and de-carboxylate amino acids (McDonald 1981). Although their nutritional requirements vary between species, in general they seem to be only slightly proteolytic.

In the clostridia, fermentation of amino acids is of three types:
(a) decarboxylation, which leads to the formation of an amine;
(b) oxidation-reduction, in which one amino acid is oxidised while another is reduced. Both oxidation and reduction result in the formation of a fatty acid; and
(c) de-amination, in which ammonia is released, producing an organic acid as the end-product. The amount of ammonia in silage is considered to be a good indicator of

the extent of proteolytic clostridial activity since only small amounts are produced by other silage micro-organisms (McDonald 1981). Apart from de-amination of amino acids, ammonia is also produced by the reduction of nitrate and nitrite. These are reduced by some types of clostridia, lactobacilli and enterobacteriaceae (Bousset-Fatianoff *et al.* 1971).

Although enterobacteriaceae are able to de-aminate and decarboxylate amino acids, they possess only weak proteolytic properties.

The N components of fresh grass, well preserved silage and poorly preserved silage are compared in Fig. 13.4.

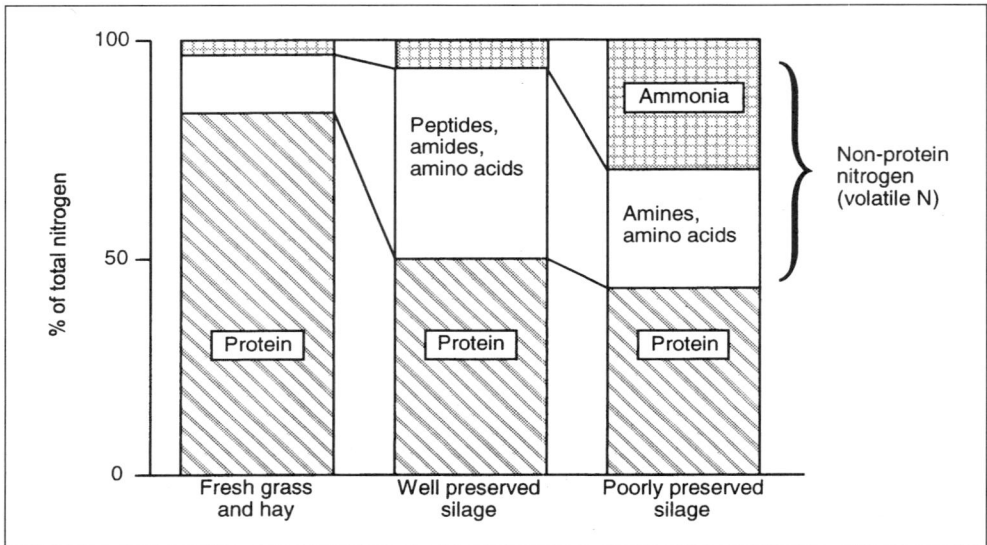

Figure 13.4 The nitrogen components of fresh grass, well preserved silage and poorly preserved silage (Wilkinson 1985).

Carotene

Carotene is a pro-vitamin which is partially destroyed during preservation by enzymic action. Losses are most rapid at pH values between 3.7 to 4.2 (Kalac & Kyzlink 1980). Little carotene should be lost where preservation is good but vitaman A may need to be supplemented when preservation is poor.

13.2.7 How to improve silage quality

A well-established method of improving silage quality is by wilting the crop before ensiling it (McDonald 1981; De Figueiredo *et al.* 1989; Rodriguez *et al.* 1989a, 1989b). This reduces microbial activity and losses from seepage. Wilting also reduces the buffering capacity of the material. However, losses of variable magnitude are bound to occur in the ensilage process. These include losses which occur in the field as well as those resulting from respiration, fermentation, effluent seepage and from aerobic deterioration.

13.2.7.1 Field losses

These result from respiratory activity, leaching of digestible nutrients by rain, shattering of the crop by machinery and a low efficiency of the pick-up operation. Such losses can be minimised by using efficient machinery, by making silage only during favourable weather conditions and by reducing the operating time.

13.2.7.2 Respiration losses

These losses occur before anaerobic conditions are established in the silo, unless the pH has been lowered by applying a strong acid. Both the enzymes in plant cells and aerobic microflora convert carbohydrates into carbon dioxide and hydrogen, with a consequent depletion of sugars and the production of heat.

Dry matter losses resulting from respiration can be reduced by quickly filling the silo and by thoroughly compacting the material. The latter can be facilitated by cutting the crop into small lengths or by bruising it (McDonald *et al.* 1964).

13.2.7.3 Fermentation losses

These losses result from the activity of anaerobic and/or facultatively anaerobic microorganisms under anaerobic conditions.

Lactic acid bacteria

The loss of dry matter through fermentation by these bacteria largely depends on whether a homofermentative or a heterofermentative population predominates. No losses are incurred when the sugars are fermented by homofermentative lactic acid bacteria, but between 4% and 6% of the dry matter is lost when the heterofermentative type ferment the carbohydrates, with the exception of pentoses where both types give total dry matter recovery. However, both types ferment organic acids and, depending on the acid fermented, dry matter losses can occur. Losses incurred when amino acids are fermented by lactic acid bacteria are small (Woolford 1984).

Enterobacteriaceae

Losses resulting from the activity of enterobacteriaceae depend on how active they are in a silage. As they are usually active only in the early stages of fermentation (Gouet & Chevalier 1966), they are not considered important. No information on the percentage of dry matter or energy lost has been published. However, in some cases, depending on the end-products produced, these losses might be considerable.

Clostridia

Silages which undergo secondary fermentation lose substantial quantities of dry matter and energy, depending on the substrate used and the reactions that occur. Crops that undergo severe secondary fermentation may lose up to 8% digestibility units in contrast to the 2% usually lost in well-preserved, wilted silages (Wilkinson 1985). Losses through the action of clostridia can be controlled by ensiling crops of high dry matter content and by creating acidic conditions.

13.2.7.4 Effluent losses

The seriousness of losses in the effluent depends on the amount of seepage, which in turn is directly related to the dry matter content of the crop being ensiled. A dry matter content greater than 25% produces no effluent (Raymond *et al.* 1986).

About 20% of a typical silage effluent is made up of nitrogenous substances, 55% of non-nitrogenous organic matter and 25% of mineral matter. Dry matter losses caused by effluent discharge range between 5% and 10% (Woolford 1984).

Apart from a loss of nutrients, silage effluent is considered to be some 20 times more polluting to rivers and streams than is cattle manure (Wilkinson 1985).

Pre-wilting of a crop before ensiling, and the use of moisture absorbents in the silos, are methods used to restrict silage effluent. Whilst pre-wilting is mainly governed by the weather, the use of absorbents is determined by their effectiveness in binding moisture. Apart from the practical aspect, the cost factor of either method should be considered in relation to its benefits.

13.2.7.5 Aerobic deterioration losses

There are two types of aerobic losses: those during storage and those during feeding. The former occur on the surface and at the sides of a silo and result from the leakage of air into the silo. Dry matter losses of this kind have been estimated to range between 23.8% and 51.5% (McDonald *et al.* 1965). Covering the silo with plastic sheeting (as in an envelope) and an even distribution of weights over the surface of the silo will reduce losses of this type. Where management is good, surface waste should not exceed 6% of the dry matter ensiled (Wilkinson 1985).

Losses during feeding commonly range between 5% and 10% (Honig & Woolford 1980). They tend to be high during hot weather. Air penetration into the mass of the silage should be minimised by uniformly removing the crop and decreasing the area of the face open to the air. In the case of bunker silos a consumption of at least 1 m of silage per day (measured horizontally) is recommended.

It is, however, difficult to eliminate all losses during fermentation. Total losses of 19% and 21% have been reported by Wilkinson (1985) from direct cut and wilted grass, respectively, when good management has been applied throughout the ensiling process.

13.2.8 Additives

Unfavourable weather often makes wilting impracticable. With an increasing demand for improved animal production, there is an increasing need for high quality feed of predictable nutritional value. Additives have been introduced in an attempt to ensure efficient preservation throughout the ensiling process. An additive should be easy to apply and non-hazardous to handle. It should restrict fermentation, and particularly clostridial activity, improve aerobic stability, be economical and have no adverse effects on human and animal health or on the environment. Since their introduction, they have gained wide acceptance. From 1949 to 1970, 178 different preparations were tested or marketed worldwide (Woolford 1984). In 1987 over 100 silage additives were reported sold within the European community alone (Adamson & Pratt 1987).

The effectiveness of an additive should be judged in terms of its chemical and micro-biological effects; its influence on the recovery of nutrients contained in the fodder; and on its effect on animal production.

Additives can be broadly classified into four categories (Woolford 1984):

(a) those producing direct acidification. Their action is mainly to lower the silage pH, with a resulting shift to lactic acid fermentation, e.g. sulphuric, hydrochloric and formic acids;

(b) those that inhibit fermentation. These either inhibit the microflora in general e.g. formaldehyde, or discourage the growth of spoilage causing micro-organisms directly, e.g. bacitracin and sodium nitrite;

(c) those that act as fermentation stimulants. These encourage fermentation by providing fermentable material, e.g. molasses and enzymes, or by establishing and promoting the dominance of lactic acid bacteria, e.g. inoculants; and

(d) those that supply additional nutrients. These are used to improve the nutritive value of the silage by rectifying specific nutrient deficiencies, e.g. starch or urea, or by protecting protein from being degraded in the rumen, e.g. formaldehyde.

No additive has yet been demonstrated to improve silage preservation under all conditions and at all times. The attributes of many commercial additives appear to have been exaggerated. Their alleged benefits are often based on obscure research data or on unreliable testimonials (Woolford 1984). One thing is certain, an additive cannot replace bad management.

13.2.9 Nutritive value of silage

The feed value of a silage depends on the amount animals are prepared to consume (intake), its digestibility, the efficiency of digestion and its protein and energy content. Raymond *et al.* (1986) consider that, under most practical feed conditions, livestock production will be determined mainly by the amount of digestible organic matter ingested. This is largely affected by the livemass of the animal and the palatability of the feed on offer. Any contamination of the silage with soil or manure can adversely affect palatability. Furthermore, animals are known to eat more short-chopped than long-chopped silage (McDonald 1981) and more silage with a high dry matter content than that with a high moisture content. The latter silage is usually associated with high protein breakdown and high contents of ammonia and volatile fatty acids. It has proved difficult to relate various chemical factors to intake and none of them have singularly been found to account for more than 40% of the variation in dry matter intake. However, recent research has been directed towards the determination of protein quality and its quantification. Jones (1997) found that protein quality was the factor responsible for the improved performance observed in animals fed silage treated with certain types of bacterial additives.

The amount of feed that a ruminant can eat is also influenced by rumen fill, which in turn is affected by digestibility. If a feed is highly digestible, the rumen empties rapidly and the animal can then eat more. The efficiency of silage digestion does not, however, appear to be very high. Diets containing a high proportion of silage also support no more than a low rate of microbial protein production. This is apparently due to the combination of poor adenosine triphosphate (ATP) yield obtained in the rumen from the silage fermentation products and the rapidity with which non-protein nitrogen is degraded to ammonia (Thomas & Thomas 1985). In addition, the efficiency with which the digested nutrients are absorbed by the animal depends on the type of fermentation in the rumen. The formation of acetate rather than lactate seems undesirable for rumen fermentation as acetate is unlikely to be available as an energy source for the micro-organisms, whereas lactate is readily metabolised. Acetate is also used less effectively than propionate (from lactate metabolism) as a source of energy for tissue growth.

Approximately 75% to 90% of the total N in plants is normally in the form of true protein which is of lower degradability in the rumen than non-protein nitrogen. Therefore a large proportion of this N would be used directly by the animal. In well-preserved silage, however, about 50% of the nitrogenous constituents are in the form of non-protein nitrogen which readily degrades to ammonia and other non-protein compounds in the rumen. This results in a reduced supply of amino acids for lean tissue growth. Supplementation of kikuyu silage with protein of low degradability has, for example, significantly improved the performance of beef weaners (De Figueredo *et al.* 1994), suggesting that high producing animals need to be supplemented with non-degradable (escape or by-pass) protein if they are to absorb sufficient protein to meet their needs. Any heating and browning of silage during preservation denatures the protein and makes it less digestible.

The amount of free sugars in silage is also lower than in the original crop. Although some dry matter loss is inevitable during ensilage, there are doubts as to the energy value of silage when compared with the original crop. However, the extent of energy losses will vary according to the type of fermentation. While all energy is retained when sugars are fermented to lactic acid by the homofermentative lactic acid bacteria, clostridia ferment sugars and lactic acid with an energy loss of 18%, despite the fact that butyric acid has a value of 24.9 MJ per kilogram, compared with 15.6 MJ for glucose. However, while the main limiting factor in forage grass silage seems to be energy, in maize silages the limiting factor is protein.

Because of the losses incurred in converting grass to silage, these feeds cannot support the same level of animal performance as the original crop (Wilkinson 1985). Also, animals fed silage are not able to select the most palatable material, as they do when grazing. They therefore eat a lower quality material than they would graze.

Silage is seldom used as the sole feed for productive stock. In the case of forage grass silage which is deficient in energy, cereal grain is usually supplemented. This grain lowers the pH in the rumen and consequently also reduces the activity of the cellulolytic microbial population which is responsible for fibre digestion. This results in a decrease in overall digestibility (Raymond *et al.* 1986) and in reduced feed intake. Feeding starch supplements leads to a pronounced increase in rumen protozoal numbers and an increase in intra-ruminal recycling of N via the protozoa (Thomas & Thomas 1985). This could offset the benefit of providing additional energy. Hence the use of free sugars, such as molasses, as an energy supplement might be a better proposition as they do not affect the protozoa in the rumen. However, the rates at which protein and sugars are degraded must be considered if N is to be retained.

13.2.10 Subtropical versus temperate grass silage

Reference has been made to the low digestibility and low carbohydrate content of subtropical forage grasses compared with temperate species. Because of this, unlike temperate grasses, they are not capable of producing silage having a digestibility greater than 50%. Forage from these subtropical grasses is often in surplus at some time during their growing season. This excess needs to be removed in order to provide good quality regrowth for grazing. Because such silage is made from surplus grass, costs are relatively low compared to maize silage or *Eragrostis* hay (Gordijn pers. comm.).

13.3 HAYLAGE

Haylage is a product intermediate in moisture content between hay and silage. The forage is chopped soon after harvest and may then be fed into a tower silo or, more appropriately, stored under vacuum in a 'Harvestore'. Here fresh material may be repeatedly fed in at the top of the Harvestore as material is removed from the base of the structure for feeding. The vacuum should be restored on each such occasion. The nutritive value is well preserved in such material but the production of haylage in this way is very capital intensive (Humphreys 1991). Haylage can be usefully produced if weather conditions should prevent complete drying of an intended hay crop when the material can be packaged into large round bales and wrapped in plastic sheeting. Fermentation then takes place within such bales (Bayer & Walters-Bayer 1998).

13.4 FOGGAGE *M.B. Hardy, P.E. Bartholomew & N.F.G. Rethman*

13.4.1 Introduction

Various terms have been used to describe foggage. These include standing-hay, autumn-saved pasture, autumn-accumulated pasture, fall-saved pasture and stock-piled pasture. In this chapter the term foggage is preferred. It is produced when a pasture is closed (or 'put-up') during the growing season to accumulate forage as a standing crop for use, *in situ*, during the dormant season when pasture growth is slow or has stopped. Herbage which happens to accumulate because of incomplete grazing of a pasture is not normally referred to as foggage.

Putting-up time is when the pasture is closed to use and forage accumulation commences. When put-up, the pasture should be cleaned off, either by mob grazing or by mowing to remove any previously accumulated poor quality material and to allow the pasture to grow out evenly for the remainder of the growing season. The pasture would normally be top-dressed with nitrogenous fertilizer when put-up so as to maximise foggage production.

13.4.2 Why foggage?

Overwintering cattle and sheep is expensive and requires major management inputs from livestock producers in the summer rainfall parts of the country. This is particularly true of the extensive grazing areas in mixed- and sourveld where the nutritive value of the veld in winter is poor. Almost all types and classes of livestock lose weight and condition if they remain on such veld during winter. Traditionally, 'trek-farming' (where animals were moved to different properties in different seasons) and hay, silage and crop (largely maize) residues have carried animals through the winter. However, with the increasing value of land and the associated need to maximise financial returns per hectare, 'trek-farming' has become less profitable. Furthermore, the cost of producing hay and silage and the difficulty of producing hay when conditions are moist during summer, point to the need for an alternative method of producing winter feed.

Foggage production is not restricted to dryland pastures and, in the summer rainfall areas, may form an important component of intensive livestock production systems based on irrigated temperate pastures as well. Whilst temperate pastures have the physiological potential to grow through winter in these areas, growth is often extremely slow or ceases altogether during the coldest months. Foggage is therefore often produced from temperate pasture species during the autumn to carry high producing animals through the winter. Such temperate species may also be used to accumulate foggage in the winter rainfall regions, but here it is the spring growth which is reserved for use during the dry summer months.

13.4.3   Species suited to foggage production

The foggaging potential of several pasture species has been evaluated (Table 13.1). In southern Africa, Smuts' fingergrass (*D. eriantha*) and kikuyu (*P. clandestinum*) are the most commonly used of the summer-growing, winter dormant species, but Guinea grass (*P. maximum*), Dallis grass (*P. dilatatum*), Rhodes grass (*C. gayana*) and Nile grass (*A. macrum*) are also used.

Temperate (spring-, winter- and autumn-growing) species suited to the production of foggage include tall fescue (*F. arundinacea*) and cocksfoot (*D. glomerata*). It may also be made from annual ryegrass (*L. multiflorum*) and perennial ryegrass (*L. perenne*), but here the material needs to be used within about three months of put-up. Delayed use results in a quality deterioration in the older leaves.

The choice of a species for foggaging depends on how well it is adapted to the local climate and soils, the availability of irrigation (for temperate species) and the livestock production system in use.

13.4.4   Production and management

The most important management practice influencing the quality and quantity of foggage produced is the put-up date.

13.4.4.1   The put-up date

Pastures must be either evenly grazed or mown to remove all standing herbage before being closed for foggage (Cooper & Morris 1973; Bartholomew *et al*. 1991). This practice ensures that only new growth is conserved so that even-aged material is produced over the whole pasture. The best time to close a pasture for use as foggage will vary according to local climatic conditions, the pasture species used and the purpose for which the foggage is required. In the summer rainfall areas in the southern hemisphere, pastures are generally put-up between mid-January and the end of March. This put-up date has a major influence on both the quantity and quality of foggage produced.

13.4.4.2   Quantity of foggage

The quantity of foggage produced is positively related to (a) the duration of suitable growing conditions following put-up; and (b) the pasture species used. If the objective is to provide a large bulk of foggage material, pastures should be closed sufficiently early to allow for a reasonable period of growth before growth slows down in the autumn. Due to the unpredictable nature of rainfall, however, both the quantity and

Table 13.1 References relating to the foggaging potential of subtropical and temperate grasses and legumes.

Species	References
Subtropical grasses	
Acroceras macrum	Rethman & De Witt (1988)
Anthephora pubescens	Van Niekerk *et al.* (1989)
Brachiaria decumbens	Filgueiras (1983)
Cenchrus ciliaris	Minon *et al.* (1988)
Chloris gayana	Minon *et al.* (1988)
	Rethman & De Witt (1991)
	Van Niekerk *et al.* (1989)
Cynodon aethiopicus	Rethman & De Witt (1988)
Cynodon dactylon	Rethman & De Witt (1988)
	Rethman & De Witt (1991)
Cynodon nlemfuensis	Postiglioni (1990)
	Omaliko (1983)
Digitaria eriantha	Dannhauser (1985, 1988)
	Rethman (1983)
	Rethman & De Witt (1991)
Digitaria scalarum	Rethman & De Witt (1988)
Eragrostis curvula	Meaker & Coetzee (1978)
Hemarthria altissima	Postiglioni (1990)
Panicum maximum	Minon *et al.* (1988)
	Omaliko (1983)
	Van Niekerk *et al.* (1989)
	Lowe (1976)
	Barnes (1966)
Paspalum notatum	Rethman & De Witt (1991)
Pennisetum clandestinum	Rethman & Gouws (1973)
	Rethman & De Witt (1988)
	Rethman & De Witt (1991)
	Rethman & De Witt (1993)
	Zacharias *et al.* (1991)
	Barnes & Dempsey (1993)
Pennisetum purpureum	Omaliko (1983)
Setaria sphacelata	Louw (1976)
Subtropical legumes	
Stylosanthes hamata	McCown *et al.* (1981)
	Wall & McCown (1989)
Stylosanthes humilis	Sturtz & Parker (1974)
Neonotonia wightii	Pizarro *et al.* (1985)
Temperate grasses	
Agropyron desertorum	Sneva *et al.* (1973)
Dactylis glomerata	Sheenan *et al.* (1985)
	Imura *et al.* (1976)
	Archer (1974)
Festuca arundinacea	Sheenan *et al.* (1985)
	Matches & Trevis (1973)
	Archer (1974)
Lolium perenne	Smith (1985)
	McCallum (1991)
Temperate legumes	
Trifolium pratense	Sheenan *et al.* (1985)

quality of foggage which accumulates will vary from year to year. The closing dates of pastures required for foggage should therefore be staggered. The amounts of N applied at closing should also be varied, based on the potential of the plants to grow for the remainder of that growing season. This will result in pastures with differing amounts of material of differing quality which can be allocated according to the needs of different classes of animals.

As a general rule, the earlier a pasture is put-up for foggage the greater is the amount of foggage produced (Table 13.2). Also, the more the pasture is used during summer for grazing or for hay or silage, the greater will be the amount of N that will be required to ensure that a reasonable amount of foggage is produced (Table 13.3).

Table 13.2 Effect of put-up date on foggage production (tons/ha) of a subtropical and a temperate pasture at Cedara (after Brockett 1983).

Species	Put-up date		
	1 January	*1 February*	*1 March*
Kikuyu	6.0	4.2	1.1
Tall fescue	3.5	2.5	0.6

Table 13.3 Expected foggage production and nitrogen requirement for Smuts' fingergrass pastures for different put-up dates at Kokstad.

Put-up	Expected foggage production (tons/ha)	Annual N requirement (kg/ha)
Mid-December	4.5 to 6.0	100
End of January	3.5 to 5.0	150 to 200
End of February	0.5 to 1.5	150

Several years of research at the Kokstad Research Station have indicated that Smuts' fingergrass pastures which are put-up after the end of January produce low foggage yields (Table 13.3). Where the foggage is being produced for cattle, pastures put-up after January will have a very low carrying capacity. However, where the foggage is being produced to feed sheep through the winter, a late put-up is necessary. This ensures a short leafy foggage which the sheep prefer. Provided moisture is not limiting, put-up dates in warmer areas may be as much as a month later than at Kokstad to achieve the levels of production indicated in Table 13.3.

What is gained in quantity through early put-up will be balanced by a loss in quality since the earlier the put-up date, the lower the quality of foggage produced. Any pasture therefore needs to be managed to achieve an appropriate balance between the quantity and quality of foggage required by the types and classes of animal to be fed.

13.4.4.3 *Quality of foggage*

In general, the earlier the pasture is put-up the lower will be its quality (Table 13.4). More specifically, foggage quality is, in most instances, inversely related to the length of the growing season following the closure of the pasture. Generally, the longer this period is, the poorer will be the quality of foggage produced, although this is not necessarily so for all species. In tall fescue (*F. arundinacea*) pastures the soluble carbohydrate content of the herbage rises in autumn. This enhances palatability and nutritive value at this time (Sheenan *et al.* 1985; Bartholomew *et al.* 1991). The suitability of the climate for plant growth is also important in determining foggage quality. For example, the quality of foggage on two pastures of the same species, closed at the same time and with the same management prior to closing, may differ significantly due to different growing conditions (e.g. moisture availability) after closing the pasture. This is so because plants which, due to unfavourable growing conditions, do not reach physiological maturity prior to the onset of the dormant season will have lower fibre and higher protein and non-structural carbohydrate contents (and thus higher quality and nutritive value) than those plants which do reach maturity.

Table 13.4 The effect of put-up date on the crude protein (% CP) and crude fibre (% CF) content of foggage produced from a subtropical and a temperate pasture species at Cedara (after Brockett 1983).

Species	% CP			% CF		
	Jan	Feb	March	Jan	Feb	March
Kikuyu	5.8	6.7	8.7	35.5	34.5	32.5
Tall fescue	7.6	9.6	11.5	32.5	29.5	25.9

The quality and acceptability of foggage also varies among different species. That produced from temperate species during autumn, for example, is comprised mainly of leaf material whereas as much as 50% of foggage produced by subtropical species at this time may comprise stem material.

Where there is a relatively short growing period after a pasture is put-up, the foggage is usually leafy and of high quality and so is suitable for young or producing livestock. A long growth period prior to the dormant season produces a high yield of poorer quality foggage, suitable for the maintenance of older, non-producing livestock.

Several workers have suggested that herbicides may be used to stop plant growth in early autumn and thus increase the winter quality of foggaged C_4 (tropical and subtropical) pastures. The herbicide Paraquat (1.1' dimethyl-4, 4' bipyridinium dichloride) has been used for this purpose (Rethman & Gouws 1973; Meaker & Coetzee 1978). Such a desiccant will artificially cure the herbage, which then retains most of the nutrients which would otherwise have leached out or have been lost through the processes of respiration and senescence. Sneva (1967) reported that the crude protein content of treated herbage declined by 10% following the application of a desiccant compared with a 50% decline in untreated material. One might nonetheless expect rain to reduce the quality of material cured in this way.

13.4.4.4 Fertilization

Of the fertilizers applied, N is the most important and, provided the other nutrients (especially K and P) are in adequate supply, has a major effect on foggage production (Hughes 1955; Rethman 1983; Bartholomew *et al.* 1991). Both the amount of N applied and its timing are important in regulating the quality and quantity of foggage produced.

Quantity

There is a reduced response to N in autumn in both temperate (Corbett 1957; Hart *et al.* 1969) and tropical pastures (Nash & Tainton 1975) relative to their response in spring. The amount of N which should be applied depends on the species concerned and the potential for growth following the closure of the pasture. In the United Kingdom the general guideline for foggaged pastures is 60 kg N/ha for cocksfoot pastures and 80 kg to 110 kg N/ha for perennial ryegrass and tall fescue pastures when they are closed in August (Cooper & Morris 1973; Clarke 1975). However, should the summer rains have been poor, or where there is a high proportion of clover in the sward, less N should be applied (Cooper & Morris 1973).

When the objective is to maximise foggage production in tropical species, there appears to be no advantage to applying more than about 40 kg to 67 kg N/ha when closing the pasture. In the Mpumalanga Highveld, applications of 0 kg, 50 kg and 100 kg N/ha, when closing kikuyu pastures in January, resulted in foggage yields of 3.2 t, 10.1 t and 11.7 t DM/ha at Nooitgedacht, and 3.2 t, 4.9 t and 6.3 t DM/ha at Athole (Rethman 1983). In the drier phase of the Highland Sourveld of KwaZulu-Natal, put-up date had a significant effect on the efficiency of use of N by *D. eriantha* pastures grown for foggage (Fig. 13.5). When the pasture was put-up at the beginning of January annual applications of N/ha had to be doubled to produce similar yields of dry matter per hectare to pastures which were put-up three weeks earlier (Hardy 1994).

A potential advantage of applying high levels of N at put-up, besides maximising foggage production, is that the pasture tends to grow more vigorously in the following spring from residual N. This effect has been observed in both temperate pasture species (Wedin *et al.* 1970) and in subtropical pasture species (Hardy 1994).

Quality

For temperate species, increasing the level of N applied at closing improves the crude protein content of the foggage (Corbett 1957; Baker *et al.* 1961), but there is some uncertainty regarding the influence of N applied in late summer/autumn on the quality of foggage in tropical pastures. Hart *et al.* (1969) reported that neither the crude protein (CP)% nor the digestibility of *Cynodon dactylon* foggage were influenced by autumn applications of N. In contrast, the CP% in both Smuts' fingergrass (Rethman & Gous 1973) and kikuyu grass foggage (Table 13.5) has been reported to increase with increased levels of N applied at put-up.

Alder (1954) and Corbett (1957) report that applications of P and K at put-up had no significant effect on the quality of foggage in temperate pastures.

13.4.5   Grazing management

Foggage is presented to animals as 'standing hay'. In many instances lodging occurs and the foggage appears to be of little value as an animal feed. This is particularly so with subtropical species. However, there are often differences in the quality of herbage in the upper and lower canopy of foggaged pastures. In areas prone to heavy frosts, the upper canopy may be frosted while leaves at ground level may remain green. Such green herbage will provide a feed of higher quality than the frosted herbage. Foggaged kikuyu pastures often exhibit this characteristic. Also, a higher quality of feed is available in the leaf than in the stem material (Rethman & De Witt 1991; Hardy 1994). Graz-

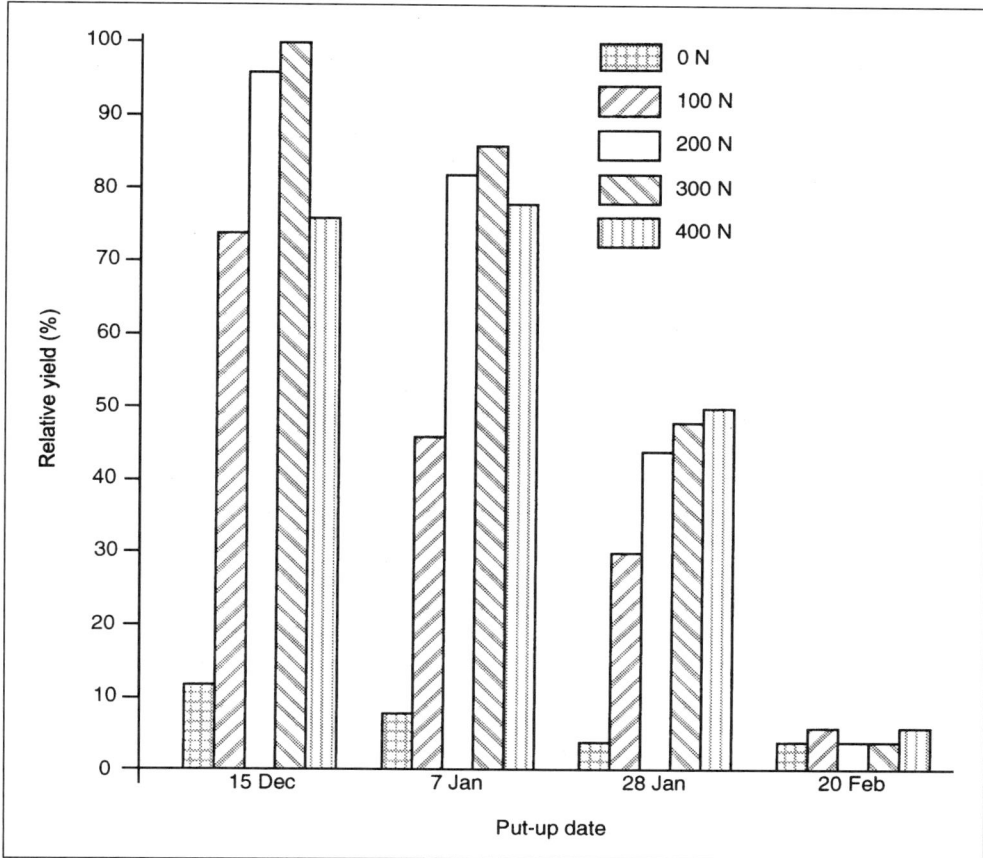

Figure 13.5 Interaction between put-up date and nitrogen on the yield of foggage from *Digitaria eriantha* at Kokstad.

Table 13.5 Per cent crude protein (% CP) in kikuyu foggage in autumn (mid-April) and early-winter (mid-June) for each of 4 levels of nitrogen (N) applied when the pasture was closed (after Rethman & Gouws 1973).

N (kg/ha)	% CP	
	mid-April	*mid-June*
68	5.9	5.6
136	7.9	7.0
204	7.2	9.5
272	10.4	8.2

ing of foggaged pastures therefore needs to be planned to ensure the most efficient use of the different types of material available on such pastures.

Most perennial pasture species can be grazed heavily during the winter without significantly influencing their future production. However, heavy grazing of foggaged subtropical pastures in autumn, before the plants become dormant, reduces the amount of photosynthetically active green leaf material and results in poor spring regrowth (Hardy 1994).

In utilising foggage, strip grazing or some other form of high intensity grazing is commonly recommended (Corbett 1957; Cooper & Morris 1973; Bartholomew *et al.* 1991). Strip grazing allows for efficient use of the available material and it allows this material to be rationed. A back-fence is not normally necessary when strip grazing dormant, subtropical pastures (Bartholomew *et al.* 1991) whereas such a fence should be used when strip grazing temperate pastures because they are normally actively growing at this time (Corbett 1957).

Continuous grazing has the advantage of requiring little capital outlay and it has a low labour requirement, but normally results in much wastage. Animal performance is normally initially higher when foggaged pastures are grazed continuously than when strip grazed since the animals will initially select the most palatable, and usually the most nutritious, components of the herbage. Thereafter performance will decline as they are forced to graze the less palatable, less nutritious residual herbage.

Lyle (1994) applied a creep grazing system involving a combination of strip and continuous grazing in a ewe/lamb production system on foggaged kikuyu pastures at Kokstad. The ewes were confined to their allocated strip while the lambs had access to the whole pasture. This practice improved lamb performance at weaning compared to either strip or continuous grazing.

Even if the foggage is strip grazed, as much as 25% to 35% of dry matter on offer is likely to be wasted. Such wastage results from trampling and fouling of the material by the animals, the decomposition of some of the herbage, and is directly linked also to the often poor acceptability and palatability of the stem material. If the foggaged material carries any dust the animals will find it unacceptable and they will tend to reject the mature dried material when new growth appears in late winter/early spring.

13.4.6 Animal performance

Selected foggage species, if appropriately managed, can provide forage capable of supporting growing animals and dry stock with little or no supplementation. Foggage produced from grasses such as tall fescue and perennial ryegrass have successfully supported ewes and lambs during the autumn and winter months, with highly acceptable lamb weaning weights and the maintenance of ewe condition (Lyle 1994). Cocksfoot pastures can also be used as foggage for autumn lambing (Lyle pers. comm.) and for cattle production systems (Hughes 1955; Corbett 1957). Mass gains of 0.2 kg to 0.6 kg per head per day have been recorded from weaner steers grazing foggaged tall fescue pastures (Louw 1991).

Animal performance on foggage produced from subtropical pastures varies widely according to put-up date, time of use of the foggage and local environmental conditions. Excellent animal performance has been obtained from kikuyu and Smuts' fingergrass pastures grazed before the grass has frosted off and become dormant. In high rainfall relatively frost-free areas, heifers on foggaged kikuyu pastures have gained 0.6 kg per head per day (Bartholomew *et al.* 1991). Rethman & De Witt (1991) reported gains of 60 g to 80 g per sheep per day on foggaged kikuyu and paspalum pastures in the Mpumalanga Highveld. In general, however, subtropical pastures which are put-up early in the growing season provide no more than maintenance diets during the winter months, whilst those put-up later in the growing season allow for modest livemass gains (Barnes 1966; Rethman & Gouws 1973; Rethman 1983; Dannhauser 1988; Rethman & De Witt 1991; Barnes & Dempsey 1993; Hardy 1994).

13.4.7 Advantages and disadvantages of foggage

Advantages

(a) There are no transport or handling costs, as with hay and silage. This represents a saving on machinery and labour;
(b) it provides out-of-season grazing;
(c) excess summer growth, which might otherwise be wasted, can be saved for winter use;
(d) production, which is based mainly on summer rain, is reasonably dependable, whereas satisfactory production of winter greenfeed (e.g. oats) is not;
(e) foggage produced from subtropical pastures will at least maintain animal mass through winter without supplementation, thus providing a saving on bought feeds. Foggaged temperate pastures can provide very good quality grazing; and
(f) in sourveld areas, foggaged subtropical pastures provide high quality grazing for weaner cattle and autumn lambing ewes during the autumn (before the plants become dormant).

Disadvantages

(a) For subtropical pastures the foggage is useful only until active growth commences in spring. Delays in removing animals from the foggaged pastures after spring growth has commenced will lead to pasture deterioration and poor animal intake;
(b) under dry-land conditions little foggage will accumulate if insufficient rain falls during the late-summer and autumn period;
(c) unseasonal winter rain may substantially reduce the quality of foggage produced from subtropical species; and
(d) losses due to weathering and trampling can be high and such pastures are highly susceptible to wild fires.

14 Fodder production planning and livestock production systems

J.R. Klug, J.M. Van Heerden & A.W. Lishman

CONTENTS

14.1 FODDER PRODUCTION PLANNING

14.1.1 Matching fodder supply to animal demand

The objective of fodder planning is to match the feed producing capability of the farm to the animal system such that the greatest margin over feed costs (the difference between feed costs and production income) is made by the livestock enterprise. Generally one should aim to produce the entire feed requirements of the livestock from farm-grown forages with the minimum supplementation at the least possible cost. This entails matching fodder supply to fodder demand, both of which can be modified in practice, at least within reasonable bounds.

The efficiency of any pastoral system is primarily dependent on the following factors:
(a) the amount, quality and seasonal distribution of the feed grown. This is first and foremost a function of soil type, the amount and seasonal distribution of rainfall, the availability of irrigation, and the average monthly temperatures – particularly during winter in the summer rainfall regions;
(b) the proportion of this feed consumed by the animal; and
(c) the efficiency with which the animal uses the food which it consumes (conversion rate).
Only the first two of these factors will be considered here.

14.1.1.1 Modification of feed supply

Within the bounds set by the natural resources of each individual farm and the economic situation, seasonal feed supply may be evened out in one or a number of the following ways:
(a) by selecting species which will provide a seasonal spread of forage;
(b) the strategic use of fertilizer to boost production when growth would normally be slow;
(c) irrigation during periods of moisture stress;
(d) the conservation of forage as hay, silage or foggage;
(e) grazing management, which will affect both the quantity and quality of forage produced; and
(f) the purchase of feeds to supplement quality (as in a dairy herd) or as a last resort, for periods of short supply. In the normal course of events, if feed must be bought in to boost the dry matter produced on the farm, then either the stocking rate is too high or the choice of enterprises is unsuitable and probably not financially viable.
It is possible to manipulate these variables to produce a variety of different fodder flow patterns at varying cost. The system that is eventually selected should largely be determined by the expected returns from the livestock enterprise.

14.1.1.2 Modification of the seasonal patterns of fodder demand by animals

The seasonal requirements for feed on any farm will depend primarily on the combination of livestock enterprises selected. If fresh milk production is selected as the major enterprise, the opportunity for manipulating seasonal demands of both quantity and quality of feed required is very restricted. Here only the size of the herd and the planned average milk yield can be used to modify the seasonal demand for forage. If beef, sheep, goats or game form the main enterprise, then there is considerable scope for seasonal manipulation of forage demand. In these livestock enterprises seasonal forage demand will depend on the following factors:
(a) age and time of year at which animals are sold/culled;
(b) date of calving, lambing or kidding;
(c) weaning date;
(d) planned ADG of calves, lambs or kids;
(e) ratio of dry animals to breeding animals; and
(f) stocking rates and species mixes.
By manipulating one or a number of these factors, seasonal forage demand can be greatly altered. This allows the fodder demand pattern to be matched to the pattern of fodder supply, determined by the nature of the resources available on the property under consideration.

14.1.1.3 Procedure used to match fodder supply and fodder demand

The procedure used to match fodder supply to fodder demand is largely one of trial and error and, although more sophisticated methods are available, the former method gives the planner the greatest degree of control over the options being tested and, therefore, the greatest opportunity of arriving at a sensible solution. The method involves four steps:

(a) Establish the current and potential long term sustainable carrying capacity of the farm. This should allow the natural resources to be maintained in a productive state. The simplest method of doing this is to use the concept of an animal unit (AU). This equates to an animal with a mass of 450 kg and a dry matter (DM) intake of 10 kg/day or 3.65 t/annum. Lactating dairy cows have an additional DM requirement of 50% or more of that of an AU, i.e. 15 kg/day or 5.5 t/annum, while other lactating females, such as beef cows, require 12.5 kg/day or 4.5 t/annum/AU. The AU equivalent of animals of different mass may also be calculated from the equation:

$$\text{AU equivalents } = \frac{M^{0.75}}{450^{0.75}} \times CF$$

where M = mass of the animal (kg);
CF = 1.50 for lactating dairy cows;
1.25 for lactating non-dairy cows; and
1.00 for all other animals.

Where veld forms a component of the system, the first step is to assess its condition and determine its current grazing capacity (refer to Morris, Hardy & Bartholomew, Chapter 7, section 7.3 in *Veld Management in South Africa*, 1999). The arable land which will be used to contribute to the livestock feed supply (pastures, hay, silage) must then be considered. To establish the number of AU's that his land can carry, divide the expected DM grazeable yield of this land by 3.65. This, added to the carrying capacity of the veld, would give the total carrying capacity of the farm.

The co-efficient of variation (CV) of mean annual rainfall gives a meaningful estimate of the amount of extra conserved fodder needed to provide for years of below average rainfall (Jones *et al.* 1989; Schulze 1997). The CV for a farm is obtained from the following general equation:

$$CV\% = -0.02261 \text{ MAP} + 42.3 \approx \frac{640}{\sqrt{MAP}}$$

where MAP = Mean annual precipitation.

If, for example, the CV of a property which receives 1 000 mm rainfall per annum is 20%, then the annual rainfall can be expected to vary between 800 mm and 1 200 mm. The carrying capacity of the property should be adjusted to accommodate this variability of annual rainfall and thus the carrying capacity should be reduced by ¼CV%, i.e. 5% in the above example. This means that in years of just below average rainfall, the farm will still produce an adequate amount of food, and in good years the surplus should be conserved to act as a fodderbank for years of severe drought. The *South African Atlas of Agrohydrology and -Climatology* (Schulze 1997) provides detailed information about rainfall distribution patterns, variability, and concentration, as well as estimates of primary production and inter-annual variation of primary production. These data were compiled for the whole of South Africa on a 1'x1' latitude by longitude grid. For KwaZulu-Natal more specific data, such as carrying capacities of veld and yield estimates of cultivated pastures, may be obtained through the Bio-Resource Programme from the Department of Agriculture at Cedara. As an example, the CV for KwaZulu-Natal varies from 15% for very wet regions to 30% for dry areas.

The above calculations may be undertaken using a variety of levels of land development and land use combinations to which a property is suited. This provides a range of carrying capacities (and/or enterprise mixes) for the farm whose economic viabilities can then be compared to ensure that the best long-term land use strategy is selected.

(b) Identify herd or flock feed requirements. Initially a stockflow is drawn up giving the month-by-month herd composition of the livestock enterprise being considered. It is convenient to describe this proposed herd in terms of a 100 parous female module and to base the assumptions of expected herd performance levels on the requirements and experience of the landowner or an experienced adviser, always bearing in mind the characteristics of the individual farm. Animals are usually grouped into age or weight classes and the number of AU equivalents in each class is determined for each month. The average herd sizes for one module of common livestock production systems are given in Table 14.1. By dividing the average monthly total AU's of the herd into the number of AU equivalents which the property can support (carrying capacity), the number of cows, and therefore total animal units, can be determined. For example, if the property is estimated to have a carrying capacity of 350 AU equivalents, and if Herd 7 from Table 14.1 is selected (210 AU), then the number of cows (and their followers) that the property can support will be:

$$\frac{350}{210} \times 100 = 167 \text{ cows} \equiv 350,7 \text{ AU}$$

This gross estimate of the form that the livestock enterprise will take should now be adjusted according to the specific dry matter, energy and protein requirements of the different classes of animals in the herd. These are usually expressed as kilograms or tons of dry matter (DM), mega or giga-joules of metabolisable energy (ME), and crude protein (CP) or digestible crude protein (DCP) in kilograms, g/kg DM, or percentage of DM. The exact structure of the herd or flock for each month of the year should be determined and a decision made on the production requirement from each class of animal. At the same time the estimated production, and so the expected gross returns from the sale of the products of the enterprise, can be estimated. Various methods can be used to estimate the amounts of DM, TDN and DCP which this herd requires. Feeding standards published by the ARC (Anon. 1980) and the NRC (Anon. 1984, 1985, 1989) are available for this purpose, as are local standards published by Stewart & Dugmore (1995) for dairy cattle and by Lesch *et al.* (1974) for beef cattle. Some generalised equations for determining DM and energy requirements are presented below, without detailed explanation, for those wishing to use them.

For lactating beef cows:

$DM = 111.47 + 0.05M + 22.7T - 4.28T^2$

For their suckling calves:

$DM = 0.182M + 29.67T - 144.27$

For non-lactating cattle:

$DM = 61 + 0.157M (2.7 + ADG)$

For lactating dairy cows:

$DM = 304 + 4.256M^{0.75} + 21.3BF - 6.14YA + 100\sqrt{YA}$ kg/month

Table 14.1 Total dry matter requirements (tons/month, concentrates & wastage included), and herd size (AU, concentrates excluded) per 100 parous female module for common livestock production systems in KwaZulu-Natal.

1	Beef, 80% calving, weaners & cull cows sold at weaning.
2	Beef, 70% calving, weaners & cull cows sold at weaning.
3	Beef, 80% calving, feedlot or onto ryegrass at weaning.
4	Beef, 70% calving, feedlot or onto ryegrass at weaning.
5	Beef, 80% calving, weaners overwintered, sold at 12 months.
6	Beef, 70% calving, weaners overwintered, sold at 12 months.
7	Beef, 80% calving, selling stores off veld at 20 months.
8	Beef, 70% calving, selling stores off veld at 20 months.
9	Beef, 80% calving, feedlot/ryegrass at 20 months for 120 days.
10	Beef, 70% calving, feedlot/ryegrass at 20 months for 120 days.
11	Beef, 80% calving, selling stores off veld at 30 months.
12	Beef, 70% calving, selling stores off veld at 30 months.
13	Beef, 100 weaners into feedlot/ryegrass for 8 months – no cows.
14	Beef, 100 stores – 20 months – into feedlot/ryegrass – no cows.
15	Dairy, Friesland, 12 ℓ/day.
16	Dairy, Friesland, 15 ℓ/day.
17	Dairy, Friesland, 18 ℓ/day.
18	Dairy, Friesland, 20 ℓ/day.
19	Sheep, spring mating, selling lambs at 35–40 kg.
20	Sheep, autumn mating, selling lambs at 35–40 kg.
21	Sheep, spring mating, remating skips in autumn, selling lambs at 35–40 kg.

Herd	Jul	Aug	Sep	Oct	Nov	Dec	Jan	Feb	Mar	Apr	May	Jun	Total	AU
1	40.3	46.7	54.0	58.7	59.5	64.0	67.0	67.6	63.4	62.9	39.8	38.8	663	163
2	40.3	45.9	52.2	56.5	56.9	61.0	63.7	63.8	60.6	60.0	39.8	38.8	639	158
3	63.1	71.1	79.3	86.3	87.8	64.0	67.0	67.6	63.4	62.9	59.3	59.4	831	198
4	58.9	66.2	73.6	80.1	81.4	61.0	63.6	63.8	60.6	60.0	59.2	55.3	784	187
5	58.3	65.5	53.8	58.5	59.4	63.8	66.9	67.6	63.3	62.8	56.0	55.4	731	177
6	55.3	61.7	52.0	56.3	56.8	60.9	63.6	63.7	60.5	59.9	53.3	52.6	697	169
7	57.0	64.3	72.0	78.1	79.3	85.2	89.2	88.8	87.3	83.7	54.7	54.1	894	210
8	54.2	60.6	67.2	72.7	73.4	78.8	82.2	81.5	80.6	77.4	52.1	51.5	832	196
9	81.6	89.4	71.1	77.0	78.1	83.9	87.7	87.3	85.6	85.1	78.1	77.4	982	230
10	74.7	81.6	66.4	71.8	72.4	77.6	81.0	80.2	79.1	78.5	71.7	71.0	906	213
11	76.5	83.8	91.0	97.8	99.7	106.3	110.4	107.9	89.6	88.6	74.0	73.0	1099	267
12	70.4	76.8	83.0	89.0	90.5	96.5	100.0	97.5	82.8	81.8	68.2	67.2	1004	247
13	27.5	29.7	31.1	34.1	35.4	38.8	0.0	0.0	0.0	0.0	20.4	23.1	265	54
14	29.0	31.2	32.5	0.0	0.0	0.0	0.0	0.0	0.0	0.0	24.6	26.1	143	29
15	70.9	69.0	71.7	78.9	76.5	76.7	76.7	70.6	78.9	74.3	76.7	71.7	893	181
16	67.4	66.0	67.7	73.4	71.1	72.1	72.1	66.0	73.4	69.9	72.1	67.7	839	170
17	65.0	63.6	65.1	71.3	69.1	69.7	69.7	64.2	71.3	67.6	69.7	65.1	812	165
18	62.5	61.8	61.6	65.7	63.7	64.7	64.7	59.2	65.7	62.8	64.7	61.6	759	155
19	9.7	9.7	7.6	6.0	4.7	4.9	4.9	6.0	7.4	6.6	7.0	8.6	83	24
20	7.8	6.9	7.5	9.4	10.7	10.8	8.1	6.2	4.7	4.9	4.9	6.2	88	25
21	9.9	8.1	6.6	5.7	5.5	5.2	6.0	7.5	6.7	7.0	8.6	10.0	87	25

$$ME = 1056 + 11.2M + 44.4M^{0.75} + 26.4YA + 3192BK \text{ MJ/month}$$

For suckling lambs:

$$DM = 1.52M + 55.80ADG - 8.81$$

For lactating ewes:

$$DM = 0.97M + 21.9$$

For non-lactating sheep:

$$DM = 0.57M + 167.9ADG - 0.43$$

where: DM = DM requirement/month/animal, kg
            M = livemass, kg
      ADG = average daily gain, kg
            T = calf age, months
          BF = butterfat % in milk

YA = milk yield, ℓ/day

BK = butterfat kg/cow/day = $\dfrac{\text{YA x BF}}{100}$

Further details may be obtained from Jones *et al.* (1989) and Stewart & Dugmore (1995).

From calculations using these standards, it is possible to determine more precisely the monthly intake requirements of DM, TDN and DCP for the livestock enterprise. To ensure that the required DM intake is achieved, it is necessary to provide for some wastage. The amount of food wasted will depend on a number of factors (e.g. animal type, feed type, management system, etc.) and typically ranges from 10–30%. The DM intake requirement must therefore be increased by this amount. This represents the required feed flow. Computer programs are available[1] to perform these calculations for the common beef, dairy and sheep production systems.

(c) The farm's forage production capability is now assessed more critically on the basis of the appropriate forage species which can be grown to suit the animal enterprises selected, and the management skills and finance which are available on the property. The objective should in general be to produce the entire feed requirements of the livestock from farm-grown forages with the minimum supplementation at the least possible cost.

A complete survey of the farm's fodder-producing potential should involve an assessment of the seasonal availability of feed from veld and from areas suited to different pasture and forage crops. Such an assessment should include an examination of the soil's potential to support different crops and pastures and of the irrigation potential of the property. On the basis of land potential and bearing in mind the feed requirements of the planned animal enterprise, suitable types of cropping and pasture programmes can be established. This requires some knowledge of growth patterns and the yield of the different pasture species. An example of the type of information which has been provided for the high rainfall areas of the KwaZulu-Natal Midlands is presented in Tables 14.2 and 14.3. These data can be used to assist in the selection of species for the forage production programme and in determining the required area of each species. Yield prediction equations are available for some of the forages, but should be used with discretion. In these equations, the following symbols are used:

D = number of defoliations per season;
F = irrigation flag: 1 = irrigated, 0 = dryland;
N = kg N/ha/year;
P_y = mm precipitation plus irrigation per year;
P_{sa} = mm precipitation plus irrigation, September to April;
S = soil factor according to Edwards & Scotney (1978);
T = average daily temperature °C, September to April;
T_{a-} = average minimum temperature °C in April;
T_{s-} = average minimum temperature °C in September;
T_{s+} = average maximum temperature °C in September;

---

[1] For more information about these, and other, computer programs contact J.R. Klug, Faculty of Science and Agriculture, University of Natal, P/Bag X01, Scottsville 3209.

Table 14.2　Possible forage yields in KwaZulu-Natal.

Crop	Harvest stage	DM %	Area 1	2	3	4	6	8	7, 9, 10	Remarks
BABALA	Dough stage	100	9–12	8–10	9–14	8–12	—	5–9	—	
	60–80 days	100	6–9	5–9	6–10	6–10	6–10	2–5	—	
	30–50 days	100	3–6	2–5	3–6	3–6	3–6		—	
JAPANESE RADISH	Full growth	100	—	—	3–5	3–5	3–5	3–5	—	Frosty areas only. Lack of cold leads to bolting.
TURNIPS	Full growth	100	—	—	3–5	3–5	3–5	3–5	—	
CABBAGE	Residues	100	5–6	5–6	5–6	5–6	5–6	5–6	—	Mainly an irrigated crop but is grown dryland in areas 3 & 4.
	Whole crop	100	10–15	10–15	10–15	10–15	10–15	10–15	—	
SOYABEANS Whole crop	Full pod pre-drying	100	6–8	6–8	6–8	6–8	6–8	6–8	—	
Residues	Post harvest	100	2–4	2–4	2–4	2–4	2–4	2–4	2–4	
COWPEAS Whole crop	Full pod	100	4–5	4–5	4–5	4–5	4–5	4–5	—	
KIDNEYBEAN Residues	Post harvest	100	2–3	2–3	2–3	2–3	2–3	2–3	—	
BUSHBEANS Residues	Post harvest	100	1–2	1–2	1–2	1–2	1–2	1–2	—	
SUNFLOWERS Whole	Well in seed	100	5–6	5–6	5–6	5–6	5–6	5–6	—	
Residues		100	2–3	2–3	2–3	2–3	2–3	2–3	—	
LUCERNE Dryland	Pre-bloom to 10% bloom	85	10–15	8–10	5–10	10–12	5–10	10–12	5–6	Soils must be well fertilized and not acid.
Irrigated	Full flower		20–30	20–30	10–15	10–15	10–15	10–30	20–30	

Possible yields (t/ha/a) under good management by bioclimatic area

Table 14.2 Continued

Crop	Harvest stage	DM %	Possible yields (t/ha/a) under good management by bioclimatic area							Remarks
			Area							
			1	2	3	4	6	8	7, 9, 10	
WHEAT										
Dryland	Pre- to early	100	1–3	1–3	1–3	1–3	1–3	1–3	1–3	
Irrigated	piping	100	5–6	5–6	5–6	5–6	5–6	5–6	5–6	
OATS										
Dryland	Pre- to early	100	2–4	2–4	2–4	2–4	2–4	2–4	2–4	
Irrigated	piping	100	5–6	5–6	5–6	5–6	5–6	5–6	5–6	
RYE										
Dryland	Pre- to early	100	—	2–4	2–4	2–4	2–4	2–4	—	
Irrigated	piping	100	—	3–6	3–6	3–6	3–6	3–6	—	
BARLEY										
Dryland	Pre- to early	100	—	1–2	1–2	1–2	1–2	2–3	—	
Irrigated	piping	100	—	2–3	2–3	2–3	2–3	2–3	—	
MAIZE										
Dryland grain	12.5% moisture	87.5	6–7	5–6	7–8	7–8	6–7	6–7	3–4	
corn + cob		87.5	7–8	6–7	8–9	8–9	7–8	7–8	4–5	
whole ear		100	7–8	6–7	8–9	8–9	7–8	7–8	4–5	
silage	blister	100	8–9	8–9	9–10	9–10	8–9	8–9	4–5	Lowest yields and quality
	milk	100	9–10	9–10	10–11	10–12	9–10	9–10	5–6	
	dough	100	12–13	12–13	12–14	12–14	11–12	11–12	5–6	
	glazed	100	13–14	13–14	14–15	14–16	13–14	13–14	6–7	Highest yields and quality
stover	after grain	100	6–8	6–8	7–9	8–10	6–8	6–8	3–4	
Irr. grain	12.5% moisture	87.5	12–14	12–14	12–14	12–14	12–14	12–14	12–14	
corn + cob		87.5	14–16	14–16	14–16	14–16	14–16	14–16	14–16	
whole ear		100	15–17	15–17	15–17	15–17	15–17			
silage	silk	100	10–12	10–12	10–12	10–12	10–12	10–12	10–12	Lowest yields and quality
	blister	100	12–14	12–14	12–14	12–14	12–14	12–14	12–14	
	milk	100	14–18	14–18	14–18	14–18	14–18	14–18	14–18	
	dough	100	20–25	20–25	20–25	20–25	20–25	20–25	20–25	
	glazed	100	25–30	25–30	25–30	25–30	25–30	25–30	25–30	Highest yields and quality
stover	after grain	100	10–14	10–14	10–14	10–14	10–14	10–14	10–14	

Table 14.2 *Continued*

Crop	Harvest stage	DM %	Possible yields (t/ha/a) under good management by bioclimatic area							Remarks
			Area							
			1	2	3	4	6	8	7, 9, 10	
GRAIN SORGHUM										
grain	12.5% moisture	87.5	6–10	6–10	6–10	6–10	6–10	6–10	4–8	
stover		100	6–10	6–10	6–10	6–10	6–10	6–10	4–8	
FODDER SORGHUMS										
Dryland		100	4–10	4–10	4–10	5–11	4–10	4–10	3–8	
JAPANESE MILLET		100	4–6	4–6	6–8	6–8	5–7	5–7	—	
Kikuyu		100	12	12	12	12	8	5	—	
Nile grass		100	12	10	5–8	6	8–10	2–8	—	
Eragrostis curvula		100	15	12	20	9–12	8–12	3–4	—	
Star grass		100	11–12	9	5	5	12	10	5	
Coastcross II		100	13	12	8	6–11	12	10	4	
Digitaria eriantha (*Smuts'*)		100	12	12	—	10–11	10	10	3–4	
Veld		—	2–3	2–3	2–3	2	1–2	1–2	1	
Midmar ryegrass dryland		100	4–8	4–8	5–10	4–8	4–8	3–7	—	
Midmar ryegrass irrigated		100	12–15	10–12	10–12	9–10	10–12	10–12	10–12	
Other ryegrass dryland cultivars		100	4–5	4–5	5–6	4–5	4–5	3–4	—	
irrigated		100	6–10	6–10	6–10	6–10	6–10	6–10	6–10	
Cocksfoot/Fescue with clover		100	—	10–11	10	10	—	—	—	Low N
Red clover No N		100	—	8	7	5	—	—	—	
White clover No N		100	8	8	7	3	—	—	—	
Cocksfoot		100	—	8	10	7	6–8	—	—	
Fescue		100	8	8	9	7	7	—	—	
Cocksfoot/Fescue irrigated		100	10–12	9–13	10–12	10–12	10	10	—	3 year mean

Table 14.3 Expected growth curves and annual DM yields of forages in KwaZulu-Natal.

Pasture Description	Bio Grp	Kg N ha/a	Jul	Aug	Sep	Oct	Nov	Dec	Jan	Feb	Mar	Apr	May	Jun	Yield t/ha
KIKUYU (Pennisetum clandestinum)															
Cedara	3d	90	—	—	0.06	0.14	0.18	0.18	0.16	0.10	0.08	0.06	0.04	—	5.0
Cedara	3d	170	—	—	0.04	0.09	0.16	0.19	0.17	0.13	0.11	0.07	0.04	—	7.5
Cedara	3d	200	—	—	0.02	0.06	0.14	0.24	0.23	0.15	0.09	0.05	0.02	—	9.5
Cedara	3d	240	—	—	0.03	0.07	0.18	0.22	0.20	0.14	0.08	0.05	0.03	—	11.8
Cedara	3d	360	—	—	0.03	0.08	0.17	0.19	0.18	0.13	0.10	0.07	0.05	—	12.6
Lidgeton	3c	0	—	—	—	0.12	0.18	0.24	0.18	0.12	0.10	0.06	—	—	1.7
Kokstad	4f	120	—	—	—	0.04	0.13	0.23	0.23	0.15	0.13	0.09	—	—	5.3
Kokstad	4f	240	—	—	—	0.07	0.14	0.20	0.20	0.15	0.14	0.10	—	—	8.1
Midlands	4e	200	—	—	—	0.07	0.13	0.25	0.27	0.17	0.08	0.03	—	—	9.2
Mooi River	4e	200	—	—	—	0.06	0.14	0.27	0.26	0.16	0.07	0.04	—	—	8.5
Karkloof	3c	200	—	—	0.02	0.06	0.13	0.24	0.23	0.15	0.10	0.05	0.02	—	10.0
Cedara	3d	200	—	—	0.02	0.06	0.13	0.24	0.23	0.15	0.10	0.05	0.02	—	9.5
Haga Haga		250	0.01	0.04	0.08	0.12	0.15	0.15	0.13	0.10	0.08	0.06	0.05	0.03	8.7
Haga Haga		450	0.01	0.04	0.08	0.12	0.15	0.15	0.13	0.10	0.08	0.06	0.05	0.03	13.9
Haga Haga		650	0.01	0.04	0.08	0.12	0.15	0.15	0.13	0.10	0.08	0.06	0.05	0.03	18.0
Bathurst		200	0.03	0.03	0.06	0.10	0.12	0.08	0.05	0.05	0.13	0.17	0.13	0.05	8.7
Bathurst, cutting 30 days		300	0.03	0.04	0.10	0.03	0.06	0.08	0.09	0.14	0.16	0.11	0.11	0.05	7.9
Bathurst, cutting 60 days		300	0.04	0.02	0.07	0.11	0.10	0.09	0.07	0.05	0.12	0.16	0.12	0.05	11.6
Alexandria, cutting 30 days		300	0.06	0.01	0.03	0.05	0.13	0.11	0.06	0.05	0.19	0.14	0.12	0.05	5.1
Alexandria, cutting 60 days		300	0.04	0.02	0.03	0.05	0.10	0.15	0.10	0.05	0.11	0.16	0.12	0.07	7.1
Tsitsikamma	3d	300	0.03	0.01	0.06	0.11	0.08	0.09	0.08	0.15	0.17	0.15	0.05	0.02	14.7
KIKUYU OVERSOWN WITH RYEGRASS (Lolium multiflorum) (W=Westerwolds. I=Italian. D=Diploid. T=Tetraploid)															
Cedara, Exalta-ID, Irr., Mar.[1]	3d	550	0.08	0.09	0.11	0.15	0.08	0.12	0.02	0.05	0.10	0.08	0.09	0.03	17.4
Cedara, Billion-WT, Irr., Mar.	3d	550	0.06	0.09	0.12	0.13	0.06	0.14	0.02	0.10	0.11	0.07	0.08	0.02	17.6
Cedara, Tetrone-IT, Irr., Mar.	3d	550	0.04	0.06	0.11	0.18	0.06	0.14	0.03	0.07	0.11	0.10	0.09	0.01	16.6
Cedara, Midmar-WD, Irr., Mar.	3d	550	0.03	0.06	0.13	0.16	0.06	0.16	0.03	0.10	0.12	0.07	0.07	0.01	17.5
Tabamhlope, Exalta-ID, Irr., Mar.	4e	450	—	0.05	0.12	0.13	0.16	0.16	0.07	0.10	0.10	0.11	—	—	19.5
Tabamhlope, Billion-WT, Irr., Mar.	4e	450	—	0.03	0.14	0.12	0.16	0.15	0.07	0.11	0.12	0.10	—	—	18.4
Tabamhlope, Tetrone-IT, Irr., Mar.	4e	450	—	0.02	0.08	0.13	0.17	0.14	0.09	0.11	0.12	0.14	—	—	18.4
Tabamhlope, Midmar-WD, Irr., Mar.	4e	450	—	0.02	0.08	0.11	0.15	0.17	0.08	0.16	0.13	0.10	—	—	17.2
RYEGRASS															
Cedara, Midmar-WD, Irr., Feb.	3d	300	0.06	0.07	0.15	0.16	0.11	0.02	—	0.01	0.07	0.13	0.15	0.03	13.5
Cedara, Midmar-WD, Irr., Mar.	3d	300	0.06	0.08	0.17	0.19	0.13	0.02	—	—	0.02	0.09	0.16	0.08	10.6
Cedara, Midmar-WD, Irr., Apr.	3d	300	0.08	0.11	0.24	0.28	0.21	0.03	—	—	—	—	—	0.05	7.5

[1] Planting month

Table 14.3 *Continued*

Pasture Description	Bio Grp	Kg N ha/a	Proportion of total production												Yield t/ha
			Jul	Aug	Sep	Oct	Nov	Dec	Jan	Feb	Mar	Apr	May	Jun	
RYEGRASS (*continued*)															
Cedara, Midmar-WD, Dry, Feb.	3d	300	—	—	0.19	0.14	0.09	0.01	—	0.01	0.08	0.20	0.23	0.05	8.6
Cedara, Midmar-WD, Dry, Mar.	3d	300	—	—	0.23	0.14	0.07	0.02	—	—	0.02	0.14	0.31	0.07	5.7
Cedara, Midmar-WD, Dry, Apr.	3d	300	—	—	0.40	0.26	0.11	0.03	—	—	—	—	0.09	0.11	3.5
Cedara, Tama, Irr., Feb.	3d	300	0.08	0.10	0.18	0.15	0.08	0.03	—	0.01	0.05	0.12	0.10	0.10	12.4
Cedara, Tama, Irr., Mar.	3d	300	0.11	0.13	0.24	0.14	0.08	0.01	—	—	0.01	0.06	0.10	0.12	9.3
Cedara, Tama, Irr., Apr.	3d	300	0.11	0.14	0.26	0.15	0.08	0.01	—	—	—	—	0.11	0.14	8.7
Cedara, Tama, Dry, Feb.	3d	300	0.03	0.03	0.10	0.05	0.02	—	—	0.02	0.10	0.34	0.21	0.10	6.2
Cedara, Tama, Dry, Mar.	3d	300	0.04	0.04	0.11	0.04	—	—	—	—	0.08	0.34	0.24	0.11	5.3
Cedara, Tama, Dry, Apr.	3d	300	0.10	0.10	0.19	0.05	—	—	—	—	—	—	0.27	0.29	2.1
Kokstad, Midmar-WD, Irr., Feb.	4f	300	0.07	0.08	0.17	0.14	0.04	—	—	0.01	0.06	0.13	0.21	0.09	10.6
Kokstad, Midmar-WD, Irr., Mar.	4f	300	0.09	0.09	0.14	0.27	0.05	0.01	—	—	0.01	0.06	0.16	0.12	8.6
Kokstad, Midmar-WD, Irr., Apr.	4f	300	0.11	0.11	0.24	0.27	0.08	0.03	—	—	—	—	0.05	0.11	7.4
Kokstad, Midmar-WD, Dry, Mar.	4f	300	—	—	0.15	0.24	0.07	—	—	—	0.02	0.09	0.25	0.18	5.5
Kokstad, Midmar-WD, Dry, Feb.	4f	300	—	—	0.12	0.15	0.05	—	—	0.01	0.09	0.20	0.26	0.12	8.4
Kokstad, Tama, Irr., Feb.	4f	300	0.07	0.09	0.23	0.20	0.08	—	—	—	0.06	0.08	0.10	0.09	8.9
Kokstad, Tama, Irr., Mar.	4f	300	0.08	0.11	0.25	0.19	0.05	—	—	—	—	0.08	0.13	0.11	7.5
Kokstad, Tama, Irr., Apr.	4f	300	0.10	0.13	0.32	0.23	0.12	—	—	—	—	—	—	0.10	6.5
Kokstad, Tama, Dry, Feb.	4f	300	—	—	0.22	0.20	0.01	—	—	0.03	0.08	0.16	0.22	0.08	7.6
Kokstad, Tama, Dry, Mar.	4f	300	—	—	0.29	0.16	—	—	—	—	—	0.05	0.21	0.29	5.6
Kokstad, Tama, Dry, Apr.	4f	300	—	—	0.38	0.21	0.12	—	—	—	—	—	0.08	0.21	4.2
Tabamhlope, Midmar-WD, Dry, Feb.	4e	200	—	—	0.17	0.22	0.13	—	—	0.01	0.07	0.11	0.20	0.09	9.1
Dundee, Midmar-WD, Dry, Mar.	8a	250	0.09	0.07	0.34	0.16	0.13	—	—	—	—	0.01	0.05	0.15	10.7
Tabamhlope, Tama, Dry, Feb.	4e	200	—	—	0.23	0.30	0.14	0.04	—	—	0.10	0.10	0.08	0.01	7.9
Dundee, Tama, Dry, Mar.	8a	250	0.17	0.13	0.32	0.14	0.01	—	—	—	—	0.01	0.06	0.16	8.7
Cedara, plus clover	3d	100	0.07	0.09	0.12	0.16	0.11	0.05	0.05	0.04	0.05	0.07	0.10	0.09	11.4
Upland, Irr., Feb.		300	0.05	0.09	0.14	0.19	0.13	0.02	—	—	0.02	0.17	0.11	0.08	10.6
Bottomland, Irr., Feb.		300	0.02	0.09	0.14	0.19	0.19	0.02	—	—	0.02	0.17	0.11	0.05	10.6
Middlerus, Irr., Mar.	6	300	0.07	0.14	0.18	0.19	0.06	—	—	—	—	0.05	0.19	0.12	10.7
Middlerus, Dry, Feb.	6	300	0.05	0.09	0.14	0.19	0.13	0.02	—	—	0.02	0.17	0.11	0.08	10.6
Döhne, Midmar-WD, Irr.		1000	0.04	0.04	0.09	0.16	0.13	0.13	0.11	0.06	0.06	0.07	0.06	0.05	12.0
Döhne, Midmar-WD, Irr.		0	0.07	0.09	0.14	0.21	0.11	0.10	—	—	—	0.08	0.10	0.10	3.8
Döhne, Midmar-WD, Irr.		300	0.07	0.09	0.14	0.21	0.11	0.10	—	—	—	0.08	0.10	0.10	7.9
Döhne, Midmar-WD, Irr.		600	0.07	0.09	0.14	0.21	0.11	0.10	—	—	—	0.08	0.10	0.10	8.8
Bathurst, Midmar-WD, Dry, cutting 30 days		300	0.04	0.13	0.08	0.45	0.10	0.20	—	—	—	—	—	—	4.4
Bathurst, Midmar-WD, Dry, cutting 60 days		300	0.05	0.04	0.20	0.34	0.22	0.10	—	—	—	—	—	—	4.6
Bathurst, Ariki, Dry, cutting 30 days		300	0.16	0.10	0.02	0.49	0.23	—	—	—	—	—	—	—	2.1
Bathurst, Ariki, Dry, cutting 60 days		300	0.08	0.07	0.06	0.29	0.50	—	—	—	—	—	—	—	2.0
Alexandria, Midmar-WD, Dry, cutting 30 days		300	0.20	0.07	0.44	0.11	0.08	0.03	—	—	—	—	—	0.07	5.0

Table 14.3 *Continued*

Pasture Description	Bio Grp	Kg N ha/a	Proportion of total production												Yield t/ha
			Jul	Aug	Sep	Oct	Nov	Dec	Jan	Feb	Mar	Apr	May	Jun	
RYEGRASS *(continued)*															
Alexandria, Midmar-WD, Dry, cutting 60 days		300	0.08	0.12	0.22	0.31	0.18	0.05	—	—	—	—	—	0.04	7.1
Alexandria, Ariki, Dry, cutting 30 days		300	—	—	0.07	0.38	0.55	—	—	—	—	—	—	—	2.7
Alexandria, Ariki, Dry, cutting 60 days		300	—	—	—	1.00	—	—	—	—	—	—	—	—	2.1
COCKSFOOT *(Dactylis glomerata)*															
Cedara, average	3d	150	—	—	0.10	0.14	0.23	0.10	0.07	0.11	0.13	0.11	0.01	—	8.3
Cedara, Irr.	3d	300	—	—	0.07	0.15	0.17	0.09	0.09	0.11	0.15	0.11	0.06	—	12.3
Cedara, Irr., plus clover	3d	100	—	0.02	0.08	0.13	0.13	0.12	0.09	0.12	0.13	0.10	0.08	—	10.1
Kokstad, Newport	4f	225	—	—	0.04	0.11	0.19	0.04	0.09	0.15	0.21	0.13	0.04	—	5.4
Bathurst, Dry, cutting 30 days		300	0.02	0.03	0.35	0.14	0.30	0.03	—	—	—	0.02	0.05	0.06	4.2
Bathurst, Dry, cutting 60 days		300	0.08	0.01	0.10	0.20	0.16	0.12	—	—	—	0.07	0.11	0.15	5.5
Alexandria, Dry, cutting 30 days		300	0.01	0.01	0.30	0.31	0.32	0.02	—	—	—	—	—	0.03	3.4
Alexandria, Dry, cutting 60 days		300	0.02	0.02	0.18	0.33	0.25	0.18	—	—	—	—	—	0.02	4.6
Döhne, Irr.		1000	0.02	0.04	0.10	0.21	0.09	0.12	0.10	0.07	0.08	0.08	0.05	0.04	10.2
TALL FESCUE *(Festuca arundinacea)*															
Cedara, Festal, Irr., infreq, defol.	3d	300	0.01	0.06	0.13	0.16	0.06	0.04	0.13	0.23	0.13	0.02	0.02	0.01	12.0
Cedara, Festal, Irr., freq. defol.	3d	300	0.03	0.09	0.14	0.10	0.03	0.01	0.14	0.19	0.16	0.07	0.03	0.01	9.4
Cedara, K31, Irr., infreq. defol.	3d	300	0.02	0.07	0.16	0.18	0.09	0.02	0.12	0.20	0.10	0.02	0.01	0.01	12.9
Cedara, K31, Irr., freq. defol.	3d	300	0.02	0.09	0.18	0.22	0.03	0.01	0.11	0.16	0.11	0.05	0.01	0.01	9.9
Cedara, K31, Dry, v. freq. defol.	3d	280	—	—	0.09	0.17	0.15	0.04	0.10	0.12	0.10	0.13	0.09	0.01	11.7
Cedara, plus clover	3d	100	—	—	0.09	0.18	0.15	0.12	0.08	0.14	0.17	0.10	—	—	11.1
Bathurst, cutting 30 days		300	0.05	0.03	0.36	0.18	0.16	—	—	—	—	0.05	0.09	0.08	4.2
Bathurst, cutting 60 days		300	0.08	0.02	0.19	0.35	—	—	—	—	—	0.11	0.12	0.13	5.5
Alexandria, cutting 30 days		300	0.05	0.05	0.36	0.32	0.14	—	—	—	—	—	—	0.08	4.1
Alexandria, cutting 60 days		300	0.06	0.06	0.30	0.52	—	—	—	—	—	—	—	0.06	5.0
OATS *(Avena sativa)*															
Cedara	3d	100	0.06	0.09	0.17	0.17	0.21	—	—	—	—	—	0.21	0.09	5.3
WHITE CLOVER *(Trifolium repens)*															
Cedara	3d	0	—	0.02	0.10	0.15	0.14	0.11	0.09	0.10	0.12	0.09	0.06	0.02	8.1
COASTCROSS II															
Kokstad, Doveton	4f	120	—	—	0.01	0.08	0.16	0.20	0.20	0.14	0.12	0.08	0.01	—	10.7
Kokstad, Arrochar	4f	120	—	—	—	0.01	0.19	0.35	0.26	0.11	0.06	0.02	—	—	8.9

Table 14.3 *Continued*

Pasture Description	Bio Grp	Kg N ha/a	Jul	Aug	Sep	Oct	Nov	Dec	Jan	Feb	Mar	Apr	May	Jun	Yield t/ha
COASTCROSS II (*continued*)															
Cedara, Hutton	3d	200	–	–	0.02	0.08	0.17	0.21	0.17	0.13	0.12	0.07	0.03	–	11.8
Wartburg, Longlands	3d	260	0.01	0.05	0.06	0.08	0.10	0.12	0.15	0.16	0.13	0.08	0.05	0.01	13.1
Umlaas Road, Glenrosa	2d	400	0.05	0.08	0.11	0.15	0.12	0.11	0.10	0.09	0.07	0.05	0.04	0.03	18.6
Bathurst, cutting 30 days		300	0.02	0.04	0.07	0.03	0.16	0.10	0.14	0.12	0.14	0.10	0.06	0.02	11.7
Bathurst, cutting 60 days		300	0.03	0.03	0.05	0.07	0.10	0.14	0.13	0.12	0.12	0.11	0.07	0.03	15.7
Alexandria, cutting 30 days		300	0.04	0.01	0.02	0.05	0.18	0.13	0.13	0.05	0.20	0.09	0.07	0.03	6.3
Alexandria, cutting 60 days		300	0.03	0.02	0.03	0.04	0.11	0.17	0.12	0.05	0.12	0.17	0.11	0.03	8.0
STAR GRASS (*Cynodon nlemfuensis*)															
Kokstad, Doveton	4f	120	–	–	–	–	–	0.19	0.41	0.20	0.14	0.06	–	–	9.3
Kokstad, Arrochar	4f	120	–	–	–	–	–	0.17	0.45	0.24	0.10	0.04	–	–	8.9
Cedara, Hutton	3d	200	–	–	–	–	0.07	0.18	0.35	0.22	0.12	0.06	–	–	10.5
Wartburg, Longlands	3d	260	–	0.02	0.06	0.08	0.14	0.15	0.15	0.14	0.13	0.08	0.04	0.01	11.4
Bathurst, cutting 30 days		300	0.02	0.03	0.06	0.04	0.15	0.09	0.13	0.12	0.16	0.11	0.08	0.01	9.7
Bathurst, cutting 60 days		300	0.03	0.02	0.04	0.07	0.09	0.10	0.12	0.13	0.13	0.14	0.09	0.04	14.3
Alexandria, cutting 30 days		300	0.02	–	0.01	0.04	0.14	0.13	0.12	0.12	0.23	0.09	0.06	0.04	4.1
Alexandria, cutting 60 days		300	0.02	0.02	0.02	0.03	0.08	0.13	0.12	0.12	0.15	0.18	0.10	0.03	5.9
NILE GRASS (*Acroceras macrum*)															
Cedara, Hutton	3d	200	–	–	–	0.11	0.17	0.19	0.20	0.16	0.08	0.05	0.04	–	8.3
Wartburg, Longlands	3d	260	0.01	0.06	0.07	0.08	0.09	0.13	0.14	0.15	0.13	0.07	0.06	0.01	10.9
Umlaas Road, Irr., Glenrosa	2d	400	0.01	0.01	0.05	0.13	0.19	0.16	0.16	0.13	0.08	0.05	0.02	0.01	17.2
ERAGROSTIS (*E. curvula*)															
Kokstad, Doveton	4f	210	–	0.02	0.05	0.09	0.14	0.15	0.16	0.15	0.10	0.09	0.05	–	10.3
Cedara, Hutton	3d	195	–	0.01	0.02	0.10	0.16	0.21	0.18	0.13	0.09	0.06	0.04	–	9.6
Cedara, Hutton	3d	290	–	0.01	0.03	0.14	0.19	0.20	0.17	0.12	0.08	0.04	0.02	–	12.2
Umlaas Road, Glenrosa	2d	244	–	0.01	0.08	0.11	0.13	0.17	0.14	0.13	0.09	0.08	0.06	–	15.0
Dundee, Longlands	8a	330	–	–	0.03	0.12	0.16	0.16	0.16	0.13	0.11	0.08	0.05	–	15.4
Dundee, Bottomland, Uitvlugt	8a	330	–	0.01	0.06	0.11	0.17	0.16	0.15	0.11	0.09	0.07	0.07	–	18.4
Tabamhlope, Farningham	4e	400	–	–	0.08	0.12	0.15	0.16	0.15	0.12	0.10	0.07	0.05	–	19.8
Bathurst, cutting 30 days		300	–	0.01	0.04	0.08	0.21	0.17	0.12	0.19	0.08	0.05	0.05	–	10.2
Bathurst, cutting 60 days		300	0.02	0.01	0.04	0.08	0.12	0.15	0.15	0.14	0.12	0.08	0.06	0.03	12.4
Alexandria, cutting 30 days		300	–	0.01	0.02	0.08	0.33	0.23	0.07	0.06	0.11	0.06	0.02	0.01	8.9
Alexandria, cutting 60 days		300	0.02	0.01	0.03	0.05	0.16	0.25	0.16	0.05	0.08	0.10	0.06	0.03	11.5

Proportion of total production

Table 14.3 *Continued*

Pasture Description	Bio Grp	Kg N ha/a	Proportion of total production												Yield t/ha
			Jul	Aug	Sep	Oct	Nov	Dec	Jan	Feb	Mar	Apr	May	Jun	
RHODES (*Chloris gayana*)															
Bathurst		200	0.03	0.01	0.05	0.09	0.10	0.13	0.12	0.12	0.11	0.11	0.08	0.05	9.0
Bathurst, cutting 30 days		300	0.01	0.01	0.08	0.07	0.16	0.11	0.12	0.13	0.14	0.10	0.06	0.01	9.6
Bathurst, cutting 60 days		300	0.03	0.01	0.05	0.09	0.11	0.13	0.12	0.12	0.11	0.10	0.08	0.05	13.5
Alexandria, cutting 30 days		300	0.01	—	0.02	0.04	0.20	0.16	0.07	0.09	0.21	0.12	0.06	0.02	6.5
Alexandria, cutting 60 days		300	0.01	0.01	0.02	0.04	0.13	0.21	0.14	0.08	0.11	0.15	0.08	0.02	9.7
DIGITARIA (*D. eriantha*)															
Kokstad, Newport	4f	120	—	—	0.04	0.11	0.14	0.16	0.18	0.15	0.12	0.08	0.02	—	10.1
SABI (*Panicum maximum*)															
Bathurst, cutting 30 days		300	0.01	0.01	0.10	0.08	0.16	0.14	0.10	0.18	0.13	0.06	0.02	0.01	10.2
Bathurst, cutting 60 days		300	0.01	0.01	0.06	0.10	0.12	0.14	0.14	0.13	0.11	0.10	0.06	0.02	11.8
Alexandria, cutting 30 days		300	0.01	—	0.02	0.05	0.21	0.17	0.07	0.17	0.20	0.06	0.03	0.01	7.6
Alexandria, cutting 60 days		300	0.02	0.01	0.03	0.05	0.14	0.23	0.15	0.08	0.10	0.11	0.06	0.02	11.7
LUCERNE (*Medicago sativa*)															
Kokstad, Newport	4f	0	0.01	—	0.05	0.12	0.15	0.10	0.09	0.12	0.14	0.11	0.08	0.03	7.4
PHALARIS (*P. tuberosa*)															
Bathurst, cutting 30 days		300	0.06	0.11	0.01	0.30	0.12	0.21	—	—	—	0.06	0.03	0.10	5.4
Bathurst, cutting 60 days		300	0.12	0.03	0.14	0.25	—	—	—	—	—	0.10	0.15	0.21	5.1
Alexandria, cutting 30 days		300	0.08	0.04	0.32	0.30	0.12	—	—	—	—	—	—	0.11	4.5
Alexandria, cutting 60 days		300	0.10	0.08	0.26	0.43	—	—	—	—	—	—	—	0.13	6.6
VELD															
Sourveld			—	—	0.05	0.14	0.18	0.18	0.14	0.14	0.09	0.08	—	—	2.2
Mixedveld			—	—	—	0.17	0.17	0.22	0.11	0.11	0.11	0.11	—	—	1.8
Sweetveld			—	—	—	0.09	0.18	0.18	0.19	0.18	0.09	0.09	—	—	1.1

Y = predicted yield, kg DM/ha/year.

Eragrostis – *Eragrostis curvula*:

Y = N(31.9 – N/57.9) + 3939 ln(P_y) – 22450

(after Hackland & Jones 1980).

Kikuyu – *Pennisetum clandestinum*:

Y = 1988 – 820T_{a-} + (0.19S + T_s)P_{sa} + N(5.1T_{s+} – 81)

Cocksfoot – *Dactylis glomerata*:

Y = 57426 + (0.29S – 16)P_{sa} – 2895T + (19.4D + 17T – 262)S + 3129F + N(1.22T – 0.039N)

Smuts' fingergrass – *Digitaria eriantha*:

Y = [$P_y^{1.5}$(N^2 + 2.5N)$^{2.5}$] / 46.65

(Dannhauser *et al.* 1987)

More yield estimate equations for pastures and forages may be obtained from Schulze (1997) and from Smith (1997), but some of these were developed for KwaZulu-Natal conditions and may not be applicable elsewhere in South Africa. Table 14.2 may assist in obtaining an estimate of yield, the seasonal distribution of which can be obtained from Table 14.3. Since management practices affect both the yield and the quality of pastures and other forages, these practices may be varied to aid in balancing the feed budget.

At this stage it should be possible to match the production of dry matter from an appropriate combination of the available fodder with the monthly requirements of the livestock enterprise. This exercise can best be undertaken in the following way (refer to Table 14.4).

1) Enter the dry matter requirements of the livestock, month by month (as a minus).
2) Calculate and add the fodderbank requirement.
3) List the pasture species which have been chosen for both grazing and for conservation. Where veld is also used, a distinction needs to be made between rested and grazed veld. The relative proportion of each will depend on the number of camps and the burning policy of the veld management system.
4) Enter the total area of the property (480 ha in Table 14.4) and the area suited to irrigation (20 ha in Table 14.4). The remaining area is apportioned to dryland cropping, dryland pastures and veld.
5) Allocate the areas of irrigable land to those forage species most likely to respond to irrigation, bearing in mind the forage production potential of these crops (refer to Tables 14.2 & 14.3) and the livestock fodder requirements in any particular month. Any areas which should logically be planted to a particular species, e.g. kikuyu in night camps, should also be apportioned at this stage.
6) Repeat the above process for the remaining available dryland areas.
7) Attempt to balance the fodder production of the listed crops against animal requirements month by month by determining the monthly and cumulative balance between the yield of grazeable material and requirement. Conserved feeds can then be apportioned in appropriate quantities to those months where there is a deficit of dry matter from grazing.
8) A gross economic analysis should be undertaken to establish the likely profitability of the forage plan. This process should be repeated again and again for different pasture and crop combinations to establish the most economic forage plan for the area.

Table 14.4 Example of a fodder production plan for a beef enterprise.

Forage description	Yield t/ha	Area ha	Jul	Aug	Sep	Oct	Nov	Dec	Jan	Feb	Mar	Apr	May	Jun	Tot DM
Irrigated ryegrass	10.6	20	13	17	36	40	28	4	—	—	4	19	34	17	212
Veld: Rested	2.2	100	—	—	11	31	40	40	31	31	20	17	—	—	221
Grazed	2.2	300	—	—	33	92	118	119	93	92	59	53	—	—	659
Coastcross II (K11)	11.8	60	—	—	14	57	120	149	120	92	85	50	21	—	708
TOTAL FORAGE PRODUCED (A)		480	13	17	94	220	306	312	244	215	168	139	55	17	1800
DM REQUIRED incl. waste (B)			−95	−107	−120	−130	−132	−142	−149	−148	−146	−140	−91	−90	−1490
SURPLUS (+) or DEFICIT (−) (C) = (A − B)			−82	−90	−26	90	174	170	95	67	22	−1	−36	−73	310
CONSERVED FODDER PRODUCED (D)															
Rested veld			—	—	11	31	40	40	31	31	20	17	—	—	221
Coastcross II hay			—	—	—	57	120	112	90	—	—	—	—	—	379
Surplus grazing			—	—	—	2	14	18	—	36	2	—	—	—	62
TOTAL CONSERVED FODDER REQUIRED (E) = (C − D)			−82	−90	−37	—	—	—	−26	—	—	−18	−36	−73	−352
FED FROM:															
Coastcross II hay			82	90	37	—	—	—	—	—	—	—	24	73	306
Surplus grazing			—	—	—	—	—	—	26	—	—	18	12	—	56
LEFT FOR FODDERBANK															
Coastcross II hay															73
Rested veld (burn if not needed that season)															221

(d) Once a reasonable balance has been reached between fodder demand and supply at a reasonable cost, the system may be examined in still greater detail. The herd or flock is separated into homogenous groups, e.g. cows, stores (2–2½ years), weaners, calves etc., and the seasonal requirements of each group determined separately for both dry matter and energy. The requirements of each group are then combined to give an overall requirement for feed during each month of the year, and feed costs determined as accurately as possible for the proposed system. These costs can then be compared to the expected returns based on planned production levels. If the margin over feed costs is unacceptable, the plan must be modified by returning to Step 3 and repeating the process, if necessary by modifying the livestock system (Step 2). In any event, it is always wise to examine the profitability of a number of possible systems and the margin over feed cost should, at least, be greater than 100%.

14.1.2   Feed budgeting

Having established a cropping and pasture plan for the property, feeds are now allocated to the separate groups of animals such that the groups do not interfere unduly with one another or with the management operations on the farm. Priorities for quality feeds are identified, and allocations made to favour the highest producers at their critical periods. Concentrate feeding can also be introduced in the following way:

(a) Determine the DM, CP and ME requirements of each group of animals for each month.
(b) Allocate feeds to each group of animals according to their requirements and the available feeds. The DM requirement should first be satisfied from named sources of feed and the ME which this feed would supply compared with the ME requirement of that group of animals. If insufficient, the roughage schedule should be altered by increasing the proportion of feed of higher ME, or by supplementation. Once the requirements for DM and ME have been satisfied, the CP provided by the ration should be checked against the CP requirement. If inadequate, feeds with a higher protein content may be incorporated in the ration or CP supplemented with a N lick or with HPC (high protein concentrate). This procedure is repeated for each group of animals. Quantities of feed required and their cost can then be determined for each such group. Provision must be made for wastage. Data from local analyses should be used in preference, but tables such as those published in Stewart & Dugmore (1995) may also be referred to.
(c) This process is repeated for each month of the year. When completed, the cumulative quantity of feed required, and its cost, can be determined.

There are invariably a number of alternative feed plans which suit any particular situation. Each should be tested before the final decision is made. Such an undertaking may be extremely tedious unless computer facilities are available to handle the calculations. A fodder production planning program, coded FODEL, is capable of undertaking the necessary calculations in seconds and is clearly an invaluable aid in formulating feed budgets.

The main weakness in the above process, particularly on undeveloped properties, is the lack of reliable information on the production levels and quality of the forage which can be expected from both veld and arable areas. The operator should therefore record yields and have samples of home-produced feed analysed to establish these norms. He can then adjust the fodder flow of the property accordingly. Unfortunately, record keeping is not popular among farmers, and pasture yields are difficult to estimate with any degree of precision. In practice, the operator generally prefers to judge the success of the feed budget at the end of each season, based on the extent to which production balanced demand. By this time it is too late to adjust the feeding programme for that season, but such information will suggest modifications for the following season and may therefore be useful. Only when production systems based predominantly on forage (as opposed to concentrates) become extremely intensive, is it likely that farmers will be prepared to spend some effort in recording the necessary data for effective feed budgeting.

14.2 PRODUCTION SYSTEMS

14.2.1 Dairy

The main differences between dairy and other livestock systems is that milk normally has to be produced uniformly throughout the year and that the industry in South Africa is heavily dependent on high quality temperate pastures produced under irrigation. A constant supply of good quality roughage is essential for profitable dairy farming, and concentrates should be used only to supplement quality deficits.

Complex interrelations exist between animal performance, feed quality and intake (refer to Chapters 5 & 7). Feed quality is itself a function of energy and protein type, and this is further complicated by the fact that in dairying, animal performance is determined by an interaction of many factors including breed, milk yield, butterfat content, lactation number, stage of lactation, stage of pregnancy, livemass and body condition. Also, since the quality of forage on offer is a function of how that pasture is managed, specific statements involving yield expectations will be deliberately avoided, both to eliminate confusion and because such detail is beyond the scope of this text.

A high proportion of cows in a dairy herd require a diet rich in energy (>10.5 MJ ME/kg DM) and good quality protein. Energy constitutes the greatest part of the cost of feeding cows, and considerable effort needs to be directed at providing energy cheaply. A protein deficiency can be corrected relatively inexpensively because, although more expensive per kg than energy, it is needed in smaller amounts. However, good quality pastures should provide sufficient energy for the low producers, but excess protein for even moderate producers. For high producers, DM intake may be too low to ensure an adequate intake of energy and high energy concentrates may be required to both supplement and substitute for the available grazing. It is convenient, in grass-based dairy systems, to plan the system separately for each of three major seasons, viz. spring/early summer, late summer/autumn, and winter in the summer rainfall regions of the country, and autumn/early winter, late winter/spring and summer in the winter rainfall regions.

14.2.1.1 Summer rainfall regions

(a) Spring–early summer. Good quality feed is normally abundant during this period. Temperate grasses (ryegrass, fescue, cocksfoot) yield well until late November, provided water is not limiting. Thereafter, midsummer temperatures depress yields, and bolting brings about a drop in quality. The tropicals green up soon after the first spring rains but only make a real contribution to the fodder flow from about early November. Unfavourable weather and the succulence of the pastures at this time of the year make haymaking difficult, but unless the surplus forage is conserved, valuable material may be wasted. The combined effects of high cost of hay- and silage-making equipment and the expense of annual forages such as maize, are leading many farmers to rely on grass silage alone. This largely eliminates the weather constraint, increases flexibility within the system and, as long as the techniques are mastered, provides a good, cheap carryover of summer surpluses into the winter, although extra concentrates would normally be needed.

(b) Late summer–autumn. Management during this period is often difficult. Tropical pastures will have matured, and as their growth rates decline the average quality of the material on offer decreases. High temperatures and dry spells inhibit growth in the temperate perennials, and temperate annuals will have only just been planted. Animal performance is therefore generally poor. No amount of compensatory feeding will restore production lost in the first stages of lactation, and since the majority of cows are calved down at this time (to promote an even production over the year), quality deficiencies can severely impact on the remainder of the lactation (Bredon & Stewart 1979). Sufficient perennial temperate pasture should therefore be established on cool aspects, preferably mixed with clover, to provide for this period. Even restricted grazing of these pastures can supplement the protein provided by the tropical pastures. Also, forage quality should be monitored regularly throughout this period and concentrate feeding adjusted accordingly. Management should aim to extend the grazing of tropical and perennial temperates as far into the autumn–winter period as possible so as to reduce the demand for irrigated annual temperate pastures. As much material as possible from the annual pastures should be carried into winter.

(c) Winter. Any green forage available during winter is usually of good quality. The limiting factor is normally the amount available. If adequate quantities of silage and foggage have been accumulated, restricted access to high quality roughages can effectively maintain milk production. The amount of pasture available at this critical period tends often to be overestimated so that animals are underfed, and so milk production declines. On the other hand, as much of an annual temperate pasture such as ryegrass should be grown as resources allow (aim for 0.4 ha/cow), because it alone has a milk production potential of 16 kg of 4% fat corrected milk per day, and with 10 kg concentrates the potential rises to 27 kg milk.

14.2.1.2 *Winter rainfall regions*

Dairying in the winter rainfall regions is largely restricted to the coastal, and therefore higher rainfall, zones. Most of what has been said about the summer rainfall regions applies also to the winter rainfall regions, but at different times of the year. Here the pastures are most active during the winter months, and the time of greatest shortage is in mid- to late summer. The predominant forage species are ryegrass and clover during winter, extending into summer if irrigation is available, with lucerne (for both grazing and hay) and kikuyu being relied upon during summer.

Feed costs make up about 70% of all variable costs in a typical dairy enterprise, and if a profit is to be made, it is essential that sufficient good quality feed be provided. This can be achieved only by careful management and, most importantly, by not overstocking.

14.2.2 Beef

The nature of any beef system is determined by factors such as the rainfall pattern, the maturity type of the animals available, the possibilities for feedlotting or finishing and the maize to beef price ratio.

The intensity of the production system is measured by the growth rate that can be maintained with available feed supplies. The most intensive system is one in which the finishing phase commences immediately after weaning. Extensive systems are those where the animals achieve the required degree of fatness only at a relatively advanced age (e.g. 30 months) with virtually no supplement. Numerous variations occur between these two extremes, with some examples being given in Table 14.1. Harwin (1989) has outlined others. Water for irrigation (or rainfall) is a vital consideration when selecting the level of intensity of a beef system. The former (irrigation) is likely to be phased out for beef operations should the proposed new water laws be implemented, with irrigation being practised only in dairy systems (Murray pers. comm.).

Primarily because of a lack of grass in the vegetation of the winter rainfall region, beef is not a major enterprise and so the production systems discussed here relate to the summer rainfall regions.

14.2.2.1 *Producing weaners*

Many beef producers throughout the country concentrate on producing weaners which are sold before winter. The major difference between extensive and intensive systems is the stocking rate used and the need for conserved fodder for the pregnant cows and replacement heifers during winter. Advisers generally recommend that farmers do not sell weaners since it is generally more profitable to keep weaners longer to gain mass off grass in summer. This will depend on the beef:maize price ratio (price/kg of A_2 class beef: price/kg of maize (Norvall pers. comm.)). The type of weaner in demand at any time also varies with this ratio. Feedlotters prefer late-maturing type weaners that can make considerable gains in body mass before achieving finished condition when the ratio exceeds 1:12, i.e. it pays them to keep animals as long as they are gaining body mass when the margin over feed costs is positive. Conversely, feedlotters select early-maturing type weaners which reach a finished condition with relatively little feed when the ratio is of the order 1:10 and the price margin (the price difference between the finished animal per kilogram and that when the animal enters the feedlot) for A_2 carcasses is positive. It now needs to be accepted, however, that the opening up of South African markets to world imports of beef may mean that the farmer will no longer be able to rely on the traditional upswing in beef prices from time to time. This may induce a move away from exotic beef breeds to the more hardy indigenous breeds (Murray pers. comm.).

14.2.2.2 *Producing finishers*

If cyclic fluctuations in the profitability of feedlotting cattle are ignored, then one of the following basic systems should be adopted in order to take the animals beyond the weaner stage.

System 1) Super A and A_1 carcasses for marketing at 13–14 months of age
This system has gained popularity in the well-watered, summer rainfall regions and in-

volves finishing weaners on cool season ryegrass pastures. Vital prerequisites for this system are suitable soils and adequate quantities of water for irrigation, although, as mentioned previously, this system is likely to be phased out as the costs of irrigation increase. Careful attention must be paid to choice of breed type and stocking intensity so that the animals are finished when pasture growth becomes limiting.

Usually the weaners are the progeny of breeding herds maintained on veld. The breeding cows are mated to calve as close to 1 September as possible. The nutritious, early spring veld grazing at this time supports both high milk production and good reconception rates. However, due to the relatively high stocking rates often used on veld, unforseen hot, dry spells during breeding can reduce subsequent calf crops. Calves should be weaned in time to allow the pregnant cows to improve their condition before winter. Ideally this would be during April but it will depend on when the majority of calves are born. When calves are well grown in summer it may be advisable to wean early so that the cows have ample time to regain condition. Crop residues, chicken litter and silage are important components of the diet of the pregnant cow, and reserves of hay are vital in bridging the winter–spring grazing gap.

System 2) Super A and A₁ carcasses for marketing at 16–18 months of age

Here weaners are overwintered on greenfeed or maize silage to maintain a growth rate of 0.5–0.6 kg per day. During summer, on pasture, the daily gains increase to between 0.75 kg and 0.9 kg so that finished carcasses are produced in January–February, and so are too late for the high priced year-end market. Production costs should, however, be lower than in System 1 so that favourable profit margins can still be obtained.

Careful attention must again be paid to the use of stocking rates which permit the animals to finish before the onset of the second winter. Provision of an energy supplement is likely to be profitable only where stocking rates are high and where this significantly increases the proportion of animals achieving the highest grades. The breeding herd is fed as for System 1.

System 3) Super A and A₁ carcasses for marketing at 19–22 months of age

Here a lower growth rate is aimed for than in the above system. The winter gain of 0.25 kg per day is based on maize residues and a protein lick. After spending the summer on veld, the animals are finished on ryegrass or in a feedlot.

System 4) Prime B and B₁ carcasses for marketing at 28–30 months of age

The first summer after weaning and the first and second winters are as for System 3. During the second summer, cultivated pastures may be used instead of good veld to achieve the desired finish.

System 5) Prime B and B₁ carcasses for marketing at 31–40 months of age

This is primarily an extensive system making almost exclusive use of veld. Strategic finishing on a small area of cultivated pasture under irrigation is sometimes practised.

The later maturing beef breeds are generally used in the intensive systems because high daily gains can be maintained for extended periods where feeding conditions are good. In contrast, earlier maturing types are normally used in extensive systems because the large framed late maturing breeds are unable to finish on veld. One exception to this general rule is where animals are finished on irrigated ryegrass during winter. Here the early maturing breeds need to be used since the large framed animals will normally not finish adequately before the end of the grazing period (November).

The feeding levels that can be maintained at different times of the year will dictate which system will be the most applicable in different regions.

14.2.2.3 Calving seasons

The determination of the time to calve down is a trade-off between conditions which maximise calf growth rate (cool season) and those which favour reconception in the breeding females (good nutrition). Thus, where calving is confined to a distinct season (usually of 3 months), the onset is planned to commence about one month before green grazing would be expected to become available. The majority of cows will then reach peak lactation when the milk requirements of the suckling calves are greatest. Cows should also have good grazing available when mating commences. Weaning should be sufficiently early to allow cows to regain the condition they lost during lactation, and should therefore take place before the onset of winter. In the drier bushveld regions where spring rains are unpredictable, calving is in summer rather than in spring, although the high temperatures in January and February and the prevalence of ectoparasites do not favour good calf performance. Since the veld is often able to sustain limited growth in calves and condition in pregnant cows during the winter, extra feed need not be provided at this time.

When considering the systems outlined above, it is clear that special attention needs to be given to fodder flow during:
(a) late summer and early winter: for pregnant cows and weaners;
(b) late winter and early spring: for cows about to calve, weaners and replacement heifers being prepared for mating;
(c) late spring: for replacement heifers to be mated three weeks before the cow herd; and
(d) winter: for young animals being finished for end-of-year marketing.

It is obvious that as one changes from intensive to extensive, the percentage of breeding cows in the herd decreases and that of older steers increases. The fodder flow must match the critical requirements of the stock for optimum performance, particularly in late summer when forage quality is generally declining and animal demands are often at their greatest (Harwin 1989).

14.2.3 Sheep

The feed, and particularly roughage, requirements of a sheep production system are extremely flexible because a complete cycle from mating to weaning/marketing can be completed in 8 to 11 months. The length of this period depends on the type of sheep, their physiological state (pregnant, lactating or dry), the feed value of the available forage and the extent to which concentrates are used.

Within the cycle from mating to weaning and sale, the feed requirements of a flock (ewes plus lambs) varies considerably (see Table 14.1). Food requirements are usually greatest during lactation, when they can rise to double those of the dry and pregnant periods.

Sheep breeds differ markedly in their feed requirements. This is primarily due to their differences in mature live mass, potential lambing and weaning percentages and their growth rates. Norms are presented in Table 14.5 for the three main types of sheep, viz. wool (e.g. Merino), dual purpose (e.g. SA Mutton Merino) and mutton types (e.g. Dorper) (Erasmus *et al.* 1982).

Table 14.5   Production norms for three major sheep breed groups.

Factor	Breed Type		
	Wool	Dual Purpose	Mutton
Conception (%)	85	90	95
Lambing (%)[1]	102	135	128
Weaning (%)[2]	87	115	122
Survival (%)[3]	85	85	90
Fecundity (%)[4]	120	150	135
Weaning age (months)	6	3–5	4–5
Wean/sale mass (kg)	25–30	35–40	35–40

[1] Total lambs born per ewe mated
[2] Total lambs weaned per ewe mated
[3] Total lambs weaned per lamb born
[4] Total lambs born per ewe lambing

To keep feed costs to a minimum, the feed requirements of a sheep enterprise should be designed to fit the forage production potential of the area. Normally the sheep type and breed are chosen to suit the forages available on the farm. The seasonal requirements of the flock may then be modified by adjusting lambing and weaning dates and percentages and retaining dry or surplus stock in times of feed abundance.

The forage production potential of a region will largely dictate which sheep production system should be followed. Two main production regions can be identified, namely the primarily summer rainfall region and the winter and all-year rainfall region. Each region can be further subdivided into extensive, semi-intensive, and intensive/irrigable areas (Erasmus *et al.* 1982).

14.2.3.1   Summer rainfall regions

In the summer rainfall regions sheep production is restricted to the grassveld (as opposed to the savanna). Here, despite the mismatch of feed supply and demand, the primary lambing season is autumn (because of the parasite problem), although the skips are often remated to lamb in spring. In the high rainfall areas, autumn lambing is generally onto irrigated ryegrass pastures. At weaning the ewes move to veld or to surplus kikuyu pastures for the summer while the lambs remain on the ryegrass until finished for the Christmas market. Hamels often run on the more extensive hilly terrain for most of the year, receiving supplements and *Eragrostis* hay during winter. In the central lower rainfall regions of the country, lambing is usually on small areas of pasture and as soon as the lambs are strong enough, they and the ewes are put onto harvested maize lands where the stover and dropped maize provide the winter diet. In summer the ewes move to veld and, depending on the availability of pastures, the lambs are either finished for the market on pasture or they move onto veld to be marketed later as mutton.

14.2.3.2   The winter and all-year rainfall region

Extensive areas

These areas are characterised by low (<300 mm/annum) and erratic rainfall. The main feed source is veld, even though its production is very seasonal because of the relatively large component of annual species. The perennials are mostly shrubs which lose most of their leaves in summer. At this time there is usually a critical shortage of green forage and so farmers often plant drought-tolerant forages which retain their leaves in summer.

In these areas, hardy, adapted mutton-types such as the Dorper are used. There is only one lambing season with mating, usually between December and February (Erasmus *et al.* 1982). Lambing thus coincides with the winter rainfall season so that lambs can be finished during the period of feed abundance and only the breeding flock needs to be carried through the summer.

Semi-intensive areas

Here the rainfall varies between about 300 mm/annum and 500 mm/annum. Annual and perennial legumes are grown in rotation with small grains. There is abundant grazing during the winter months (Van Heerden & Tainton 1987; Van Heerden 1990). The main mating season is therefore in November and December, with lambing in autumn to early winter (Erasmus *et al.* 1982). Skips are often re-mated in spring. Stocking rates of three to five ewes and their lambs per hectare are recommended for the period from lambing to weaning in the main breeding flock, depending on the productive potential of the pasture and the timing, amount and distribution of winter rainfall. Considering a simple scenario of a farming system where half the farm is planted to cereal crops and the other half to an annual legume pasture with a stocking rate of four ewes (and their lambs) per hectare on the winter pastures, the overall carrying capacity of the farm would be somewhat less than two ewes per hectare (Hardy pers. comm.).

Very little, if any, green forage is normally available during the often very dry summers. Plant residues from the medic and other pastures form an important part of the fodder flow during this period. The dry stems, and pods in the case of the medic pastures, provide sufficient nutrients to maintain mass, and even allow for mass gain in dry animals. More than 2 000 kg/ha of dried medic pods may be left on the surface of the soil from a productive medic pasture. It is important to appreciate, however, that at least 500 kg/ha of pods should be left on the soil surface to ensure that sufficient seed is added to the seed bank (Hardy pers. comm.).

Wooled and dual purpose sheep breeds predominate. Lambs from the dual purpose breeds are marketed directly after weaning at four to five months of age. In wooled breeds, lambs may be weaned from six months of age onwards, but this is usually delayed until they are 12 to 18 months old.

Intensive areas

These areas largely comprise the all-year rainfall regions, where there is ample surface water for irrigation (Van Heerden *et al.* 1989). Grazing is based on intensively irrigated grass/clover or pure grass pastures. The seasonal production of forage is linked primarily to temperature, with the peak production period in spring and early summer (September to January) (Van Heerden & Tainton 1989; Van Heerden *et al.* 1989). During this period forage yields are usually at least double those of mid-winter.

Slaughter lamb production systems are mainly used with dual purpose breeds, although problems with internal parasites and hence poor growth rates amongst the lambs can be expected. Such systems use a primary spring and a secondary autumn lambing season, with mating in April and November. A general practice is to mate every eight months and thus produce three lamb crops in two years. However, this system does not always fit the seasonal production pattern of the pastures, and considerable feed supplementation is required.

14.3 CONCLUSION

The primary aim of any farmer must surely be to generate a stable, sustainable income from his property. To do this he must develop a stable marketing strategy. This means choosing a livestock production system that is consistent with the forage producing capabilities of the farm. Precautions must be included to accommodate destabilising factors which can be expected (such as drought and changing price structures), and if planning has been based on sound principles, even unexpected disturbances should not have a major impact on the system.

Glossary

W.S.W. Trollope, L.A. Trollope & O.J.H. Bosch

Since Booysen (1967) published a set of terms describing grazing and grazing management in southern Africa, several important developments have occurred in management and general terrestrial ecology. All these and other developments have necessitated a revision of and addition to the original set of terms proposed by Booysen (1967). Consequently, the Grassland Society of Southern Africa appointed a committee to undertake this task. The procedure that was adopted in the selection of the terms was that they should have an explanatory role rather than a prescriptive one. This is because there are particular terms that have been recommended for official use in preference to others. While recognising the desirability of using only recommended terms, the problem exists where non-recommended terms appear and are widely used in the more important role of disseminating knowledge about pasture management. Therefore, these terms have also been included in an effort to avoid confusion in the ongoing debate on pasture matters. However, only those terms that are both scientifically and conceptually meaningful have been included.

It has also been necessary to define new terms that have been recently added to the pasture vocabulary. It is recognised that these definitions may not receive the full approval of everyone in the pure and applied biological fields. Nevertheless, great care and thought has been taken and given in developing these definitions, and a plea is made to assess them in terms of their explanatory function rather than their grammatical niceties.

Abiotic	Non-living, basic elements and compounds of the environment.
Acceptability	Attractiveness of feed to animals as determined by factors of the forage and the environment.
Agro-ecological unit	An area in which the climate, landscape, soil and vegetation are homogeneous to the extent that the adaptability and response of any particular plant species would not change markedly from place to place within the unit.
Animal unit (AU)	An animal with a mass of 450 kg and which gains 0.5 kg per day on forage with a digestible energy percentage of 55%.
Anthesis	Stage in floral development when pollen is shed.
Animal unit equivalent	$M^{0.75}/97.7$ where M is the mass of the animal under consideration.
Apical dominance	Inhibiting effect of a terminal bud upon the development of lateral buds.
Apomixis	Formation of viable embryos without actual union of male and female gametes.
Area selective grazing/ browsing	Habit of grazing/browsing animals to graze/browse certain areas of the veld/pasture in preference to others.
Artificial pasture	An area that has been artificially established to selected forage plants.
Aspect	Predominant direction of slope of the land.
Auricle	Earlike appendage sometimes present at the collar of the leaf sheath in grasses.

Average daily gain (ADG)	Average gain in live mass per animal per day for a specified period – kg/d.
Awn	Bristle-like projection arising from the seed of a grass.
Axillary bud	New bud arising in the axil of a leaf.
Axillary tiller	New tiller arising in the axil of a leaf.
Biomass	Total amount of living material (plant and animal) present in a particular area at any given time – kg/ha.
Biome	Major regional ecological community of plants and animals.
Biota	All the species of plants and animals occurring within an area.
Biotic	Living components of the ecosystem.
Browse	That portion of the woody vegetation that is available for consumption by animals.
Browsing	Utilisation of woody vegetation by animals.
Browsing capacity	*See* Grazing capacity.
Browser	An animal that utilises browse.
Camp	Smallest unit to which grazing and/or browsing management is applied.
Canopy	Cover of leaves and branches formed by the tops or crowns of plants.
Canopy cover	Proportion of the ground area covered by the vertical projection of the canopy – %.
Carrying capacity	Potential of an area to support livestock through grazing and/or browsing and/or fodder production over an extended number of years without deterioration to the overall ecosystem – ha/AU or AU/ha.
Community	An assemblage of plants growing together and interacting among themselves in a specific location.
Continuous grazing/ browsing	Type of management whereby animals are placed in a camp when the forage becomes ready for grazing/browsing at the start of the growing season and they or their replacements are left in that camp for the entire grazeable/browseable period of each year.
Crop growth rate	Rate of increase in dry mass per unit area of all or part of a sward – g/m²/d.
Crown cover	*See* Canopy cover.
Culm	Flowering stem of a grass plant.
Cultivated pasture	Pasture which has been established by conventional means involving soil disturbance, removal of existing vegetation and seedbed preparation.
Density	Number of plants per unit area – No/ha.
Digestibility	Proportion of a feed that has the potential to be ingested by animals.
Diversity	An expression of the variety of species that exists in a community.
Dominance	Degree of influence that a plant species exerts over a community as measured by its mass or basal area per unit area of the ground surface or by the proportion it forms of the total cover, mass, or basal area of the community.
Dystrophic	Habitat which is both low in basic nutrients and toxic substances.
Ecology	Study of the interrelationships between organisms, and between them and their environment.

Ecosystem	Biological system comprising both living organisms and the abiotic components of the environment.
Ecotype	Plant type or strain within a species resulting from long-term exposure to a particular environment.
Eutrophic	Habitat rich in nutrients.
Fodder	Livestock feed that includes forage, hay and silage.
Fodder bank	Fodder reserve intended for use during periods of scarcity.
Fodder flow	Supply of fodder available to livestock throughout the year expressed on a monthly basis – kg/month.
Fodder production capacity	Potential of a farming unit to produce livestock fodder from arable areas without deterioration to the edaphic environment – ha/AU or AU/ha.
Fodder unit	An animal that has a daily intake of 10 kg of fodder containing at least 66 per cent total digestible nutrients.
Forage	That portion of a living plant that is available for consumption by animals.
Forage factor	Index of the sustained forage production potential of a plant species.
Forage flow	Supply of forage (grazing/browse) available to livestock throughout the year expressed on a monthly basis – kg/month.
Forb	Non-graminaceous, herbaceous plant.
Genotype	Genetic constitution of an organism.
Grazer	An animal that utilises grazing.
Grazing (n)	That portion of the herbaceous vegetation that is available for consumption by animals.
Grazing (v)	Utilisation of herbaceous vegetation by animals.
Grazing/browsing capacity	Productivity of the grazeable/browseable portion of a homogeneous unit of vegetation expressed as the area of land required to maintain a single animal unit over an extended number of years without deterioration to vegetation or soil – ha/AU or AU/ha.
Herbage	Leaves, stems and other succulent parts of herbaceous plants.
Inflorescence	Flowering part of a plant.
Intake	Mass of forage consumed by an animal per day expressed on a dry matter basis – kg/d.
Intensification	Process of increasing the agricultural production per unit area of land.
Internode	Intervening section of the grass stem between the nodes.
Introduced pasture	Pasture which has been established without soil disturbance or complete removal of existing vegetation.
Large stock unit (LSU)	*See* Animal unit.
Leaf area index	Ratio of leaf area to ground surface area.
Leaf blade	Upper portion of the grass leaf extending beyond the collar region.
Leaf sheath	Lower portion of the grass leaf below the collar region, normally enclosing the stem.
Ley	The pasture phase of a crop rotation specifically designed to improve soil productivity.
Ligule	Outgrowth on the inner surface of the grass leaf at the junction of the blade and the sheath.
Livestock unit	*See* Animal unit.

Mature livestock unit (MLU)	*See* Animal unit.
Meristem	Area of rapidly dividing cells.
Metabolic mass	Mass of an animal raised to the power three quarters.
Nett assimilation rate	Dry mass increment of a plant per unit leaf area per unit time – $g/m^2/d$.
Nett primary production	Amount of organic matter produced in the plant in excess of that used during respiration.
Node	Joint in the grass stem from which leaves and buds develop.
Nutritive value	Concentration of nutrients in a feed.
Overgrazing	Excessive defoliation of the grass sward by animals to the detriment of the condition of the veld or pasture.
Overseeding	Distribution of seed over the soil surface in the process of establishing or renovating a pasture.
Overstocking	When the stocking rate exceeds the carrying capacity of the veld or pasture.
Paddock	*See* Camp.
Palatability	Attractiveness of feed to animals as determined by specific factors of the forage.
Period of absence	Length of time between successive periods of occupation during which the veld or pasture is allowed time to grow and attain a suitable stage for grazing and/or browsing.
Period of occupation	Length of time which a particular camp is being utilised without interruption.
Permanent pasture	Pasture which has been established with a view to providing forage for an indefinite but extended number of years.
Phenotype	Externally obvious characteristics manifested by an organism as contrasted with the set of genes possessed by it.
Phytomass	Total mass of plants, including dead attached parts, per unit area – kg/ha.
Potential carrying capacity	Potential of a farming unit to support livestock through grazing and/or browsing and/or fodder production when all the factors which affect its productivity are at an optimum level – ha/AU or AU/ha.
Potential fodder production capacity	Fodder production capacity of an arable area when all the factors which affect its productivity are at an optimum level for producing livestock fodder – ha/AU or AU/ha.
Potential grazing/ browsing capacity	Grazing/browsing capacity of the vegetation when all the factors which affect its productivity are at an optimum level for grazing/browsing purposes – ha/AU or AU/ha.
Preferred species	Plant species that are preferred and utilised first by animals.
Primary production	Total amount of organic matter formed in the plant including that used during respiration.
Radical veld improvement	Maximisation of the potential herbage production of the veld through either veld reinforcement or replacement.
Regrowth rate	Rate of increase in dry mass of the regrowth per unit area of veld or pasture – $g/m^2/d$.
Relative growth rate	Rate of increase in dry mass of a plant per unit plant mass per unit time – $g/g/d$.
Resting intensity	Mean annual number of days over all periods of rest given to each camp during a complete cycle of rotational resting.

Rhizobia	Species of nitrogen-fixing bacteria that live in symbiotic relationship with leguminous plants within nodules on their roots.
Rhizome	Underground stem.
Rotational grazing/ browsing	Type of management which requires the grazing/browsing allotted to a group or groups of animals for the entire grazeable/browseable period, to be subdivided into at least one (usually more) camp more than the number of animal groups. It involves successive grazing/browsing of the camps by the animals in a rotation so that at any time the animals are concentrated on as small a part of the grazing/browsing available to them during the entire grazeable/browseable period, as fencing will permit.
Rotational resting	Type of management where the pasture is subdivided into at least one more camp than there are groups of animals and involves the successive resting of the camp from grazing and/or browsing for a specific purpose aimed at the restoration of vigour and productivity rather than merely the regrowth of vegetative material for grazing and/or browsing.
Roughage	Plant materials that are relatively high in fibre and low in nutrients.
Scarification	Mechanical scarring of the seed coat of 'hard' or impermeable seed to permit the rapid imbibition of water for initiating germination.
Senesce	To age.
Selective grazing/ browsing	Selective utilisation of the grazing/browse by animals.
Silage	Forage preserved in a succulent condition by partial fermentation under anaerobic conditions.
Small stock unit (SSU)	An animal which is equivalent to the $\frac{1}{6}$ of an Animal unit, e.g. goat or sheep.
Sod	Top layers of soil permeated and held together by roots and rhizomes of grasses and other herbaceous plants.
Sod seeding	Mechanical placement of seed directly into a grass sod.
Species composition	Relative proportion of different plant species occurring in a specific area.
Species selective grazing/browsing	Habit of grazing/browsing animals to graze/browse certain species of the vegetation in preference to others.
Standing crop	Total amount of above-ground plant material per unit area.
Stocking density	Concentration of livestock on the veld and/or pasture at any instant in time – AU/ha.
Stocking intensity	Mathematical expression reflecting simultaneously both the degree of concentration of livestock and the length of the period of occupation – AU/ha/day.
Stocking pressure	Amount of available forage that has been allocated per animal unit during a short period of occupation – kg/AU.
Stocking rate	Area of land in the system of management which the operator has allotted to each animal unit in the system, and is expressed per length of the grazeable and/browseable period of the year – ha/AU or AU/ha.
Stolon	Horizontal stem, which grows along the surface of the soil, and roots at the nodes.

Stubble	Basal portion of herbaceous plants remaining after the upper portion has been defoliated.
Sward	Above-ground parts of a population of herbaceous plants characterised by a relatively short growth habit.
Tiller	Basic unit of the grass plant.
Tillering	Vegetative reproduction in grasses comprising the development of new tillers.
Total annual yield	Total annual production of dry matter by all species of a plant community.
Trampling	Effect of hoof-action by ungulates on herbaceous plants and the soil surface.
Veld reinforcement	Introduction of more productive pasture plants into the veld and/or the addition of plant nutrients through veld fertilization.
Veld replacement	Complete replacement of veld with more desirable pasture plants.
Xerophyte	Plant capable of surviving periods of prolonged moisture deficiency.
Zero grazing/browsing	Type of management whereby herbaceous forage is cut and fed green to livestock.

References

Acocks, J.P.H., 1988. Veld types of South Africa. *Mem. Bot. Surv. S. Afr.* No. 40, Government Printer, Pretoria: 128.

Adams, F., 1980. Interactions of phosphorus with other elements in soils and plants. In: *The role of phosphorus in agriculture*, eds. F.E. Khasawneh, E.C. Sample & E.J. Kamprath, Amer. Soc. Agron., Madison, Wisconsin: 655–680.

Adams, F. & Martin, J.B., 1984. Liming effects on nitrogen use and efficiency. In: *Nitrogen in crop production*, ed. R.D. Hauck, Amer. Soc. Agron., Madison, Wisconsin: 417–426.

Adamson, A.H. & Pratt, C., 1987. An approval scheme for silage additives in the United Kingdom. In: *Developments in silage*, eds. J.M. Wilkinson & B.A. Stark, Chalcombe Publications, 13 Highwoods Drive, Marlow Bottom, Marlow, Bucks. Sl 7 3PU: 63–69.

Aitken, Y., 1974. *Flowering time, climate and genotype*. Melbourne University Press, Melbourne.

Alberda, Th., 1968. Some aspects of nitrogen in plants: more specifically in grasses. *Dutch Nitrogenous Rev. Stikstof* 12: 97–103.

Alder, F.E., 1954. Some effects of fertilizers on the output of leys with particular reference to extension of the grazing season. *J. Br. Grassld Soc.* 9(1): 29–33.

Allen, O.N. & Allen, E.K., 1981. The genus Rhizobium. In: *The leguminosae: a source book of characteristics, uses and nodulation*, Macmillan, London.

Allen, W.M., Moore, P.R. & Sansom, B.F., 1981. Control release glasses for selenium supplementation. In: *Trace element metabolism in man and animals*, eds. J.M. Gawthorne, J.McC. Howell & C.L. White, Aust. Acad. Sci., Canberra: 195–198.

Ammerman, C.B., 1965. *Feedstuffs* 37: 18.

Ammerman, C.B., Kawashima, T. & Henry, P.R., 1989. Inter-relationships of copper, zinc and iron in animal nutrition. In: *Recent progress on mineral nutrition and mineral requirements in ruminants*. Proc. Symp., Kyoto Shomado Insatsu Co., Kyoto, Japan: 47–55.

Andrew, C.S. & Johansen, C., 1978. Differences between pasture species in their requirements for nitrogen and phosphorus. In: *Plant relations in pastures*, ed. J.R. Wilson, CSIRO, East Melbourne, Australia: 111–127.

Andrew, C.S. & Robins, M.F., 1969. The effect of phosphorus on the growth and chemical composition of some tropical pasture legumes. II. Nitrogen, calcium, magnesium, potassium and sodium contents. *Aust. J. Agric. Res.* 20: 675–685.

Anon., undated (a). Inoculation and pelleting of legume seed. Unpublished report. Pasture Science Section, Natal Agric. Res. Inst., Dept. Agric. Tech. Serv., Pietermaritzburg.

Anon., undated (b). The construction and use of bench terraces to conserve steep land as developed for pineapple culture in the Umkomaas valley. Dept of Agriculture, Natal Region.

Anon., 1977. Farm planning manual: second draft. Natal Agric. Res. Inst., Dept. Agric. Tech. Serv., Pietermaritzburg.

Anon., 1980. *The nutrient requirements of ruminant livestock*. Agricultural Research Council, Commonwealth Agricultural Bureau, London.

Anon., 1983. *Energy allowances and feeding systems for ruminants: use of metabolisable energy system for sheep (section iv)*. Ministry of Agric. & Food, Her Majesty's Stationery Office, London: 51–65.

Anon., 1984. *Nutrient requirements of domestic animals: nutrient requirements of beef cattle*. 6th ed. Nat. Acad. Sci., Washington DC.

Anon., 1985. *Nutrient requirements of domestic animals: nutrient requirements of sheep*. 6th ed. Nat. Acad. Sci., Washington DC.

Anon., 1986. Regional development programme. Department of Agriculture, Transvaal Region, Pretoria.

Anon., 1989. *Nutrient requirements of domestic animals: nutrient requirements of dairy cattle*. 6th ed. Nat. Acad. Sci., Washington DC.

Anon., 1990. Gebruik Joodse soutbos gereeld. *Landb. Wkbld.* March 23: 52–55.

Anon., 1999. *Western Cape agricultural conditions*. Department of Economic Affairs, Agriculture and Tourism, Elsenberg 4(2): 7–9.

Anslow, R.C., 1966. The rate of appearance of leaves on Graminea. *Herb. Abst.* 36: 149–155.

ARC., 1980. *The nutrient requirements of ruminant livestock*. Commonwealth Agricultural Bureau, Farnham Royal.

Archibald, K.P., 1990. Ekonomie. In: *The Kynoch pasture handbook*, eds. E.B. Dickinson, G.F.S. Hyam & W.A.S. Breytenbach, Keyser & Versveld, Houghton.

Ariovich, D. & Cresswell, C.F., 1983. The effect of nitrogen on starch accumulation in two varieties of *Panicum maximum* Jacq. *Plant, Cell & Environment* 6: 657–664.

Arman, P. & Hopcraft, D., 1975. Nutritional studies on East African herbivores: 1. Digestibilities of dry matter, crude fibre and crude protein in antelope, cattle and sheep. *Br. J. Nutr.* 33: 255–264.

Arnold, G.W., 1964. Some principles in the investigation of selective grazing. *Proc. Aust. Soc. Anim. Prod.* 5: 258–271.

Arnold, G.W., 1981. Grazing behaviour. In: *Grazing animals:* World Animal Science, B1, ed. F.W. Morley, Elsevier, Amsterdam.

Arnold, G.W. & Dudzinski, M.L., 1978. *Ethology of free-ranging domestic animals.* Elsevier, New York.

Arnold, G.W. & Hill, J.L., 1972. Chemical factors affecting selection of food plants by ruminants. In: *Phytochemical ecology,* Proc. Phytochem. Symp., ed. J.B. Harborne, Academic Press, London: 71–101.

Arp, W.J., Darke, B.G., Pockman, W.T., Curtis, P.S. & Whigman, D.F., 1993. Interactions between C_3 and C_4 salt marsh plant species during four years of exposure to elevated atmospheric CO_2. *Vegetatio* 104/105: 133–143.

Atkinson, C.J. & Farrar, J.F., 1983. Allocation of photosynthetically fixed carbon in *Festuca ovina* L and *Nardis stricta. New Phytol.* 95: 519–531.

Aucamp, J.D. & Cloete, J.G., 1970. The utilisation of oldman saltbush. *Farming S. Afr.* 46: 3–7.

Aucamp, J.D. & De Kock, G.C., 1970. *Doringlose turksvye: die boer se voorsorg teen droogte.* Bull. No. 37, Dept. Agric. Tech. Serv., Pretoria.

Bahrani, J., Beaty, E.R. & Tan, K.H., 1983. Relationship between carbohydrate, nitrogen contents and regrowth of tall fescue tillers. *J. Range Manage.* 36: 234–235.

Baker, H.K., Chard, J.R.A. & Hughes, G.P., 1961. The production and utilization of winter grass at various centres in England and Wales. *J. Br. Grassld Soc.* 16(3): 185–189.

Ball, P.R. & Field, T.R.O., 1987. Nitrogen cycling in intensively-managed grasslands: a New Zealand viewpoint. In: *Nitrogen cycling in temperate agricultural systems,* Vol. 1, eds. P.E. Bacon, J. Evans, R.R. Storrier & A.C. Taylor, Aust. Soc. Soil Sci.: 91–112.

Ball, P.R. & Ryden, J.C., 1984. Nitrogen relationships in intensively-managed temperate grasslands. *Plant Soil* 76: 23–33.

Barber, S.A., 1980. Soil-plant interactions in the phosphorus nutrition of plants. In: *The role of phosphorus in agriculture,* eds. F.E. Khasawneh, E.C. Sample & E.J. Kamprath, Amer. Soc. Agron., Madison, Wisconsin: 591–616.

Barnard, S.A., 1985. *Oumansoutbos in winterreënstreek.* Weiding No. 140/1986. Dept. Agric. & Water Supply, Pretoria.

Barneix, A.J., James, D.M., Watson, E.F. & Hewett, E.J., 1984. Some effects of nitrate abundance and starvation on metabolism and accumulation of nitrogen in barley. *Planta* 162: 469–474.

Barnes, D.L., 1966. Studies in the use of Sabi *Panicum* for foggage. *Rhod. Agric. J.* 63: 144–154.

Barnes, D.L. & Dempsey, C.P., 1993. Grazing trials with sheep on kikuyu (*Pennisetum clandestinum* Chiov.) foggage in the eastern Transvaal highveld. *Afr. J. Range & For. Sci.* 10(1): 66–71.

Barnes, D.L., Rethman, N.F.G., Beukes, B.H. & Kotze, G.D., 1984. Veld composition in relation to grazing capacity. *J. Grassld Soc. Sth. Afr.* 1: 16–19.

Barnes, R.F., 1973. Laboratory methods of evaluating feed value of herbage. In: *Chemistry and biochemistry of herbage.* Vol. 3, eds. G.W. Butler & R.W. Bailey, Academic Press, London & New York: 179–214.

Barnes, R.F. Personal communication. Department of Agriculture, PO Box 3, Ermelo 2350.

Bartholomew, P.E., 1985. Beef production from kikuyu and Italian ryegrass. Ph.D. thesis. University of Natal, Pietermaritzburg.

Bartholomew, P.E., 1991. *A stocking rate model.* Natal Pastures Extension Pamphlet. Dept. Agric. Dev., Natal Region, Pietermaritzburg.

Bartholomew, P.E. & Booysen, P.deV., 1969. The influence of clipping frequency on reserve carbohydrates and regrowth of *Eragrostis curvula. Proc. Grassld Soc. Sth. Afr.* 4: 35–43.

Bartholomew, P.E. & Cross, G.W., 1973. *Eragrostis curvula.* Unpublished report. Pasture Enterprise Team, Natal Agric. Res. Inst., Dept. Agric. Tech. Serv., Pietermaritzburg.

Bartholomew, P.E. & Miles, N., 1984. Factors affecting the nitrogen requirement of pasture grasses in Natal. *Proc. Nitrogen Symposium.* Dept. Agric. Tech. Comm. No. 187: 122–126.

Bartholomew, P.E., Du Plessis, T.M. & MacDonald, C.I., 1991. Production and utilization of pasture foggage. *Cultivated Pastures in Natal.* Department of Agriculture, Government Printer, Pretoria.

Bartholomew, P.E., Louw, B.P. & MacDonald, C.I., 1994. Beef off annual ryegrass and annual ryegrass/clover: a natural progression. Presented at a series of seminars on intensive pasture production held under the auspices of the Grassld Soc. Zimbabwe. Natal Agric. Res. Inst., Dept. Agric. Tech. Serv., Pietermaritzburg.

Baurenfiend, J.C. & De Ritter, E., 1983. Nutritional supplements for animals: vitamins. In: *Handbook of nutritional supplements, Vol. II, Agricultural use,* ed. M. Rechigl Jr., CRC Press, Florida: 3–48.

Bayer, W. & Waters-Bayer, A., 1998. *Forage husbandry.* The Tropical Agriculturalist Series, Macmillan, London & Basingstoke.

Baylor, J.E., 1974. Satisfying the nutritional requirements of grass-legume mixtures. In: *Forage fertilization,* ed. D.A. Mays, Amer. Soc. Agron., Madison, Wisconsin: 171–188.

Beard, J.B., 1973. *Turfgrass: science and culture.* Prentice-Hall, Engelwood Cliffs, N.J.: 658 pp.

Beck, T., 1978. *The microbiology of silage fermentation.* Bayerische Landesanstalt für Bodenkultur, Pflanzenbau und Pflanzenschutz, München, German Federal Republic: 61–115.

Beckerling, A.C., 1989. Personal communication. Dept. Agric., Private Bag X804, Potchefstroom 2520.

Beede, D.K., 1992. The DCAD concept: transition rations for dry pregnant cows. *Feedstuffs* Dec. 28: 12–16.

Beekman, J.H. & Prins, H.H.T., 1989. Feeding strategies of sedentary large herbivores in East Africa, with emphasis on the African buffalo, *Syncerus caffer*. *Afr. J. Ecol.* 27: 129–147.

Bell, F.R., 1984. Aspects of ingestive behaviour in cattle. *J. Anim. Sci.* 59: 1369–1372.

Bell, R.H.V., 1970. The use of the herb layer by grazing ungulates in the Serengeti. In: *Animal populations in relation to their food resources*, ed. A. Watson, Blackwell, Oxford: 111–124.

Bell, R.H.V., 1971. A grazing ecosystem in the Serengeti. *Sci. Am.* 255: 86–93.

Bembridge, T.L., 1963. Protein supplementary feeding of breeding stock. *Rhod. Agric. J.* 60: 98–103.

Beringer, H., 1988. Potassium, sodium and magnesium requirements of grazing ruminants. *J. Sci. Food Agric.* 43: 323.

Beukes, D.J. & Kotze, A., 1988. Produksie praktyke vir graansorghum op turfgronde. Hoëveldfokus 2/88, Dept. Agric. Dev., Private Bag X804, Potchefstroom.

Bhattacharya, A.S. & Fontenot, J.P., 1966. Protein and energy value of peanut hull and wood-shaving poultry litters. *J. Anim. Sci.* 25: 367–371.

Bishop, E.J.B. & Kotze, J.J.J., 1965. Good strategy with beef cows: give supplementary feed just after calving. *Farming S. Afr.* 41: 6.

Bjorkman, O., 1973. Comparative studies on photosynthesis in higher plants. In: *Photophysiology Vol. III*, ed. A.C. Giese, Academic Press, New York: 1–63.

Black, A.S., Sherlock, R.R. & Smith, N.P., 1987. Effect of timing of simulated rainfall on ammonia volatilization from urea, applied to soil of varying moisture content. *J. Soil Sci.* 38: 679–687.

Black, A.S., Sherlock, R.R., Smith, N.P., Cameron, K.C. & Goh, K.M., 1985. Effects of form of nitrogen, season, and urea application rate on ammonia volatilization from pastures. *N. Z. J. Agric. Res.* 28: 469–474.

Black, J.N., 1957. The influence of varying light intensity on the growth of herbage plants. *Herb. Abstr.* 27: 89–98.

Blackman, G.E., 1956. Influence of temperature and light on leaf growth. In: *The growth of leaves*, ed. F.L. Milthorpe, Butterworths, London: 151–169.

Blair, R., 1975. Utilising wastes in animal feeds: a European overview. *Feedstuffs* June 30: 16.

Blankenship, L.A. & Qvortrup, S.A., 1974. Resource management on a Kenya game ranch. *J. Sth. Afr. Wildl. Mgmt. Ass.* 4: 185–190.

Blaxter, K.L., 1962. *The energy metabolism of ruminants*. Hutchinson, London.

Bonsma, H.C. & Maré, G.S., 1942. *Cactus and Oldman-saltbush as feed for sheep*. Dept. Agric. & For. Ext. Series No. 36, Pretoria.

Booysen, P.deV., 1954. An investigation into the effects of certain fertilisers on the yield and protein content and botanical composition of the veld.

M.Sc. Agric. thesis. University of Natal, Pietermaritzburg.

Booysen, P.deV., 1966. The physiological approach to research in pasture utilisation. *Proc. Grassld Soc. Sth. Afr.* 1: 77–85.

Booysen, P.deV., 1972. Pastoral productivity and intensification. *Proc. Grassld Soc. Sth. Afr.* 7: 51–55.

Booysen, P.deV., 1981. Radical veld improvement. In: *Veld and pasture management in South Africa*, ed. N.M. Tainton, Shuter & Shooter and University of Natal Press, Pietermaritzburg: 57–90.

Booysen, P.deV., Tainton, N.M. & Scott, J.D., 1963. Shoot apex development in grasses and its importance in grassland management. *Herb. Abstr.* 33: 209–213.

Bornemissza, G.F. & Williams, C.H., 1970. An effect of dung beetle activity on plant yield. *Pedobiologia* 10: 1–7.

Bosch, O.J.H., 1999. The ecology of the main grazing lands of South Africa. In: *Veld Management in South Africa*, ed. N.M. Tainton, University of Natal Press, Pietermaritzburg: 23–53.

Bousset-Fatianoff, N., Gouet, P., Bousset, J. & Comtrepois, M., 1971. Nitrate in fresh and ensiled forages. 2. Origin of nitrates, catabolism and effect of conservation treatments on nitrate degradation in silage. *Anales Biol. Anim. Biochem. Biophys.* 11: 715–723.

Bowen, G.D., 1961. The toxicity of legume seed diffusates towards rhizobia and other bacteria. *Plant & Soil* 15: 155–165.

Boyazoglu, P.A., 1973. Mineral imbalances of ruminants in southern Africa. *S. Afr. J. Anim. Sci.* 3: 149–152.

Bradshaw, A.D., 1980. Mineral nutrition. In: *Amenity grassland: an ecological perspective*, eds. I.H. Rorison & R. Hunt, John Wiley & Sons, New York: 101–118.

Bransby, D.I., 1981. The value of veld and pasture as animal feed. In: *Veld and pasture management in South Africa*, ed. N.M. Tainton, Shuter & Shooter and University of Natal Press, Pietermaritzburg: 173–214.

Bredon, R.M. & Stewart, P.G., 1979. *Guide to balanced feeding and management of dairy cattle*. Dept. Agric. Tech. Serv., Natal Region, Private Bag X9059, Pietermaritzburg 3200.

Bredon, R.M., Stewart, P.G. & Dugmore, T.J., 1987. *Nutritive value and chemical composition of commonly used South African farm feeds*. Natal Regional Dept. Agric. & Water Supply, Pietermaritzburg.

Brierem, K. & Ulvesli, O., 1960. Ensiling methods. *Herb. Abstr.* 30: 1–8.

Briggs, M.H., 1983. Nutritional supplements for animals: nonprotein nitrogen other than amino acids – origin, manufacture, composition, uses, dosage, effect. In: *Handbook of nutritional supplements, Vol. II, Agricultural use*, ed. M. Rechigl Jr., CRC Press, Florida: 87–97.

Briske, D.D. & Richards, J.H., 1995. Plant responses to defoliation: a physiological, morphological and demographic evaluation. In: *Wildland*

Plants: physiological ecology and developmental morphology, eds. D.J. Bedunah & R.E Sosebee, Soc. Range Manage., Denver, Colorado: 635–710.

Briske, D.D. & Woie, B.M., 1984. Plant response to defoliation: morphological considerations and allocation of priorities. *Proc. 2nd Int. Rangeland Congress*. Adelaide, Australia: 425–427.

Brockett, G.M., 1978. An evaluation of grazing management techniques for the characterisation of seasonal production potential of a ladino clover pasture. M.Sc. Agric. thesis. University of Natal, Pietermaritzburg.

Brockett, G.M., 1983. Pasture foggage as winter feed. *Arena* 6(2): 7–8.

Brockett, G.M. Personal communication. Nitrochem, 23 Plant Road, Howick 3290.

Brockett, G.M., Tainton, N.M., Booysen, P.deV. & Bransby, D.I., 1979. A comparison of grazing management techniques with sheep on ladino clover pastures. *Proc. Grassld Soc. Sth. Afr.* 14: 65–69.

Brockwell, J. & Phillips, L.J., 1970. Studies on seed pelleting as an aid to legume inoculation. 3. Survival of *Rhizobium* applied to seed sown in hot, dry soil. *Aust. J. Exp. Agric. Anim. Husb.* 10: 739–744.

Brockwell, J., Bottomby, P.J. & Thies, J., 1995. Manipulation of rhizobia microflora for improving legume productivity and soil fertility: a critical assessment. *Plant & Soil* 174: 143–180.

Brockwell, J., Gault, R.R., Chase, D.L., Hely, F.W., Zorin, H. & Corbin, E.J., 1980. An appraisal of practical alternatives to legume seed inoculation: field experiments on seedbed inoculation with solid and liquid inoculants. *Aust. J. Agric. Res.* 31: 47–60.

Bromfield, S.M., Cumming, R.W., David, D.J. & Williams, C.H., 1987. Long-term effects of incorporated lime and topdressed lime on the pH in the surface and subsurface of pasture soils. *Aust. J. Exp. Agric.* 27: 533–538.

Brougham, R.W., 1958. Interception of light by the foliage of pure and mixed stands of pasture plants. *Aust. J. Agric. Res.* 9: 39–52.

Brougham, R.W., 1960. The relationship between the critical leaf area, total chlorophyll content and maximum growth-rate in some pasture and crop plants. *Ann. Bot.* 24: 463–474.

Brougham, R.W., 1970. Frequency and intensity of grazing and their effects on pasture production. *Proc. N. Z. Grassld Assn.* 32: 137–144.

Brown, R.H. & Ashley, D.A., 1974. Fertiliser effects on photosynthesis, organic reserves and regrowth mechanisms of forages. In: *Forage Fertilisation,* ed. D.A. Mays, Amer. Soc. Agron., Madison, Wisconsin: 455–480.

Brown, R.H. & Blaser, R.E., 1965. Relationships between reserve carbohydrate accumulation and growth rate in Orchard grass and tall fescue. *Crop Sci.* 5: 577–582.

Brown, R.H. & Blaser, R.E., 1968. Leaf area index in pasture growth. *Herb. Abstr.* 38: 1–9.

Brown, R.W., 1995. The water relations of range plants: adaptations to water deficits. In: *Wildland plants: physiological ecology and developmental morphology*, eds. D.J. Bedunah & R.E. Sosebee, Soc. Range Manage., Denver, Colorado: 291–413.

Bucher, H.P., Machler, F. & Nöesberger, J., 1987a. Sink control of assimilate partitioning in meadow fescue (*Festuca pratensis* Huds). *J. Plant Physiol.* 129: 469–477.

Bucher, H.P., Machler, F. & Nöesberger, J., 1987b. Storage and remobilisation of carbohydrates in meadow fescue (*Festuca pratensis* Huds). *J. Plant Physiol.* 130: 101–109.

Butler, G.W. & Bailey, R.W., 1973. *Chemistry and Biochemistry of herbage.* Academic Press, London.

Buwai, M. & Trlica, M.J., 1977. Multiple defoliation effects on herbage yield, vigour and total nonstructural carbohydrate of five range species. *J. Range Manage.* 30: 164–171.

Buxton, D.R., 1989. Major edaphic and climatic stresses in the United States. In: *Persistence of forage legumes*, eds. G.C. Marten, A.G. Matches, R.F. Barnes, R.W. Brougham, R.J. Clements & G.W. Sheath, Proceedings of a trilateral workshop, Honolulu, Hawaii, July 1988, Madison, Wisconsin: 217–232.

Calder, C.M., 1966. Inflorenscence induction and initiation in the Graminea. In: *The growth of cereals and grasses*, eds. F.L. Milthorpe & K.D. Ivins, Butterworths, London: 59–73.

Caldwell, M.M., Osmond, C.B. & Not, D.L., 1977. C_4 pathway photosynthesis at low temperatures in cold-tolerant *Atriplex* species. *Plant Physiol.* 60: 157–164.

Caldwell, M.M., Richards, J.H., Johnson, D.A., Nowak, R.S. & Dzurec, R.S., 1981. Coping with herbivory: photosynthetic capacity and resource allocation in two semi-arid bunch grasses. *Oecologia* 50: 14–24.

Caloin, M., El Khodre, A. & Atry, M., 1980. Effects of nitrate concentration on the root shoot ratio in *Dactylis glomerata* L and on the kinetics of growth in the vegetative phase. *Ann. Bot.* 46: 165–173.

Carpenter, J.A., Boyce, K.G., Cameron, D.G., Collins, W.J., Read, J.N., Ross, B. & Williams, A., 1990. Pastures. In: *The manual of Australian agriculture*, ed. R.L. Reid, Butterworths, Sydney: 228–280.

Carter, E.D., 1987. Establishment and natural regeneration of annual pastures. In: *Temperate pastures, their production, use and management*, eds. J.L. Wheeler, C.J. Pearson & G.E. Robards, Australian Wool Corporation/CSIRO, Australia.

Carter, E.D. & Lake, A., 1985. Seed, seedlings and species dynamics of grazed annual pastures in South Australia. *Proc. XV Int. Grassld Congress*, Kyoto, Japan.

Cate, R.B. & Nelson, L.A., 1971. A simple statistical procedure for partitioning soil test correlative data into two classes. *Soil Sci. Soc. Amer. Proc.* 35: 658–660.

Chang, J., 1968. *Climate and agriculture: an ecological survey.* Aldine Publishing Company, Chicago.

Chapin, S.F. (III) & Slack, M., 1979. Effect of defoliation on root growth, phosphate absorption and respiration in nutrient-limited tundra graminoids. *Oecologia* 42: 67–79.

Chapman, G.O., 1996. *The biology of grasses.* Wallingford, Oxford.

Chatel, D.L. & Parker, C.A., 1973. Survival of field grown rhizobia over the dry summer period in western Australia. *Soil Biol. Biochem.* 5: 415–423.

Chatterton, N.J., Akao, S., Calson, G.E. & Hungerford, W.E., 1977. Physiological components of yield and tolerance to frequent harvests in alfalfa. *Crop Sci.* 17: 918–923.

Chiy, C.P. & Phillips, C.J.C., 1991. The effects of sodium chloride application to pasture, or its direct supplementation, on dairy cow production and grazing preference. *Grass & Forage Science* 46: 325–331.

Chu, A.C.P. & Robertson, A.G., 1974. The effects of shading and defoliation on nodulation and nitrogen fixation by white clover. *Plant Soil* 41: 509–519.

Chung, H.H. & Trlica, M.J., 1980. ^{14}C distribution and utilisation in blue gramma as affected by temperature, water potential and defoliation regimes. *Oecologia* 47: 190–195.

Clark, J.H. & Bartyh, K.M., 1970. Use of laboratory methods to estimate the performance of beef cattle fed mixed rations containing silage. *J. Anim. Sci.* 30: 268–273.

Clarke, A.L., 1980. Crop rotations in dryland farming systems. In: *Proceedings of an international congress on dryland farming.* Adelaide, South Australia.

Clarke, N.A., 1975. Forage crops. *The Agronomist.* University of Maryland, USA.

Clatworthy, J.N., 1973. Growth of *Stylosanthes* spp. as spaced plants at eight sites in Rhodesia. *Proc. Grassld Soc. Sth. Afr.* 8: 65–71.

Conolly, J., 1976. Some comments on the shape of the gain-stocking rate curve. *J. Agric. Sci. Camb.* 86: 103–109.

Coombe, J.B. & Tribe, D.E., 1962. The feeding of urea supplements to sheep and cattle: the results of penned feeding and grazing experiments. *J. Agric. Sci.* 59: 125–141.

Coombs, J., 1985. Carbon metabolism. In: *Techniques in bioproductivity and photosynthesis.* 2nd ed., eds. J. Coombs, D.O. Hall, S.P. Long & J.M.O. Scurlock, Pergamon Press, New York: 139–157.

Cooper, M. & Owen-Smith, N., 1985. Condensed tannins deter feeding by browsing ruminants in a South African savanna. *Oecologia* 67: 142–146.

Cooper, M.McG. & Morris, D.W., 1973. *Grass farming.* 3rd ed. Farming Press Ltd, Ipswich: 98–103.

Cooper, S.M. & Owen-Smith, N., 1986. Effects of plant spinescence on large mammalian herbivores. *Oecologia* 68: 446–455.

Cooper, S.M., Owen-Smith, N. & Bryant, J.P., 1988. Foliage acceptability to browsing ruminants in relation to seasonal changes in leaf chemistry of woody plants in South African savanna. *Oecologia* 75: 336–342.

Coppock, C.E., Everett, R.W. & Merrill, W.G., 1972. Effect of ration on free choice of calcium-phosphorus supplements by dairy cattle. *J. Dairy Sci.* 55: 245–256.

Corbett, J.L., 1957. Studies on the extension of the grazing season. Part I. *J. Br. Grassld Soc.* 12(2): 81–146.

Cornforth, I.S. & Sinclair, A.G., 1984. *Fertilizer recommendations for pastures and crops in New Zealand.* 2nd rev. ed., Ministry of Agric. & Fisheries, Wellington.

Cossens, G.G., 1984. Physical conditions of the soil influencing pasture production. In: *Pasture: the export earner.* New Zealand Inst. of Agric. Sci., Dunmore Press, Palmerston North: 111–115.

Costigan, O. & Ellis, K.J., 1980. Retention of copper oxide needles in cattle. *Anim. Prod. Aust.* 13: 451.

Coyne, P.I., Trlica, M.J. & Ownesby, C.E., 1995. Carbon and nitrogen dynamics in range plants. In: *Wildland plants: physiological ecology and developmental morphology,* eds. D.J. Bedunah & R.E. Sosebee, Soc. Range Manage., Denver, Colorado: 59–167.

Crabtree, J.R. & Williams, G.L., 1971a. The voluntary intake and utilisation of roughage-concentrate diets by sheep. 1. Concentrate supplements for hay and straw. *Anim. Prod.* 13: 71–82.

Crabtree, J.R. & Williams, G.L., 1971b. The voluntary intake and utilisation of roughage-concentrate diets by sheep. 2. Barley and soya bean meal supplementation of hay diets. *Anim. Prod.* 13: 83–92.

Crawley, M.J., 1983. *Herbivory: the dynamics of animal-plant interactions.* University of California Press, Berkley.

Cresswell, C.F., Ferrar, P., Grunow, J.O., Grossman, D., Rutherford, M.C. & Van Wyk, J.J.P., 1982. Phytomass, seasonal phenology and photosynthetic studies. In: *Ecology of tropical savannas,* eds. B.J. Huntley & B.H. Walker, Ecological Studies 42, Springer-Verlag, Berlin, Heildelberg, New York, Tokyo: 476–497.

Crider, F.J., 1955. Root-growth stoppage resulting from defoliation of grass. *USDA Tech. Bull.* No. 1102: 2–23.

Cronje, P.B., 1983. Protein degradability of several South African feedstuffs by the artificial fibre bag technique. *S. Afr. J. Anim. Sci.* 13: 225–228.

Cross, G.W. & Theron, E.P., 1970. Effectiveness of paraquat on the ngongoni veld of the Natal Mistbelt. *Proc. Grassld Soc. Sth. Afr.* 5: 101–105.

Crowder, L.V. & Chheda, H.R., 1982. *Tropical Grassland Husbandry.* Longman, New York: 225–278.

Curll, M.L. & Jones, R.M., 1989. The plant-animal interface and legume persistence: an Australian perspective. In: *Persistence of forage legumes,* eds. G.C. Marten, A.G. Matches, R.F. Barnes, R.W. Brougham, R.J. Clements & G.W. Sheath, Proceedings of a trilateral workshop, Honolulu, Hawaii, July 1988, Madison, Wisconsin: 339–359.

Daer, T. & Willard, E.E., 1981. Total nonstructural carbohydrate trends in bluebush wheatgrass related to growth and phenology. *J. Range Manage.* 34: 377–379.

Dahl, B.E., 1995. Developmental morphology of plants. In: *Wildland plants: physiological ecology and developmental morphology,* eds. D.J. Bedunah & R.E. Sosebee, Soc. Range Manage., Denver, Colorado: 22–58.

Daines, T., 1980. The use of grazing pattern in the management of Döhne Sourveld. *Proc. Grassld Soc. Sth. Afr.* 15: 185–188.

Daitz, J., 1954. *Available carbohydrate reserves in the roots of* Themeda triandra *from a seasonal burn experiment at Bethal.* Rep. Frankenwald Fld. Res. Stn., University of the Witwatersrand: 27–29.

Danckwerts, J.E., 1984. Towards improved livestock production off sweet grassveld. Ph.D. thesis. University of Natal, Pietermaritzburg.

Danckwerts, J.E., 1988. Growth and desiccation of *Themeda triandra* and *Sporobolus fimbriatus* in relation to diminishing moisture availability. *J. Grassld Soc. Sth. Afr.* 5(2): 96–102.

Danckwerts, J.E., 1989. Animal performance. In: *Veld management in the Eastern Cape,* eds. J.E. Danckwerts & W.R. Teague, Dept. Agric. & Water Supply, Eastern Cape Region, Private Bag X15, Stutterheim 4930: 47–60.

Danckwerts, J.E., 1993. Reserve carbon and photosynthesis: their role in regrowth of *Themeda triandra,* a widely distributed subtropical graminaceous species. *Funct. Ecolo.* 7: 634–641.

Danckwerts, J.E. & Gordon, A.J., 1987. Long-term partitioning, storage and re-mobilisation of ^{14}C assimilation by *Lolium perenne* (cv. Melle). *Ann. Bot.* 59: 55–66.

Danckwerts, J.E. & King, P.G., 1984. Conservative stocking or maximum profit: a grazing management dilemma? *J. Grassld Soc. Sth. Afr.* 1: 25–28.

Danckwerts, J.E., Aucamp, A.J. & Barnard, H.J., 1983. Herbaceous species preference by cattle in the False Thornveld of the eastern Cape. *Proc. Grassld Soc. Sth. Afr.* 18: 89–94.

Dannhauser, C.S., 1988. A review of foggage in the Central Grassveld area with special reference to *Digitaria eriantha. J. Grassld Soc. Sth. Afr.* 5: 193–196.

Dannhauser, C.S., Van Rensburg, W.L.J., Opperman, D.P.J. & Van Rooyen, P.J., 1987. Die produksie van smutvingergras in die Wes-Transvaal. *J. Grassld Soc. Sth. Afr.* 4(4): 148–151.

Daphne, C., 1993. Response of *Themeda triandra* to defoliation. M.Sc. Agric thesis. University of Natal, Pietermaritzburg.

Date, R.A., 1970. Microbiological problems in the inoculation and nodulation of legumes. *Plant & Soil* 32: 703–725.

Date, R.A. & Brockwell, J., 1978. *Rhizobium* strain competition and host interaction for nodulation. In: *Plant relations in pastures,* ed. J.R. Wilson, CSIRO, Australia: 202–216.

Davidson, J.L. & Milthorpe, F.L., 1966. The effects of defoliation on the carbon balance in *Dactylis glomerata. Ann. Bot.* 30: 185–198.

Davies, B.E. & Jones, L.H.P., 1988. Micronutrients and toxic elements. In: *Russell's soil conditions and plant growth,* 11th ed., ed. A. Wild, Longman, England: 780–814.

Davies, R.A. & Skinner, J.D., 1986. Diet selection by springbok and Merino sheep during a Karoo drought. *Trans. R. Soc. S. Afr.* 46: 165–176.

Decker, A.M., Taylor, T.H. & Willard, C.J., 1982. Establishment of new seedlings. In: *Forages,* eds. M.E. Heath, D.S. Metcalf & R.F. Barnes, Iowa State University Press, Iowa: 384–395.

De Figueiredo, M., 1987. Factors affecting the quality of *Pennisetum clandestinum* (Kikuyu grass) silage. Ph.D. thesis. University of Natal, Pietermaritzburg.

De Figueiredo, M., Marais, P. & Tainton, N.M., 1989. Effect of wilting on biochemical and microbiological changes in kikuyu (*Pennisetum clandestinum* Hochst) silage. *J. Grassld Soc. Sth. Afr.* 6(2): 99–103.

De Figueiredo, M., Stewart, I.B. & Botha, W.A., 1994. Performance of beef weaners on kikuyu silage supplemented with maize meal alone and in combination with protein of low degradability. *J. Sci. Food Agric.* 66: 133–137.

Deinum, B.E., 1984. Chemical composition and nutritive value of herbage in relation to climate. In: *Proc. X General Meeting European Grassland Federation: the impact of climate on grass production and quality,* eds. H. Riley & A.O. Skelvag, As, Norway: 338–350.

De Kock, G.C., 1980a. Drought resistant fodder crops in South Africa. In: *Browse in Africa,* ed. H.N. Le Houerou, Symp. on browse in Africa, Int. Livestock Centre for Africa, Addis Ababa, Ethopia.

De Kock, G.C., 1980b. Droogtevoergewasse. *Proc. Grassld Soc. Sth. Afr.* 2: 147–156.

Dennison, C. & Phillips, A.M., 1983. Balancing the duodenal amino acid supply in ruminants with practical feed ingredients. *S. Afr. J. Anim. Sci.* 13: 229–235.

Denton, D.A. & Sabine, J.R., 1963. The behaviour of Na deficient sheep. *Behaviour* 20: 264–276.

Deregibus, V.A., Trlica, M.J. & Jameson, D.A., 1982. Organic reserves in herbage plants: their relationship to grassland management. In: *Handbook of agricultural productivity. Vol. 1, Plant Productivity,* ed. M.J. Rechigl Jr., CRC Press, Florida: 315–344.

Devlin, R.M. & Witham, F.W., 1983. *Plant Physiology.* 4th ed. Willard Grant Press, Boston: 577 pp.

De Waal, H.O., 1990. Animal production from native pasture (veld) in the Free State region: a perspective of the grazing ruminant. *S. Afr. J. Anim. Sci.* 20: 1–9.

De Waal, H.O., Baard, M.A. & Engels, E.A.N., 1989a. Effects of sodium chloride on sheep. 1. Diet composition, body mass change and wool production of young merino wethers grazing native pasture. *S. Afr. J. Anim. Sci.* 19: 27–33.

De Waal, H.O., Baard, M.A. & Engels, E.A.N., 1989b. Effects of sodium chloride on sheep. 2. Voluntary feed intake and changes in certain rumen parameters of young merino wethers grazing native pasture. *S. Afr. J. Anim. Sci.* 19: 34–42.

Dewald, C.L. & Sims, P.L., 1981. Seasonal vegetative establishment and shoot reserves of eastern gamagrass. *J. Range Manage.* 34: 300–304.

Dewar, A., McDonald, P. & Whittenbury, R., 1963. The hydrolysis of grass hemicellulose during ensilage. *J. Sci. Food Agric.* 14: 411–417.

Donald, C.M., 1963. Competition among crop and pasture plants. *Adv. Agron.* 15: 1–118.

Donald, C.M. & Black, J.N., 1958. The significance of leaf area in pasture growth. *Herb. Abstr.* 28: 1–6.

Downtown, W.J.S., 1971. Adaptive and evolutionary aspect of C₄ photosynthesis. In: *Photosynthesis and photorespiration*, eds. M.D. Hatch, C.B. Osmond & R.O. Slatyer, Wiley-Interscience, New York: 3–17.

Doyle, P.T., 1987. Supplements other than forages. In: *The nutrition of herbivores*, eds. J.B. Hacker & J.H. Ternouth, Academic Press, Sydney: 429–464.

Drewes, R.H., 1995. Personal communication. Dept. Agric., Nelspruit.

Dugmore, T.J., Lesch, S.F. & Walsh, K.P., 1987. The effects of magnesium oxide supplementation on the fertility of dairy cows grazing fertilised pastures. *S. Afr. J. Anim. Sci.* 17: 183–185.

Duke, J.A., 1981. *Handbook of legumes of world economic importance*. Plenum Press, New York & London.

Duncan, P., 1991. *Horses and grasses: A study of horses and their impact on the Camargue*. Springer-Verlag: 275 pp.

Du Pisani, L.G., 1992. Simulasiestudies met *Cenchrus ciliaris* L. cv. Molopo. Ph.D. thesis. University of the Orange Free State, Bloemfontein.

Du Pisani, L.G., Van Niekerk, P.A., De Waal, H.O. & Knight, I.W., 1987. Die evaluasie van *Cenchrus ciliaris* cv. Molopo vir speenkalfproduksie in die sentrale grasveld. *Proc. Grassld Soc. Sth. Afr.* 4(2): 55–58.

Du Plessis, T.M., 1978. An investigation into the validity of some of the pasture and veld norms used in budget feed programmes in the Underberg district. M.Sc. Agric. thesis. University of the Orange Free State, Bloemfontein.

Du Toit, E.W., 1992. An investigation into the dynamics of tannin production in southern African grasses. M.Sc. thesis. University of Pretoria, Pretoria.

Du Toit, P.J., Louw, J.G. & Malan, A.I., 1940. A study of the mineral content and feeding value of natural pastures in the Union of South Africa. *Onderstepoort J. Vet. Res.* 14: 123–129.

Eckard, R.J., 1989. The response of Italian ryegrass to applied nitrogen in the Natal Midlands. *J. Grassld Soc. Sth. Afr.* 6: 19–22.

Eckard, R.J., 1990. The effect of source of nitrogen on the dry matter yield, nitrogen and nitrate-N content of *Lolium multiflorum. J. Grassld Soc. Sth. Afr.* 7: 208–209.

Eckard, R.J. & Dugmore, T.J., 1993. Livestock health and production as influenced by nitrogen fertiliser management. In: *Pastures in Natal*, Natal Region, Dept. Agric., Cedara.

Eckard, R.J. & Franks, D.R., 1998. Strategic nitrogen fertiliser use on perennial ryegrass and white clover pasture in north-western Tasmania. *Australian Journal of Experimental Agriculture* 38: 155–160.

Edmeades, D.C. Personal communication. Agricultural Research Ruakuru, Private Bag 3123, Hamilton, New Zealand.

Edmeades, D.C., Wheeler, D.M. & Pringle, R.M., 1990. Effects of liming on soil phosphorus availability and utilization. In: *Phosphorus requirements for sustainable agriculture in Asia and Oceania*. International Rice Res. Institute, Manila, Phillipines: 255–267.

Edwards, P.J., 1970. Radical veld improvement in South Africa, with special reference to the Highland Sourveld of Natal. Ph.D. thesis. University of Natal, Pietermaritzburg.

Edwards, P.J., 1978. Methods of veld reinforcement, their action and adaptability to various sites. *Proc. Grassld Soc. Sth. Afr.* 13: 71–74.

Edwards, P.J. & Booysen, P.deV., 1972. The future for radical veld improvement in South Africa. *Proc. Grassld Soc. Sth. Afr.* 7: 61–66.

Edwards, P.J. & Mappledoram, B.D., 1979. The effect of stocking rate on beef production from star grass (*Cynodon nlemfuensis* Harl.). *Proc. Grassld Soc. Sth. Afr.* 14: 101–105.

Edwards, P.J. & Scotney, D.M., 1978. Site assessment for pasture production. *Proc. Grassld Soc. Sth. Afr.* 13: 65–70.

Ellis, R.P., 1977. Distribution of the Krantz syndrome in southern African *Eragrostideae* and *Panicoideae* according to bundle sheath anatomy and cytology. *Agroplantae* 9: 73–110.

Ellis, R.P., Vogel, J.C. & Fuls, A., 1980. Photosynthetic pathways and the geographical distribution of grasses in South West Africa/Namibia. *S. Afr. J. Sci.* 76: 307–314.

Engels, E.A.N., De Waal, H.O., Biel, L.C. & Malan, A., 1981. Practical implications of the effect of drying and treatment on nitrogen content and *in vitro* digestibility of samples collected by oesophageally fistulated animals. *S. Afr. J. Anim. Sci.* 11: 116–119.

Erasmus, L.S., De Villiers, T.T. & Du Plessis, J.J., 1982. Die kleinveebedryf in die Winterreënstreek. *Winterreën Spes. Uitgawe* 5: 2–7.

Erasmus, L.J., Prinsloo, J. & Meissner, H.H., 1988. The establishment of a protein degradability data base for dairy cattle using the nylon bag technique. 1. Protein sources. *S. Afr. J. Anim. Sci.* 18: 23–29.

Etzel, M.G., Volenec, J.J. & Vorst, J.J., 1988. Leaf morphology, shoot growth, and gas exchange of multifoliate alfalfa phenotypes. *Crop Sci.* 28: 263–269.

Evans, M.W., 1958. Growth and development in certain economic grasses. *Ohio Agr. Exp. Sta. Agron. Series* 147: 1–23.

Everson, C.S., 1999. Veld burning: Grassveld. In: *Veld management in South Africa*, ed. N.M. Tainton, University of Natal Press, Pietermaritzburg: 228–235.

Fair, J., 1986. *Guide to profitable pastures.* M & J Publications, Harrismith, South Africa.

Farina, M.P.W., 1983. Lime and liming: quality and requirement. *Mielies/Maize* June: 29–32.

Farina, M.P.W. & Johnston, M.A., 1987. *Fertilization guidelines for maize production in Natal.* Farming in South Africa leaflet series, Maize F.2.4.

Farina, M.P.W., Cross, G.W. & Channon, P., 1972. The influence of sulphur on the yield of a grass-clover pasture fertilized with different sources of phosphorus. *Fert. Soc. Sth Afr. J.* 1: 1–3.

Farina, M.P.W., Channon, P., Thibaud, G.R. & Phipson, J.D., 1992. Soil and plant potassium optima for maize on a kaolinitic clay soil. *S. Afr. J. Plant Soil* 9: 193–200.

Farquhar, G.D. & Richards, R.A., 1984. Isotopic composition of plant carbon correlates with water-use efficiency of wheat genotypes. *Aust. J. Plant Physiol.* 11: 539–552.

Farquhar, G.D., Ehleringer, J.R. & Hubick, K.T., 1989. Carbon isotope discrimination and photosynthesis. *Ann. Rev. Plant Physiol. & Plant Molec. Biol.* II: 539–557.

Fey, M.V., Manson, A.D. & Schutte, R., 1990. Acidification of the pedosphere. *S. Afr. J. Sci.* 86: 403–406.

Field, C.R., 1976. Palatability factors and nutritive value of the food of buffaloes (*Syncerus caffer*) in Uganda. *East Afr. Wildl. J.* 14: 181–201.

Field-Dodgson, J., 1973. *Producing seed of weeping lovegrass.* Herbage Seed Production Series. No. 1, Dept. Pasture Science, University of Natal, Pietermaritzburg.

Field-Dodgson, J., 1974. *Producing Italian ryegrass seed.* Herbage Seed Production Series. No. 2, Dept. Pasture Science, University of Natal, Pietermaritzburg.

Finch, V.A. & Western, D., 1977. Cattle colours in pastoral herds: natural selection or social preference? *Ecol.* 58: 1384–1392.

Finck, A., 1982. *Fertilizers and fertilization.* Verlag Chemie, New York.

Fitter, A.H. & Hay, R.K.M., 1981. *Environmental physiology of plants.* Academic Press, London: 355 pp.

Fitzgerald, R.D., 1976. Effects of stocking rate, lambing time and pasture management on wool and lamb production on annual Subterranean clover pastures. *Aust. J. Agric. Res.* 27: 261–265.

Forde, M.B., Hay, M.J.M. & Brock, J.L., 1989. Development and growth characteristics of temperate perennial legumes. In: *Persistence of Forage legumes,* eds. G.C. Marten, A.G. Matches, R.F. Barnes, R.W. Brougham, R.J. Clements & G.W. Sheath, Proc. trilateral workshop, Honolulu, Hawaii, July 1988, Madison, Wisconsin: 91–109.

Fox, R.L., Nishimoto, R.K., Thompson, J.R. & De La Pena, R.S., 1974. Comparative external phosphorus requirements of plants growing in tropical soils. *Trans. 10th Int. Congr. Soil Sci.*, Moscow: 232–239.

Frame, J., 1992. *Improved grassland management.* Farming Press, Ipswich, UK.

Frame, J. & Newbould, P., 1986. Agronomy of white clover. *Advances in Agronomy* 40: 1–88.

Frame, J., Charlton, J.F.L. & Laidlaw A.S., 1998. *Temperate forage legume.* CAB, Wallingford, UK.

Frame, J., Harkness, R.D. & Talbot, M., 1989. The effect of cutting frequency and fertilizer nitrogen rate on herbage productivity from perennial ryegrass. *Res. Dev. Agric.* 6: 99–105.

Fritz, J.C., 1983. Nutritional supplements for animals: trace mineral elements. In: *Handbook of nutritional supplements Vol. II, Agricultural use,* ed. M. Rechigl Jr., CRC Press, Florida: 67–85.

Gammon, D.I. & Roberts, B.R., 1980. Grazing behaviour of cattle during continuous and rotational grazing of the Matopos sandveld in Rhodesia. *Rhod. J. Agric. Res.* 18: 13–27.

Gartrell, J.W. & Bolland, M.D.A., 1987. Phosphorus nutrition of pastures. In: *Temperate pastures: their production, use and management,* eds. J.L. Wheeler, C.J. Pearson & G.E. Robards, CSIRO, Australia: 127–136.

Gault, R.R. & Brockwell, J., 1980. Studies on seed pelleting as an aid to legume inoculation. 5. Effects of incorporation of molybdenum compounds in the seed pellets on inoculant survival, seedling nodulation and plant growth of lucerne and subterranean clover. *Aust. J. Exp. Agric. Anim. Husb.* 20: 63–71.

Gibb, M.J. & Treacher, T.T., 1983. The performance of lactating ewes offered diets containing different proportions of fresh perennial ryegrass and white clover. *Anim. Prod.* 37: 433–440.

Gibb, M.J. & Treacher, T.T., 1984. The performance of weaned lambs offered diets containing different proportions of fresh perennial ryegrass and white clover. *Anim. Prod.* 38: 413–420.

Gibson, T., Stirlin, A.C., Kedie, R.M. & Roenberger, R.F., 1961. Bacteriological changes in silage as affected by laceration of fresh grass. *J. Appl. Bact.* 24: 60–70.

Gifford, R.M. & Marshall, C., 1973. Photosynthesis and assimilate distribution in *Lolium multiflorum* LAM following differential tiller defoliation. *Aust. J. Biol. Sci.* 26: 517–526.

Gillard, P. 1967. Coprophagous beetles in pasture ecosystems. *J. Aust. Inst. Agric. Sci.* 33: 30–34.

Goh, K.M. & Nguyen, M.L., 1990. Effects of grazing animals on the plant availability of sulphur fertilizers in grazed pastures. *Proc. N. Z. Grassld Ass.* 52: 181–185.

Gold, W.G. & Caldwell, M.M., 1989. The effects of

spatial pattern of defoliation on regrowth of a tussock grass. II. Canopy gas exchange. *Oecologia* 81: 437–442.

Golding, E.J., 1985. Sources of supplemental protein or energy. (b) Biuret. In: *Nutrition of grazing ruminants in warm climates*, ed. L.R. McDowell, Academic Press, Orlando: 140.

Goodrich, R.D., 1978. *Sulphur in ruminant nutrition*. National Feed Ingredients Assn, West Des Moines, Iowa.

Goodrich, R.D. & Garrett, J.E., 1986. Sulphur in livestock nutrition. In: *Sulphur in agriculture*, ed. M.A. Tabatabai, Amer. Soc. Agron., Madison, Wisconsin: 617–634.

Gordijn, R.J., 1989. Personal communication. Macaulay Land Use Research Institute, Cragiebuckler, Aberdeen, AN9 2QJ, Scotland, UK.

Gordon, A.J., Ryle, G.J.A. & Powell, C.E., 1977. The strategy of carbon utilisation in uniculm barley. *J. Exp. Bot.* 28(197): 1258–1269.

Gordon, I.J. & Lascano, C., 1993. Foraging strategies of ruminant livestock on intensively managed grasslands: potential and constraints. *Proc. XVII Int. Grassld Congr.*, New Zealand & Australia: 681–690.

Gordon, J.G., Tribe, D.E. & Graham, T.C., 1954. The feeding behaviour of phosphorus-deficient cattle and sheep. *Br. J. Anim. Behaviour* 2: 72–74.

Gouet, P. & Chevalier, R., 1966. The evolution of gram-negative microflora in direct-harvested and wilted alfalfa silages. *Proc. 10th Int. Grassld Congr.*, Kelsinki, Finland: 533–536.

Graven, E.H., 1967. Some edaphic considerations regarding the introduction of legumes into veld. *Proc. Grassld Soc. Sth. Afr.* 2: 63–69.

Green, G.C., Elwin, R.L., Mattershead, B.E., Keogh, R.G. & Lynch, J.J., 1984. Long-term effect of early experience to supplementary feeding in sheep. *Proc. Aust. Soc. Anim. Prod.* 15: 373–375.

Greenhill, L., 1964. Buffering capacity of pasture plants with respect to ensilage. *Aust. J. Agric. Res.* 15: 511–519.

Groenewald, J.W. & Boyazoglu, P.A., 1980. *Animal nutrition: concepts and application*. J.L. van Schaik, Pretoria.

Grunow, J.O., 1979. Personal communication. Dept. Plant Production, University of Pretoria, Pretoria.

Grunow, J.O., 1980. Feed and habitat preferences among some large herbivores on African veld. *Proc. Grassld Soc. Sth. Afr.* 15: 141–146.

Grunow, J.O. & Pieterse, P.A., 1984. A comparison of highveld fodders and fodder flows for growing out long yearling steers. *J. Grassld Soc. Sth. Afr.* 3: 25–29.

Grunow, J.O., Pienaar, A.J. & Breytenbach, C., 1970. Long term nitrogen application to veld in South Africa. *Proc. Grassld Soc. Sth. Afr.* 5: 75–90.

Gulbransen, B., 1974. Utilisation of grain supplements by roughage fed cattle. *Proc. Aust. Soc. Anim. Prod.* 10: 74–77.

Gutierrez, M., Gracen, V.E. & Edwards, G.E., 1974. Biochemical and cytological relationships in C_4 plants. *Planta* 119: 179–300.

Hacker, J.B. & Minson, D.J., 1981. The digestibility of plant parts. *Herb. Abstr.* 51: 459–482.

Hackland, N.G.E. & Jones, R.I., 1980. Predicting seasonal production of *Eragrostis curvula*. *Proc. Grassld Soc. Sth. Afr.* 15: 85–89.

Hall, T.D., Meredith, D. & Murray, S.M., 1940. Fertilising natural veld and its effect on sward, chemical composition, carrying capacity and leaf production. *S. Afr. J. Sci.* 37: 11.

Hardy, M.B., 1994. Annual reports. Facet No. N5413/30/5/14. Cedara Agricultural Development Institute, Department of Agriculture.

Hardy, M.B., 1999. Personal communication. Private Bag X1, Elsenberg 7607.

Hardy, M.B. & Mentis, M.T., 1986. Grazing dynamics in sour grassveld. *S. Afr. J. Sci.* 82: 566–572.

Harkin, J.M., 1973. Lignin. In: *Chemistry and biochemistry of herbage Vol. III*, eds. G.W. Butler & R.W. Bailey, Academic Press, London: 323–373.

Harris, W. & Brougham, R.W., 1968. Some factors affecting change in botanical composition in a ryegrass-white clover pasture under continuous grazing. *N. Z. J. Agric. Res.* 11: 15–38.

Harris, W. & Thomas, V.J., 1972. Competition among pasture plants. 2. Effects of frequency and height of cutting on competition between *Agrostis tenuis* and two ryegrass cultivars. *N. Z. J. Agric. Res.* 15: 19–32.

Hart, A.L., 1987. Physiology. In: *White Clover*, eds. M.J. Baker & W.M. Williams, CAB International, The Cambrian News Ltd, Aberystwyth: 132–133.

Hart, R.H., 1978. Stocking rate theory and its application to grazing on rangeland. In: *Proc. 1st Int. Range. Congr.*, ed. D.N. Hyder, Soc. Range Manage., Denver, Colorado: 547–550.

Hart, R.H., Monson, W.G. & Lowrey, R.S., 1969. Autumn-saved coastal Bemuda grass (*Cynodon dactylon* L Pers.): effects of age & fertility on quality. *Agron. J.* 61(6): 940–941.

Harwin, G.O., 1989. Strategies for beef production in South Africa: a selection of lectures. Stockowners Co-operative Limited.

Hassan, N.I., Abd-Elaziz, & El-Tabbakh, A.E., 1979. Evaluation of some forages introduced to newly-reclaimed areas in Egypt. *World Rev. Anim. Prod.* XV(2): 31–36.

Hatch, D.J. & MacDuff, J.H., 1991. Concurrent rates of N_2 fixation, nitrate and ammonium uptake by white clover in response to different root temperatures. *Ann. Bot.* 67: 265–274.

Hatch, M.D. & Slack, C.R., 1966. Photoynthesis by sugarcane leaves. A new carboxylation reaction and the pathway of sugar formation. *Biochem. J.* 101: 103–111.

Hatch, M.D., Slack, C.R. & Johnson, H.S., 1967. Further studies on a new pathway of photosynthetic carbon dioxide fixation in sugarcane and its occurrence in other plant species. *Biochem. J.* 102: 417–422.

Hattersley, P.W. & Watson, L., 1976. C_4 grasses: an anatomical criterion for distinguishing betweeen NADP malic enzyme species and PCK or NAD malic enzyme species. *Aust. J. Bot.* 24: 297–308.

Hay, R.K.M. & Walker, A.J., 1989. *An introduction to the physiology of crop yield*. Longman Scientific and Technical, Essex, England: 292 pp.

Haynes, R.J., 1983. Soil acidification induced by leguminous crops. *Grass Forage Sci*. 38: 1–11.

Haynes, R.J., 1984. Lime and phosphate in the soil–plant system. *Adv. Agron*. 37: 249–315.

Haynes, R.J., 1986. *Mineral nitrogen in the plant–soil system*. Academic Press, London.

Haynes, R.J. & Goh, K.M., 1978. Ammonium and nitrate nutrition in plants. *Biol. Rev*. 53: 465–510.

Heady, H.F., 1964. Palatability of herbage and animal preference. *J. Range Manage*. 17: 76–82.

Heady, H.F., 1975. *Rangeland management*. McGraw Hill Book Co., New York.

Heany, D.P., 1970. Voluntary intake as a component of an index of forage quality. *Proc. Nat. Conf. Forage Qual. Eval. Util*. Lincoln, N.E.

Hefer, G.D., 1989. The effects of liquid urea ammonium nitrate fertilizer on dryland kikuyu and coastcross II pastures. M.Sc. Agric. thesis. University of Natal, Pietermaritzburg.

Heitschmidt, R.K. & Taylor, C.A., 1991. Livestock production. In: *Grazing management: an ecological perspective*, eds. R.K. Heitschmidt & J.W. Stuth, Timber Press, Portland, Oregon: 161–178.

Hely, F.W. & Brockwell, J., 1962. An exploratory survey of the ecology of *Rhizobium meliloti* in inland New South Wales and Queensland. *Aust. J. Agric. Res*. 13: 864–879.

Hely, F.W., Hutchings, R.J. & Zorin, M., 1976. Legume inoculation by spraying suspensions of nodule bacteria into soil beneath seed. *J. Aust. Inst. Agric. Sci*. 42: 241–244.

Helyar, K., 1991. Do perennial pastures reduce soil acidity? *Proc. 6th Ann. Conf. Grassld Soc. NSW*: 64–71.

Hennessy, D.W., Williamson, P.J., Nolan, J.V., Kempton, T.J. & Leng, R.A., 1983. The role of energy and protein rich supplements in the subtropics for young cattle consuming basal diets that are low in digestible energy and protein. *J. Agric. Sci*. 100: 657–666.

Henzell, E.F. & Ross, P.J., 1973. The nitrogen cycle of pasture ecosystems. In: *Chemistry and biochemistry of herbage*, eds. C.W. Butler & R.W. Bailey, Academic Press, New York: 227–246.

Herridge, D.F., 1982. Crop rotations involving legumes. In: *Nitrogen fixation in legumes*, ed. J.M. Vincent, Academic Press, New York.

Herriott, J.B.D. & Wells, D., 1960. Clover nitrogen and sward productivity. *J. Brit. Grassld Soc*. 15: 63–69.

Hewitt, E.J., 1975. Assimilatory nitrate-nitrite reduction. *Ann. Rev. Plant Physiol*. 26: 73–100.

Higgins, S.P., 1992. Selenium availability in soils and pastures of the Natal Midlands. M.Sc. Agric. thesis. University of Natal, Pietermaritzburg.

Hill, P.R., Scotney, D.M. & Wilby, A.F., 1981. Wetland development ridge and furrow system. Natal Farming guide (Section 7). Natal Region, Dept. Agric. & Fisheries.

Hobson, V., Grabbelaar, P.D., Wentzel, D. & Koen, A., 1986. Effect of level of supplementary feeding on mohair production and reproductive performance of Angora ewes grazing *Atriplex nummularia*. *S. Afr. J. Anim. Sci*. 2: 95–96.

Hodgson, J. & Grant, S.A., 1985. The grazing ecology of hill and upland swards. In: *Hill and upland livestock production*, eds. T.J. Maxwell & R.G. Gunn, Br. Soc. Anim. Prod. Occ. Publ. No. 10: 77–84.

Hofmann, R.R., 1989. Evolutionary steps of ecophysiological adaption and diversification of ruminants: a comparative view of their digestive systems. *Oecologia* 78: 443–457.

Hofmann, R.R. & Stewart, D.R.M., 1972. Grazer or browser: a classification based on the stomach structure and feeding habits of East African ruminants. *Mammalia* 36: 226–240.

Holmes, C.W. & Wilson, G.F., 1987. *Milk production from pasture*. Butterworths, Wellington, New Zealand.

Holmes, W., 1989. *Grass: its production and utilisation*. 2nd ed. Blackwell Scientific Publications, Oxford: 7–42.

Honig, H. & Woodford, M.K., 1980. Changes in silage on exposure to air. *Occ. Symp. Brit. Grassld Soc*. 11: 76–87.

Hoppe, P.P., 1977. Comparison of voluntary food and water consumption in Kirk's dik-dik and suni. *East Afr. Wildl. J*. 15: 41–48.

Hoppe, P.P., Qvortrup, S.A. & Woodford, M.H., 1977. Rumen fermentation and food selection in East African Zebu cattle, wildebeest, Coke's hartebeest and topi. *J. Zool. London* 181: 1–9.

Horn, G., 1988. Experiences with 'Magnesia-Kainet' on grassland. *J. Sci. Food Agric*. 43: 324.

Hudson, R.J. & Christofferson, R.J., 1985. Maintenance metabolism. In: *Bioenergetics of wild herbivores*, eds. R.J. Hudson & R.G. White, CRC Press, Florida: 121–142.

Hughes, G.P., 1955. The production and utilization of winter grass. *J. Agric. Sci*. 45(2): 179–201.

Humphrey, J., 1978. A planned comparison of five warm season grasses at Noble Foundation. In: *Proc. summer grass congress on management of warm season grasses for utilisation with beef animals*, ed. C.A. Griffiths, Ardmore, Oklahoma, USA.

Hunt, L.A., 1964. Some implications of death and decay in pasture production. *J. Br. Grassld Soc*. 20: 27–31.

Huston, J.E. & Eng, K.G., 1974. Comparative aspects of ammonium sulphate and urea on toxicity, blood ammonium and blood urea in sheep. *Feedstuffs* June 3: 22–23.

Hyam, G.F.S. & Clayton, J.H., 1968. The response of *Eragrostis curvula* to nitrogen fertilizer. *Proc. Grassld Soc. Sth. Afr*. 3: 29–31.

Hyder, D.N. & Sneva, F.A., 1959. Morphological factors affecting the grazing management of crested wheatgrass. *Crop Sci*. 3: 267–271.

Hylton, L.O., Ulrich, A. & Cornelius, D.R., 1967. Potassium and sodium interrelations in growth and mineral content of Italian ryegrass. *Agron. J.* 59: 311–314.

Illius, A.W. & Gordon, I.J., 1987. The allometry of food intake in grazing ruminants. *J. Anim. Ecol.* 56: 989–999.

I'Ons, J.H., 1967. Veld improvement in Swaziland through the introduction of a tropical legume. *Proc. Grassld Soc. Sth. Afr.* 2: 71–73.

I'Ons, J.H., 1969. The effect of introducing Stylo (*Stylosanthes guyanensis*) into veld on animal production in the middleveld of Swaziland. *Proc. Grassld Soc. Sth. Afr.* 4: 113–115.

I'Ons, J.H., 1973. Effect of fertilising and reinforcing Tall Grassveld on a duplex soil. *Proc. Grassld Soc. Sth. Afr.* 8: 61–63.

Jackman, R.H. & Mouat, M.C.H., 1972. Competition between grass and clover for phosphate. II. Effect of root activity, efficiency of response to phosphate and soil moisture. *N. Z. J. Agric. Res.* 15: 667–675.

Jacobs, G.A. Personal communication. PO Box 125, Dundee 3000.

Jacobs, G.A., 1977. Waarde van droogtevoergewasse vir kleinveeproduksie in die extensiewe streke. *Boerdery Keur.* Promedia Publ., Pretoria: 102–105.

Jacobs, G.A. & Smit, C.J., 1977. Benutting van vier *Atriplex* spesies deur skape. *Agroanimal* 9: 37–43.

Jacoby, P.W., 1989. *A glossary of terms used in range management.* Glossary revision special committees, Soc. Range Manage., Denver, Colorado: 12 pp.

Jenkins, K.J., Hidiroglou, M., Wanthy, J.M. & Proulx, J.E., 1974. Prevention of nutritional muscular dystrophy in calves and lambs by selenium and vitamin E additions to the maternal mineral supplement. *Can. J. Anim. Sci.* 54: 49–50.

Jewell, P.A. & Nicholson, M.J., 1989. Strategies for water economy among cattle pastoralists and in wild ruminants. In: *The biology of large African mammals in their environments*, eds. P.A. Jewell & G.M. Maloiy, Clarendon Press, Oxford: 73–87.

Jewiss, O.R., 1966. Morphological and physiological aspects of growth of grasses during the vegetative phase. In: *The growth of cereals and grasses*, eds. F.L. Milthorpe & J.D. Ivins, Butterworths, London: 39–54.

Johansen, C., Kerridge, P.C., Luck, P.E., Cook, B.G., Lowe, K.F. & Ostrowski, H., 1977. The residual effect of molybdenum fertilizer on growth of tropical pasture legumes in a subtropical environment. *Aust. J. Exp. Agric. Anim. Husb.* 17: 961–968.

Johnson, D.A., 1980. Improvement of perennial herbaceous plants for drought-stressed western rangelands. In: *Adaptations of plants to water and temperature stress*, eds. N.C. Turner & P.J. Kramer, John Wiley & Sons, New York: 419–433.

Johnson, D.A., Asay, K.H., Tieszen, L.L., Ehleringer, J.R. & Jefferson, P.G., 1990. Carbon isotope discrimination: potential in screening cool-season grasses for water limited environments. *Crop Sci.* 30: 338–343.

Johnson, H.S. & Hatch, M.D., 1969. The C$_4$ dicarboxylic acid pathway of photosynthesis. *Biochem. J.* 114: 127–134.

Johnson, M.A., 1989. Personal communication. Dept. Agronomy, University of Natal, Pietermaritzburg.

Jones, C.A., 1985. *Grasses and cereals: growth, development and stress response.* John Wiley & Sons, New York: 419 pp.

Jones, D.I.H. & Wilson, A.D., 1987. Nutritive quality of forage. In: *The nutrition of herbivores*, eds. J.B. Hacker & J.H. Ternouth, Academic Press, Australia: 65–89.

Jones, R., 1997. Silage losses and animal performance. *Effective farming.* October 1997. Effective Farming Publications Pty Ltd, PO Box 1649, Pietermaritzburg: 422–426.

Jones, R.I. & Arnott, J., 1977. *Natal farming guide: Section F. Fodder program planning.* Natal Agric. Res. Inst., Dept. Agric. Tech. Serv., Pietermaritzburg.

Jones, R.I., Arnott, J.K. & Klug, J.R., 1989. *Fodder production planning.* Dept. Agric., Natal Region, Private Bag X9059, Pietermaritzburg 3200.

Jones, R.J. & Sandland, R.L., 1974. The relation between animal gain and stocking rate: derivation of the relation from results of grazing trials. *J. Agric. Sci. Camb.* 83: 335–342.

Jones, R.M. & Carter, E.D., 1989. Demography of pasture legumes. In: *Persistence of forage legumes*, eds. G.C. Marten, A.G. Matches, R.F. Barnes, R.W. Brougham, R.J. Clements & G.W. Sheath, Proceedings of a trilateral workshop, Honolulu, Hawaii, July 1988, Madison, Wisconsin: 139–158.

Jones, W.F. & Watson, V.H., 1991. Response of hybrid bermudagrass to sulphur application. *Commun. Soil Sci. Plant Anal.* 22: 505–515.

Jordaan, G., undated. Die gebruik van droogtebestaande voergewasse. Unpublished report. Institute for Range and Forage, Pretoria.

Juritz, C.F., 1920. Turksvijg as veevoer. Dept. Landbouw Wetenschap. Pamphlet No. 16, Pretoria.

Kalac, P. & Kyzlink, V., 1980. The enzymic nature of the degradation of beta-carotene in red clover in an acid medium during ensiling. *Anim. Feed Sci. Techn.* 4: 81–89.

Kamprath, E.J. & Foy, C.D., 1985. Lime–fertilizer–plant interactions in acid soils. In: *Fertilizer technology and use*, ed. O.P. Engelstad, Soil Science Society of America, Madison, Wisconsin: 91–152.

Kautz, J.E. & Van Dyne, G.M., 1978. Comparative analyses of diets of bison, cattle, sheep and pronghorn antelope on short grass prairie in northeastern Colorado, USA. *Proc. 1st Int. Rangeland Congress*, Denver, Colorado: 438–443.

Kearl, L.C., 1982. *Nutrient requirements of ruminants in developing countries.* International Feedstuffs Institute, Utah State University, Logan.

Keeling, C.D., Bacastow, R.B. & Whorf, T., 1982.

Measurement of the concentration of CO_2 at Mauna Loa Observatory, Hawaii. In: *Carbon dioxide review*, ed. W.C. Clark, Clarendon Press, New York: 337–385.

Kemp, A. & Geurink, J.H., 1978. Grassland farming and minerals in cattle. *Neth. J. Agric. Sci.* 26: 161–169.

Kemp, A. & 't Hart, M.L., 1957. Grass tetany in grazing milking cows. *Neth. J. Agric. Sci.* 5: 4–17.

Kerridge, P.C., 1978. Fertilization of acid tropical soils in relation to pasture legumes. In: *Mineral nutrition of legumes in tropical and subtropical soils*, eds. C.S. Andrew & E.J. Kamprath, Commonwealth Scientific and Industrial Research Organization, Melbourne: 395–415.

Kibe, K., Ewart, J.M. & McDonald, P., 1977. Chemical studies with silage microorganisms in artificial media and sterile herbages. *J. Sci. Food Agric.* 28: 355–364.

Kleiber, M., 1961. *The Fire of Life*. John Wiley & Sons, New York.

Kleiber, M., 1975. The fire of life: an introduction to animal energetics. ed. R.E. Krieger, Huntington, New York.

Klug, J.R. & Webster, R.M., 1993. *Natal farming guide: Section F. Fodder programme planning*. 7th rev. ed. Natal Agric. Res. Inst., Dept. Agric., University of Natal, Pietermaritzburg.

Knott, P., Algar, B., Zervas, G. & Telfer, S.B., 1985. Mass: a medium for providing animals with supplementary trace elements. In: *Trace elements in man and animals*, eds. C.F. Mills, I. Bremmer & J.K. Chesters, Farnham Royal, UK.

Kondos, A.C. & Mutch, B., 1975. Biuret and urea in maintenance and production diets of cattle. *J. Agric. Sci.* 85: 359–368.

Kotzé, T.N., 1999. Seedling dynamics and subsequent production of annual *Medicago* spp. as affected by pasture utilization, seedbed preparation and soil type. Ph.D. thesis. University of Stellenbosch, Stellenbosch.

Krause, V. & Klopfenstein, T., 1978. *In vitro* studies of dried alfalfa and complementary effects of dehydrated alfalfa and urea in ruminant rations. *J. Anim. Sci.* 46: 499–504.

Kreft, H.W., 1963. Feeding urea to cattle. *Proc. S. Afr. Soc. Anim. Prod.* 2: 43–44.

Kreft, H.W., 1966. Urea and biuret as nitrogen supplements for cattle. *Proc. S. Afr. Soc. Anim. Prod.* 5: 66–70.

Kresge, C.B. & Younts, S.E., 1963. Response of orchardgrass to potassium and nitrogen fertilization on a Wickham silt loam. *Agron. J.* 55: 161–164.

Kretschemer, A., 1989. Tropical forage legumes development, diversity and methodology for determining persistence. In: *Persistence of Forage legumes*, eds. G.C. Marten, A.G. Matches, R.F. Barnes, R.W. Brougham, R.J. Clements & G.W. Sheath, Proceedings of a trilateral workshop, Honolulu, Hawaii, July 1988, Madison, Wisconsin: 117–138.

Krog, M.M., Theron, E.P. & Andrews, C., 1969. An experimental sodseeder. *Proc. Grassld Soc. Sth. Afr.* 4: 126–130.

Ladd, J.N., Oades, J.M. & Amato, M., 1981. Distribution and recovery of nitrogen from legume residues decomposing in soils sown to wheat in the field. *Soil Biol. Biochem.* 13: 251–256.

Lamprecht, M.P., 1957. Droogtebestande voergewasse. *Hulpboek vir boere in Suid-Afrika* 3: 693–700.

Langenhoven, J.D., 1983. Personal communication. Dept. Agric., Private Bag, Elsenberg, 7607.

Langer, R.H.M., 1972. *How grasses grow. Studies in Biology No. 34,* Edward Arnold, London.

Langlands, P., 1965. Diurnal variation in the diet selected by free-ranging sheep. *Nature* 207: 666–667.

Lascano, C., 1979. Determinants of grazed forage voluntary intake. Ph.D. dissertation, Texas A & M University, College Station, Texas.

Le Mare, P.H., 1991. Rock phosphates in agriculture. *Expl. Agric.* 27: 413–422.

Lesch, S.P., Jones, R.I., Louw, B.P., Archibald, K.P. & Kaiser, H.W., 1974. *Guide to beef production in Natal.* Dept. Agric. Tech. Serv., Pietermaritzburg.

Lishman, A.W., Lyle, A.D., Smith, V.W. & Botha, W.A., 1984. Conception rate of beef cows and growth of suckling calves as influenced by date of calving and supplementary feeding. *S. Afr. J. Anim. Sci.* 14: 10–19.

Little, D.A., 1982. Utilisation of minerals. In: *Nutritional limits to animal production from pastures*, ed. J.B. Hacker, Commonwealth Agric. Bureau, Farnham Royal, UK: 259–279.

Loosli, J.K., 1978. Mineral problems as related to tropical climates. In: *Proceedings of the Latin American symposium on mineral nutrition research with grazing ruminants*, eds. J.H. Conrad & L.R. McDowell, University of Florida, Gainsville: 5–9.

Louw, B.P., 1991. Annual reports. Facet No. N5311/41/2/3. Cedara Agricultural Development Institute, Private Bag X9059, Pietermaritzburg 3200.

Lowther, W.L., 1987. Application of molybdenum to inoculated, lime-coated white clover seed. *New Zeal. J. Exp. Agric.* 15: 271–275.

Ludlow, M.M., 1976. Ecophysiology of C_4 grasses. In: *Water and plant life: problems and modern approaches*, eds. O.L. Lange, L. Kappen & E.D Schulze, Ecological Studies No. 19, Springer-Verlag, Berlin: 364–386.

Ludlow, M.M., 1980. Adaptive significance of stomatal responses to water stress. In: *Adaptation of plants to water and high temperature stress*, eds. N.C. Turner & P.J. Kramer, John Wiley & Sons, New York: 123–138.

Ludlow, M.M. & Charles-Edwards, D.A., 1980. Analysis of regrowth of a tropical grass/legume sward subjected to different frequencies and intensities of defoliation. *Aust. J. Agric. Res.* 31: 673–692.

Lyle, A.D., 1995. Personal communication. Kokstad Agricultural Research Station, Dept. Agric., Private Bag X501, Kokstad 4700.

Lyle, A.D., 1994. Annual reports. Facet No. N5312/ 05/3/3. Cedara Agricultural Development Institute, Private Bag X9059, Pietermaritzburg 3200.

MacDonald, C.I., 1991a. Establishment with vegetative material. In: *Pastures in Natal: agricultural production guidelines for Natal*, Cedara Agricultural Development Institute, Pietermaritzburg.

MacDonald, C.I., 1991b. Establishment of seeded pasture species. In: *Pastures in Natal: agricultural production guidelines for Natal*, Cedara Agricultural Development Institute, Pietermaritzburg.

MacDuff, J.H., Jarvis, S.C. & Mosquera, A., 1989. Nitrate nutrition of grasses from steady-state supplies in flowing solution culture following nitrate deprivation and/or defoliation. II. Assimilation of NO_3^- and short term effects on NO_3^- uptake. *J. Exp. Bot.* 40: 977–984.

MacFarlane, V. & Howard, B., 1970. Water in the ecology of ruminants. In: *Physiology of digestion and metabolism in the ruminant*, ed. A.J. Phillipson, Orvil Press, Newcastle: 362–374.

MacFarlane, W.V. & Howard, B., 1972. Comparative water and energy economy of wild and domestic mammals. In: *Comparative physiology of desert animals*, ed. G.M.O. Maloiy, Symp. Zool. Soc., London No. 31: 261–296.

MacKenzie, D.J. & Wylam, C.B., 1957. Analytical studies on the carbohydrates of grasses and clovers. 8. Changes in carbohydrate composition during the growth of perennial ryegrass. *J. Sci. Food. Agric.* 8: 38–45.

MacLaurin, A.R., Tainton, N.M. & Bransby, D.I., 1981. *Leucaena leucocephala* (Lam.) de Wit as a forage plant: a review. *Proc. Grassld Soc. Sth. Afr.* 16: 63–69.

MacVicar, C.N., De Villiers, J.M., Loxton, R.F., Verster, E., Lambrechts, J.J.N., Merryweather, F.R., Le Roux, J. & Von M. Harmse, H.J., 1977. *Soil classification: a binomial system for South Africa*. 1st ed. Soils and Irrigation Research Institute, Dept. ATS, Pretoria.

Maeda, S. & Yonetani, T., 1978. Optimum cutting stage of forage plants: II. Seasonal changes in CGR and average productivity in Italian ryegrass populations. *J. Japan Grassld Sci.* 24: 10–16.

t'Mannetje, L., 1984. Nutritive value of tropical and subtropical pastures, with special reference to protein and energy deficiency in relation to animal production. In: *Herbivore nutrition in the subtropics and tropics*, eds. F.M.C. Gilchrist & R.I. Mackie, The Science Press, Craighall: 51–66.

Manson, A.D., 1995. The response of Italian ryegrass to sodium, lime and potassium on an acidic Natal soil. *S. A. J. Plant Soil* 12: 117–123.

Manson, A.D., 1999. Personal communication. Dept. Soil Science, CADA, Private Bag X9059, Pietermaritzburg.

Mappledoram, B.D., 1989. The reinforcement of Highland Sourveld with cocksfoot. M.Sc. Agric. thesis. University of Natal, Pietermaritzburg.

Marais, P.G., 1990. Personal communication. Private Bag 529, Middelburg, 5900.

Marshall, C. & Sagar, G.R., 1965. The influence of defoliation on distribution of assimilates in *Lolium multiflorum* Lam. *Ann. Bot.* 29: 365–372.

Marshall, C. & Sagar, G.R., 1968. The distribution of assimilated carbon in *Lolium multiflorum* Lam. following defoliation. *Ann. Bot.* 32: 715–719.

Marshall, K.C., 1964. Survival of root-nodule bacteria in dry soils exposed to high temperatures. *Aust. J. Agric. Res.* 15: 273–281.

Marshall, K.C., 1975. Clay mineralogy in relation to survival of soil bacteria. *Ann. Rev. Phytopathol.* 13: 357–373.

Marten, G.C., 1978. The animal-plant complex in forage palatability phenomena. *J. Anim. Sci.* 46: 1470–1477.

Martens, D.C. & Westermann, D.T., 1991. Fertilizer applications for correcting micronutrient deficiencies. In: *Micro-nutrients in agriculture*, eds. J.J. Mortvedt, F.R. Cox, L.M. Shuman & R.M. Welch, 2nd ed., Soil Science Society of America, Madison, Wisconsin: 549–592.

Matches, A.G., 1966. Influence of intact tillers and height of stubble on regrowth responses of tall fescue (*Festuca arundinacea* Schreb). *Crop Sci.* 6: 484–487.

May, L.H., 1960. The utilisation of carbohydrate reserves in pasture plants. *Herb. Abstr.* 30: 239–245.

Mays, D.A., Wilkinson, S.R. & Cole, C.V., 1980. Phosphorus nutrition of forages. In: *The role of phosphorus in agriculture*, eds. F.E. Khasawneh, E.C. Sample & E.J. Kamprath, Amer. Soc. Agron., Madison, Wisconsin: 805–846.

McAllan, A.B. & Phipps, R.H., 1977. The effect of sample date and plant density on the carbohydrate content of forage maize and the changes that occur in silage. *J. Agric. Sci.* 89: 589–597.

McDonald, P., 1981. *The biochemistry of silage*. John Wiley & Sons, New York.

McDonald, P. & Henderson, A.R., 1962. Buffering capacity of herbage samples as a factor in ensilage. *J. Sci. Food Agric.* 13: 395–400.

McDonald, P., Edwards, R.A. & Greenhalgh, J.F.D., 1973. *Animal nutrition*. 2nd ed. Longman, London, 479 pp.

McDonald, P., Stirling, A.C., Henderson, A.R. & Whittenbury, R., 1964. Fermentation studies on inoculated herbages. *J. Sci. Food Agric.* 15: 429–436.

McDonald, P., Stirling, A.C., Henderson, A.R. & Wittenbury, R., 1965. Fermentation studies on Red Clover. *J. Sci. Food Agric.* 16: 549–557.

McDowell, L.R., 1985a. Iron, manganese and zinc. In: *Nutrition of grazing ruminants in warm climates*, ed. L.R. McDowell, Academic Press, Orlando: 291–315.

McDowell, L.R., 1985b. Free choice mineral supplementation and methods of mineral evaluation. In: *Nutrition of grazing ruminants in warm climates*, ed. L.R. McDowell, Academic Press, Orlando: 383–408.

McDowell, L.R., Conrad, J.H., Ellis, G.L. & Loosli, J.K., 1983. *Minerals for grazing ruminants in tropical regions.* University of Florida, Gainsville and US Agency for International Development.

Mcewen, J., Darby, R.J., Hewitt, M.V. & Yeoman, D.P., 1989. Effects of field beans, fallow, lupin, oats, oilseed rape, peas, ryegrass, sunflowers and wheat, on nitrogen residues in the soil and on the growth of a subsequent wheat crop. *J. Agric. Sci.* 115: 209–219.

McKenzie, F.R., 1996. The effect of grazing frequency and intensity on the root development of *Lolium perenne* L. under subtropical conditions. *S. A. J. Plant Soil* 13(1): 22–26.

McKenzie, F.R., 1997. The influence of grazing management on weed invasion of *Lolium perenne* pastures under subtropical conditions in South Africa. *Trop. Grassld* 31: 24–30.

McKenzie, F.R. & Tainton, N.M., 1996. Effect of grazing frequency and intensity on *Lolium perenne* L. pastures under subtropical conditions: herbage quality. *Afr. J. Range & For. Sci.* 13: 6–8.

McKinney, G.T., 1974. Management of lucerne for sheep grazing in the southern Tablelands of New South Wales. *Aust. J. Exp. Agric. Anim. Husb.* 14: 726–734.

McNaughton, S.J., 1984. Grazing lawns: animals in herds, plant form and coevolution. *Amer. Nat.* 124: 863–886.

Meaker, H.J. & Coetzee, T.P.N., 1978. Effectiveness of paraquat on the foggage value of *Eragrostis curvula* for beef cattle. *S. Afr. J. Anim. Sci.* 8(2): 153–154.

Meissner, H.H., 1982. Theory and application of a method to calculate forage intake of wild southern African ungulates for purposes of estimating carrying capacity. *S. Afr. J. Wildl. Res.* 12: 41–47.

Meissner, H.H., Koster, H.H., Nieuwoudt, S.H. & Coertze, R.J., 1991. The effect of energy supplementation on intake and digestion of early and mid season ryegrass and *Panicum*/Smuts' finger hay, and *in sacco* disappearance of various forage species. *S. Afr. J. Anim. Sci.* 21: 33–42.

Meissner, H.H., Van Niekerk, W.A., Spreeth, E.B. & Koster, H.H., 1989. Voluntary intake of several planted pastures by sheep and an assessment of NDF and IVDOM as possible predictors of intake. *J. Grassld Soc. Sth. Afr.* 6: 156–162.

Mengel, K. & Kirkby, E.A., 1987. *Principles of plant nutrition.* International Potash Inst., Worblaufen-Bern, Switzerland.

Menke, J.W. & Trlica, M.J., 1981. Carbohydrate reserve, phenology and growth cycles in nine Colorado range species. *J. Range Manage.* 36: 70–84.

Mentis, M.T., 1977. Stocking rates and carrying capacity for ungulates on African rangelands. *S. Afr. J. Wildl. Res.* 7: 89–98.

Mentis, M.T., 1981. Acceptability and palatability. In: *Veld and pasture management in South Africa*, ed. N.M. Tainton, Shuter & Shooter and University of Natal Press, Pietermaritzburg: 186–191.

Mentis, M.T., 1984. Optimising stocking rate under commercial and subsistence pastoralism. *J. Grassld Soc. Sth. Afr.* 1: 20–24.

Mertens, D.R., 1983. Using neutral detergent fibre to formulate dairy rations and estimate the net energy content of forages. *Proc. Cornell Nutri. Cont.*: 60–68.

Metcalf, D.S., 1973. Forage statistics. In: *Forages,* eds. M.E. Heath, D.S. Metcalf & R.F. Barnes, 3rd ed. Iowa State University Press: 64–77.

Meyer, J.H., 1985. Sulphur availability in soils of the South African sugar industry. *Proc. S. Afr. Sug. Technol. Ass.* June: 190–194.

Miles, N., 1986. Pasture responses to lime and phosphorus on acid soils in Natal. Ph.D. thesis. University of Natal, Pietermaritzburg.

Miles, N., 1988. Response of pasture species to applied P on highly weathered soils. In: *Proc. phosphorus symposium.* Dept. Agric. & Water Supply, Pretoria: 159–163.

Miles, N., 1991. Personal communication. Department of Agricultural Development, Private Bag X9059, Pietermaritzburg 3200.

Miles, N., 1997. Responses of productive and unproductive kikuyu pastures to top-dressed nitrogen and phosphorus fertiliser. *Afr. J. Range & For. Sci.* 14(1): 1–6.

Miles, N., Bartholomew, P.E., Bennet, R.J. & Wood, S.M., 1985a. Observations on the potassium nutrition of cultivated pastures in the Natal Midlands. In: *Proceedings: Potassium Symposium.* Tech. Comm. No. 187. Dept. Agric., Pretoria: 193–197.

Miles, N., Bartholomew, P.E. & MacDonald, C.I., 1985b. The influence of lime and phosphorus on the growth of white clover on highly weathered Natal soils. *S. Afr. J. Plant Soil* 2: 67–71.

Miles, N., Eckard, R.J. & De Villiers, J.M., 1991. Responses of Italian ryegrass to phosphorus on highly weathered soils. *J. Grassld Soc. Sth Afr.* 8: 86–91.

Millar, K.R. & Meads, W.J., 1987. Blood selenium levels in sheep transferred from selenium topdressed to selenium deficient pasture and vice versa. *N. Z. J. Agric. Res.* 30: 177–181.

Millard, P., 1988. The accumulation and storage of nitrogen by herbaceous plants. *Plant, Cell & Environment* 11: 1–8.

Millard, P., Thomas, R.J. & Buckland, S.T., 1990. Partitioning and remobilisation of nitrogen during regrowth of defoliated *Lolium perenne* L. *J. Exp. Bot.* 41: 941–947.

Miller, W.J., 1979. *Dairy cattle feeding and nutrition.* Academic Press, New York.

Minson, D.J., 1971. The nutritive value of tropical pastures. *J. Aust. Inst. Agric. Sci.* 37: 255–263.

Minson, D.J., 1977. The chemical composition and nutritive value of tropical legumes. In: *Tropical forage legumes, FAO Plant Production Series 2,* ed. P.J. Skerman, FAO, Rome: 186–218.

Minson, D.J., Raymond, W.F. & Harris, C.E., 1960. Studies in the digestibility of herbage. Vlll. The digestibility of S.37 Cocksfoot, S.23 Ryegrass, and S.24 Ryegrass. *J. Br. Grassld Soc.* 15: 174–180.

Moen, A.N., 1973. *Wildlife ecology.* W.H. Freeman, San Francisco.

Morley, F.H.W., Bennett, D. & McKinney, G.T., 1969. The effects of rotational grazing with breeding ewes on *Phalaris*-Subterranean clover pastures. *Aust. J. Exp. Agric. Anim. Husb.* 9: 74–84.

Morris, E.J., 1984. Degradation of the intact plant cell wall of subtropical and tropical herbage by human bacteria. In: *Herbivore nutrition in the subtropics and tropics,* eds. F.M.C. Gilchrist & R.I. Mackie, The Science Press, Craighall: 378–396.

Morrison, J. & Russell, R.D., 1980. Fertilizer recommendations for grassland. *Chemistry and Industry.* 686–688.

Morrison, J., Jackson, M.V. & Sparrow, P.E., 1980. The response of perennial ryegrass to fertilizer nitrogen in relation to climate and soil. Tech. Rep. No. 27. Rothamsted Experimental Station, Hurley.

Mulholland, J.G., Coombe, J.B. & McManns, W.R., 1976. Effect of starch on the utilisation by sheep of a straw diet supplemented with urea and minerals. *Aust. J. Agric. Res.* 27: 139–153.

Muller, J.D., Schaffer, L.V., Ham, L.C. & Owens, M.J., 1977. Cafeteria style free-choice mineral feeders for lactating dairy cows. *J. Dairy Sci.* 60: 1574–1582.

Mundy, G.N., 1983. Effects of potassium and sodium concentrations on growth and cation accumulation in pasture species grown in sand culture. *Aust. J. Agric. Res.* 34: 469–481.

Murphy, M.D., 1980. Essential micronutrients III: sulphur. In: *Applied soil trace elements,* ed. B.E. Davies, John Wiley, New York: 235–258.

Murphy, J.S. & Briske, D.D., 1992. Regulation of tillering by apical dominance: chronology, interpretive value and current perspectives. *J. Range Manage.* 45: 419–429.

Murray, R.N., 1975. Veld reinforcement for increased meat production. Part 2, Keynote symposium. *Proc. Grassld Soc. Sth. Afr.* 10: 155–158.

Murray, R.N., 2000. Personal communication. PO Box 288. Kokstad 4700.

Nash, M.J., 1959. Partial wilting of grass crops for silage. I. Field trials. *J. Br. Grassld Soc.* 14: 65–73.

Nash, R. & Tainton, N.M., 1975. Seasonal response of *Eragrostis curvula* to nitrogen. *Proc. Grassld Soc. Sth. Afr.* 10: 91–94.

Nel, J. & Van Niekerk, B.D.H., 1970. The value of protein and energy-rich supplements in the maintenance of merino sheep grazing sour grassveld. *Proc. S. Afr. Soc. Anim. Prod.* 9: 155–158.

Nelson, C.J. & Smith, D., 1968. Growth of birdsfoot trefoil and alfalfa. III. Changes in carbohydrate reserves and growth analysis under field conditions. *Crop Sci.* 8: 25–28.

Norman, M.J. & Green, J.O., 1958. Local influence of cattle dung and urine upon yield and botanical composition of a permanent pasture. *J. Br. Grassld Soc.* 13: 39–45.

Norris, D.O., 1967. The intelligent use of inoculants and lime pelleting for tropical legumes. *Trop. Grassld* 1: 107–121.

Norvall, F. 2000. Personal communication. Stockowners Co-operative Limited, Pietermaritzburg.

NRC., 1976. *Nutrient requirements of domestic animals, No. 4. Nutrient requirements of beef cattle.* 5th rev. ed. National Academy of Sciences, National Research Council, Washington DC.

NRC., 1978. *Nutrient requirements of dairy cattle.* 5th rev. ed. National Academy of Sciences, National Research Council, Washington DC.

NRC., 1984. *Nutrient requirements of beef cattle.* 6th rev. ed. National Academy of Sciences, National Research Council, Washington DC.

NRC., 1985. *Nutrient requirements of sheep.* 6th rev. ed. National Academy of Sciences, National Research Council, Washington DC.

Nursey, W.R.E., 1971. Starch deposits in *Themeda triandra* Forsk. *Proc. Grassld Soc. Sth. Afr.* 6: 157–160.

Obioha, F.C., Clanton, D.D., Ritterhouse, L.R. & Streeter, C.L., 1970. Sources of variation in chemical composition of forage ingested by oesophageal fistulated cattle. *J. Range Manage.* 23: 133–136.

Odberg, F.O. & Francis-Smith, K., 1977. Studies on the formation of ungrazed eliminative areas in fields used by horses. *Appl. Anim. Ethol.* 3: 27–34.

Olson, B.E. & Richards, J.H., 1988a. Annual replacement of the tillers of *Agropyron desertorum* following grazing. *Oecologia* 76: 1–6.

Olson, B.E. & Richards, J.H., 1988b. Tussock regrowth after grazing: intercalary meristems and axillary bud activity of *Agropyron desertorum.* *Oikos* 51: 374–382.

Oltjen, R., Slyter, L.L., Kozak, A. & Williams, E.E., 1968. Evaluation of urea, biuret, urea phosphate and uric acid as NPN sources for cattle. *J. Nutr.* 94: 193–202.

Oltjen, R.R. & Dinius, D.A., 1976. Processed poultry waste compared with uric acid, sodium urate, urea and biuret as nitrogen supplements for beef cattle fed forage diets. *J. Anim. Sci.* 43: 201–208.

Oltjen, R.R., Burns, W.C. & Ammerman, C.B., 1974. Biuret versus urea and cottonseed meal for wintering and finishing steers. *J. Agric. Sci.* 38: 975–983.

Oparka, K.J., Marshall, B. & MacKerran, D.K.L., 1986. Carbon partitioning in a potato crop in response to applied nitrogen. In: *Phloem transport,* eds. J. Cronshaw, W.J. Lucas, & R.T. Giaquinta, Allan Liss, New York: 577–587.

Opperman, D.P.J. & Human, J.J., 1976. Die invloed van ontblaring en vogstremming op die groeikragtigheid van *Themeda triandra* Forsk. onder gekontroleerde toestande. *Proc. Grassld Soc. Sth. Afr.* 12: 65–71.

O'Reagain, P.J. & Mentis, M.T., 1989a. The effect of plant structure on the acceptability of different grass species to cattle. *J. Grassld Soc. Sth. Afr.* 6: 163–170.

O'Reagain, P.J. & Mentis, M.T., 1989b. Sequence and process of species selection by cattle in relation to optimal foraging theory on an old land in the Natal Sour Sandveld. *J. Grassld Soc. Sth. Afr.* 6: 71–76.

O'Reagain, P.J. & Turner, J.R. 1992. An evaluation of the empirical basis for grazing management recommendations for rangeland in southern Africa. *J. Grassld Soc. Sth. Afr.* 9: 38–49.

Osmond, C.B., Winter, K. & Powles, S.B., 1980. Adaptive significance of carbon dioxide cycling during photosynthesis in water-stressed plants. In: *Adaptations of plants to water and high temperature stress*, eds. N. Turner & P.J. Kramer, John Wiley & Sons, New York: 139–154.

Osbourn, D.F., 1982. The feeding of grass and grass products. In: *Grass: its production and utilisation*, ed. W. Holmes, Blackwell Sci. Publs., Oxford: 70–124.

Osbourn, D.F., Thomson, D.J. & Terry, R.A., 1966. The relationship between voluntary intake and digestibility of forage crops, using sheep. In: *Proc. X Int. Grassld Congress*, Helsinki, Finland: 363–367.

Ourry, A., Boucaud, J. & Salette, J., 1990. Partitioning and remobilisation of nitrogen during regrowth in nitrogen-deficient ryegrass. *Crop Sci.* 30: 1251–1254.

Owens, F.N. & Bergen, W.G., 1983. Nitrogen metabolism of ruminant animals: historical perspective, current understanding and future implications. *J. Anim. Sci.* Suppl. 2: 498–518.

Owensby, C.E., Coyne, P.I., Ham, J.M., Auen, L.M. & Knapp, A.K., 1993. Biomass production in tallgrass prairie ecosystem exposed to ambient CO_2. *Ecol. Appl.* 3: 644–653.

Owen-Smith, N., 1985. Niche separation among African ungulates. In: *Species and speciation*, ed. E.S. Vrba, Transvaal Mus. Monograph 4: 167–171.

Owen-Smith, N., 1988. *Megaherbivores: the influence of very large body size on ecology*. Cambridge University Press, Cambridge.

Owen-Smith, N., 1991. Veld condition and animal performance: application of an optimal foraging model. *J. Grassld Soc. Sth. Afr.* 8: 77–82.

Owen-Smith, N., 1992. Grazers and browsers: ecological and social contrasts among African ruminants. In: *Ongules/Ungulates 91*, eds. F. Spitz, G. Janeau, G. Gonzales & S. Aulagnier, SFEPM-IRGM, France: 175–182.

Owen-Smith, N. Personal communication. Zoology Department, University of the Witwatersrand, Johannesburg.

Owen-Smith, N. & Cooper, S.M., 1985. Comparative consumption of vegetation components by kudus, impalas and goats in relation to their commercial potential as browsers in savanna regions. *S. Afr. J. Sci.* 81: 72–76.

Owen-Smith, N. & Cumming, D.H.M., 1993. Comparative foraging strategies of grazing ungulates in African savanna grasslands. *Proc XVII Int. Grassld Congress*, New Zealand & Australia: 691–698.

Owen-Smith, N. & Novelli, P., 1982. What should a clever ungulate eat? *Amer. Nat.* 119: 151–178.

Ozanne, P.G., 1980. Phosphate nutrition of plants: a general treatise. In: *The role of phosphorus in agriculture*, eds. F.E. Khasawneh, E.C. Sample & E.J. Kamprath, Amer. Soc. Agron., Madison, Wisconsin: 559–590.

Ozanne, P.G. & Sewell, P.C., 1980. Increasing moisture availability by improved distribution of plant nutrients. In: *Agrochemicals in Soils*, eds. A. Banin & U. Kafkafi, Pergamon Press, Oxford: 267–275.

Ozanne, P.G., Asher, C.J. & Kirton, D.J., 1965. Root distribution in a deep sand and its relationship to the uptake of added potassium by pasture plants. *Aust. J. Agric. Res.* 16: 785–800.

Page, B.R. & Walker, B.H., 1978. Feeding niches of four large herbivores in the Hluhluwe Game Reserve, Natal. *Proc. Grassld Soc. Sth. Afr.* 13: 117–122.

Paulson, G.D., Broderick, G.A., Baumann, C.A. & Pope, A.L., 1968. Effect of feeding sheep selenium fortified trace mineralised salt: effect of tocopherol. *J. Anim. Sci.* 27: 195–202.

Payne, W.J.A. & MacFarlane, J.S., 1963. A brief study of cattle browsing behaviour in a semi-arid area in Tanganyika. *East Afr. Agric. For. J.* 29: 131–133.

Payne, W.J.A., Laing, W.E. & Raivoka, E.N., 1951. Grazing behaviour of dairy cattle in the tropics. *Nature* 167: 610–611.

Pearcy, R.W., Bjorkman, O., Caldwell, M.M., Keeley, J.E., Monson, R.K. & Strain, B.R., 1987. Carbon gains by plants in natural environments. *BioScience* 37: 21–29.

Peddie, G.M., Tainton, N.M. & Hardy, M.B., 1996. The effect of past grazing intensity on the vigour of *Themeda triandra* and *Tristachya leucothrix*. *Afr. J. Range & For. Sci.* 12: 111–115.

Penning, P.D., 1986. Some effects of sward condition on grazing behaviour and intake by sheep. In: *Grazing research in northern latitudes*, ed. O. Gudmundsson, Plenum Press, London: 219–226.

Peterson, R.G., Lucas, H.L. & Mott, G.O., 1965. Relationship between rate of stocking and per animal and per acre performance on pasture. *Agron. J.* 57: 27–30.

Peterson, T.A. & Varvel, G.E., 1989. Crop yield as affected by rotation and nitrogen rate: 1. Soyabean. *Agron. J.* 81: 727–731.

Phillips, R.L., 1994. The influence of management on tiller dynamics, dry matter yield and certain economic aspects of Tall fescue (*Festuca arundinacea* Schreb.) pastures. M.Sc. Agric. thesis. University of Natal, Pietermaritzburg.

Pieterse, P.J.S., 1966. *Supplementary feeding of ruminants during the winter months in South Africa*. Tech. Comm. No. 48. Dept. Agric. Tech. Serv., S. Afr.

Pieterse, P.J.S. & Preller, J.H., 1965. Voorlopige resultate met byvoedings aan vleisbeeste op somerveld. *Proc. S. Afr. Soc. Anim. Prod.* 4: 123–126.

Playne, M.J. & McDonald, P., 1966. The buffering constituents of herbage and of silage. *J. Sci. Food Agric.* 17: 264–268.

Pond, K.R., Ellis, W.C. & Akin, D.E., 1984. Ingestive mastication and fragmentation of forages. *J. Anim. Sci.* 58: 1567–1574.

Prins, W.H. & Den Boer, D.J., 1985. Aspects of nitrogen and phosphate fertilisation on grassland. *Fert. & Agric.* 90: 27–36.

Provenza, F.D. & Balph, D.F., 1990. An assessment of five explanations for diet selection by ruminants facing five foraging challenges. In: *Behavioural mechanisms of food selection*, ed. R.N. Hughes, Springer-Verlag, Berlin.

Provenza, F.D., Burritt, E.A., Clausen, T.P., Bryant, J.P., Reichardt, P.B. & Dustel, R.A., 1990. Conditional flavour aversion: a mechanism for goats to avoid condensed tannins in blackbrush. *Amer. Nat.* 136: 810–828.

Raymond, F., Redman, P. & Waltham, R., 1986. *Forage conservation and feeding.* Farming Press Ltd, Wharfedale Road, Ipswich, Suffolk.

Raymond, W.F., 1969. The nutritive value of forage crops. *Adv. Agron.* 21: 1–108.

Read, M.V.P., Engels, E.A.N. & Smith, W.A., 1986a. Phosphorus and the grazing ruminant. 1. The effect of supplementary P on sheep at Armoedsvlakte. *S. Afr. J. Anim. Sci.* 16: 1–6.

Read, M.V.P., Engels, E.A.N. & Smith, W.A., 1986b. Phosphorus and the grazing ruminant. 2. The effect of supplementary P on cattle at Glen and Armoedsvlakte. *S. Afr. J. Anim. Prod.* 16: 7–12.

Reason, G.K., McGuigan, K.R. & Waugh, P.D., 1989. Some consequences to dairy cattle of the addition of nitrogen fertilisers to dryland grass pastures in South-East Queensland. In: *Recent advances in animal nutrition in Australia*, ed. D.J. Farrell, University of New England, Armidale, Australia: 31–36.

Rees, W.A., 1974. Preliminary studies into bush utilisation by cattle in Zambia. *J. Appl. Ecol.* 11: 207–214.

Reid, R.L. & Jung, G.A., 1974. Effects of elements other than nitrogen on the nutritive value of forage. In: *Forage fertilisation*, ed. D.A. Mays, Amer. Soc. Agron., Madison, Wisconsin: 105–108.

Rethman, N.G. & Beukes, H., 1973. Overseeding of *Eragrostis curvula* on North-eastern Sandy Highveld. *Proc. Grassld Soc. Sth. Afr.* 8: 57–59.

Rethman, N.G.F., 1971. Elevation of shoot apices of two ecotypes of *Themeda triandra* on the Transvaal highveld. *Proc. Grassld Soc. Sth. Afr.* 6: 86–92.

Rethman, N.G.F., 1983. Planted pastures for foggage. *Agrivaal* 5(12): 5–12.

Rethman, N.G.F., 1987. The effect of form and level of nitrogen fertilization on the yield of *Digitaria eriantha* Steud. *J. Grassld Soc. Sth. Afr.* 4: 105–108.

Rethman, N.G.F. & Beukes, B.H., 1988. The influence of nitrogen fertilisation, spring burning and height of stubble on hay and seed production of *Eragrostis curvula*. *J. Grassld Soc. Sth. Afr.* 5: 208–212.

Rethman, N.G.F. & De Witt, C.C., 1991. The value of subtropical grass pastures for use as foggage on the eastern Transvaal Highveld. *J. Grassld Soc. Sth. Afr.* 8(1): 19–21.

Rethman, N.G.F. & Gous, C.I., 1973. Foggage value of kikuyu (*Pennisetum clandestinum* Hochst. ex Chiov.). *Proc. Grassld Soc. Sth. Afr.* 8: 101–105.

Rhoades, D.F., 1979. Evolution of plant chemical defence against herbivores. In: *Herbivores: their interactions with secondary plant metabolites*, eds. G.A. Rosenthal & D.H. Janzen, Academic Press, New York.

Richards, J.H., 1984. Root growth response to defoliation in two *Agropyron* bunchgrasses: field observations with an improved root periscope. *Oecologia* 64: 21–25.

Richards, J.H. & Caldwell, M.M., 1985. Soluble carbohydrates, concurrent photosynthesis and efficiency in regrowth following defoliation. A field study with *Agropyron* species. *J. Appl. Ecol.* 22: 907–920.

Richards, J.H., Mueller, R.J. & Mott, J.J., 1988. Tillering in tussock grasses in relation to defoliation and apical bud removal. *Ann. Bot.* 62: 173–179.

Rimmer, D.L., Shiel, R.S., Syers, J.K. & Wilkinson, M., 1990. Effects of soil application of selenium on pasture composition. *J. Sci. Food Agric.* 51: 122–126.

Robbins, G.B. & Faulkner, G.B., 1983. Productivity of six ryegrass (*Lolium* spp.) cultivars grown as irrigated annuals in the Burnett district of southeast Queensland. *Trop. Grassld* 17: 49–54.

Robinson, J.J., 1990. Nutrition in the reproduction of farm animals. *Nutrition Res. Rev.* 3: 253–276.

Robson, M.J. & Parsons, A.J., 1977. Nitrogen deficiency in small closed communities of S24 ryegrass. I. Photosynthesis, respiration, dry matter production and partitioning. *Ann. Bot.* 42: 1184–1197.

Rodriguez, J.A., Poppe, S. & Meier, H., 1989a. The influence of wilting on the quality of tropical grass silage in Cuba. 3. Pasto estrella jamaicano (*Cynodon nlemfuensis* cv. *jamaecano*) Arch. *Anim. Nutr.* 39 (10): 843–850.

Rodriguez, J.A., Poppe, S. & Meier, H., 1989b. The influence of wilting on the quality of tropical grass silage in Cuba. 4. Bermuda grass cross No. 1 (*Cynodon dactylon*). Arch. *Anim. Nutr.* 39 (10): 851–858.

Rogosa, M., 1974. Genus l. Lactobacillus. In: *Bergey's manual of determinative bacteriology*, Willams & Wilkins Co., Baltimore.

Rohweder, D.A., Barnes, R.F. & Jorgensen, N.A., 1978. Proposed hay grading standards based on laboratory analysis for evaluating quality. *J. Anim. Sci.* 47: 747–759.

Römheld, V. & Marschner, H., 1991. Function of micronutrients in plants. In: *Micronutrients in agriculture*, eds. J.J. Mortvedt, F.R. Cox, L.M. Shuman & R.M. Welch, 2nd ed. Soil Science Society of America, Madison, Wisconsin: 297–328.

Roux, E.R., 1954. The nitrogen sensitivity of *Eragrostis curvula* and *Trachypogon plumosis* in relation to grassveld succession. *S. Afr. J. Sci.* 50: 173–176.

Rovira, A.D., 1961. *Rhizobium* numbers in the rhizospheres of red clover and paspalum in relation to soil treatment and the numbers of bacteria and fungi. *Aust. J. Agric. Res.* 12: 77–83.

Rowell, D.L., 1988. Soil acidity and alkalinity. In: *Russell's soil conditions and plant growth*, ed. A. Wild, John Wiley & Sons, New York: 844–898.

Ryden, J.C., 1984. The flow of nitrogen in grassland. *Proc. Fertilizer Soc.* No. 229: 1–44.

Ryle, G.J.A. & Powell, C.E., 1974. The utilisation of recently assimilated carbon in graminaceous plants. *Ann. Appl. Biol.* 77: 145–158.

Ryle, G.J.A. & Powell, C.E., 1975. Defoliation and regrowth in the graminaceous plant: the role of current assimilate. *Ann. Bot.* 39: 297–310.

Sagar, G.R. & Marshall, C., 1966. The grass plant as an integrated unit: some studies of assimilate distribution in *Lolium multiflorum* Lam. *Proc. 9th Int. Grassld Congress:* 493–497.

Salette, J., 1988. Nitrogen research: results and questions. In: *Nitrogen and water use by grassland*, ed. R.J. Wilkins, AFRC Inst. for Grassland and Animal Production, Hurley: 43–57.

Salisbury, F.B. & Ross, C.W., 1985. *Plant physiology.* 3rd ed. Wadsworth Publishing Co., Belmont: 540 pp.

Sanchez, P.A., 1976. *Properties and management of soils in the tropics.* Wiley-Interscience, New York.

Sanchez, P.A. & Salinas, J.G. 1981. Low input technology for managing Oxisols and Ultisols in tropical America. *Adv. Agron.* 34: 279–406.

Sandland, R.L. & Jones, R.J., 1975. The relation between animal gain and stocking rate in grazing trials: an examination of published theoretical models. *J. Agric. Sci. Camb.* 85: 123–128.

Santhirasegaram, K. & Black, J.N., 1965. Agronomic practices aimed at reducing competition between cover crops and undersown pasture. *Herb. Abstr.* 34: 221–225.

Saunders, W.M.H., Sherrell, C.G. & Gravett, I.M., 1987. A new approach to the interpretation of soil tests for phosphate response by grazed pasture. *N. Z. J. Agric. Res.* 30: 67–77.

Schields, R.G., Campbell, D.R., Hughes, D.M. & Dillingham, D.A., 1982. Researchers study vitamin A stability in feeds. *Feedstuffs* 54(47): 22.

Schulze, R.E., 1997. *South African Atlas of Agrohydrology and -Climatology.* Report TT82/96. Water Research Commission, Pretoria.

Scotcher, J.S.B., 1979. Personal communication, Natal Parks Board, PO Box 662, Pietermaritzburg 3200.

Senock, R.S., Sisson, W.B. & Donart, G.B., 1991. Compensatory photosynthesis of *Sporobolus flexuosus* (Thurb.) rydb following simulated herbivory in the northern Chihuahuan desert. *Bot. Gaz.* 152: 275–281.

Sharman, B.C., 1947. The biology and developmental morphology of the shoot apex in the Graminea. *New Phytol.* 46: 20–34.

Sheenan, W., Fontenot, J.P. & Blaser, P.E., 1985. *In vitro* dry matter digestibility and chemical composition of autumn-accumulated tall fescue, orchard grass and red clover. *Grass & Forage Science* 40(3): 317–322.

Shewmaker, G.E., Mayland, H.F., Rosenau, R.C. & Asay, K.H., 1989. Silicain C_3-grasses: effects on forage quality and sheep preference. *J. Range Manage.* 42: 122–127.

Shirley, R.L., 1978. Sulphur in ruminant nutrition. In: *Proceedings of the Latin American symposium on mineral nutrition research with grazing ruminants*, eds. J.H. Conrad & L.R. McDowell, University of Florida, Gainsville: 66–72.

Shirley, R.L., 1986. *Nitrogen and energy nutrition of ruminants.* Academic Press, Orlando: 154.

Silsbury, J.H., 1970. Leaf growth in pasture grasses. *Trop. Grassld* 4: 17–36.

Simpson, J.R., 1987. Nitrogen nutrition of pastures. In: *Temperate pastures: their production, use and management*, eds. J.L. Wheeler, C.J. Pearson & G.E. Robards, CSIRO, Australia: 143–154.

Skerman, P.J., 1977. *Tropical forage legumes.* FAO, Rome.

Slayter, R.O., 1967. *Plant water relationships.* Academic Press, New York: 366 pp.

Smith, A., 1988. The development of a white clover for use in the eastern high potential areas of South Africa. Ph.D. Agric. thesis. University of Natal, Pietermaritzburg.

Smith, C.A., 1959. Studies on the Northern Rhodesia *Hyparrhenia* veld. I. The grazing behaviour of indigenous cattle grazed at light and heavy stocking rates. *J. Agric. Sci.* 52: 369–375.

Smith, C.A., 1962. The utilisation of *Hyparrhenia* veld for the nutrition of cattle in the dry season. III. Studies on the digestibility and the produce of mature veld and veld hay, and the effect of feeding supplementary protein and urea. *J. Agric. Sci.* 58: 173–178.

Smith, D., 1962. Carbohydrate root reserves in alfalfa, red clover and birdsfoot trefoil under several management schedules. *Crop Sci.* 2: 75–78.

Smith, G.S., Cornforth, I.S. & Henderson, H.V., 1985. Critical leaf concentrations for deficiencies of nitrogen, potassium, phosphorus, sulphur and magnesium in perennial ryegrass. *New Phytol.* 101: 393–409.

Smith, G.S., Middleton, K.R. & Edmonds, A.S., 1978. A classification of pasture and fodder plants according to their ability to translocate sodium from their roots into areal parts. *N. Z. J. Exp. Agric.* 6: 183–188.

Smith, H.R.H., 1987. Identification of companion legumes for Midmar Italian ryegrass. *J. Grassld Soc. Sth. Afr.* 4: 100–104.

Smith, J.M.B., 1997. Crop, pasture and timber yield estimates for KwaZulu-Natal. Cedara Report No. N/A/97/9.

Smith, L.D.S. & Hobbs, G., 1974. Genus lll. Clostridium. In: *Bergey's Manual of determinative bacteriology*, Williams & Wilkins Co., Baltimore.

Smith, L.H., 1962. Theoretical carbohydrate requirement for alfalfa silage production. *Agron. J.* 54: 291–293.

Smith, L.H. & Marten, G.C., 1970. Foliar regrowth of alfalfa utilising ¹⁴C-labelled carbohydrates stored in roots. *Crop Sci.* 10: 146–150.

Sneva, F.A., 1967. Chemical curing of range grasses with Paraquat. *J. Range Manage.* 6: 389–394.

Spears, J.W., 1991. Advances in mineral nutrition in grazing ruminants. *Proc. Grazing Livestock Nutrition Conf.*, Oklahoma State University: pp 138–149.

Spedding, C.R.W. & Diekmahns, F.L., 1972. Grasses and legumes in British Agriculture. *Bull. Commonw. Agric. Bur.*, No. 49, Farnham Royal, UK.

Steinke, T.D., 1975. Effect of height of cut on translocation of ¹⁴C-labelled assimilates in *Eragrostis curvula* (Schrad) Nees. *Proc. Grassld Soc. Sth. Afr.* 10: 41–47.

Steinke, T.D. & Booysen, P.deV., 1968. The regrowth and utilisation of carbohydrate reserves of *Eragrostis curvula* after different frequencies of defoliation. *Proc. Grassld Soc. Sth. Afr.* 3: 105–110.

Stephens, D.W. & Krebs, J.R., 1986. *Foraging theory*. Princeton University Press, Princeton.

Stewart, P.G. & Dugmore, T.J., 1995. Nutritional requirements of dairy cattle. In: *Dairying in KwaZulu-Natal*, ed. T.J. Dugmore, KwaZulu-Natal Dept. Agric., Private Bag X9059, Pietermaritzburg 3200.

Steynberg, H. & De Kock, G.C., 1987. Aangeplante weidings in die veeproduksiestelsels van die Karoo en ariede gebiede. *Karoo Agric.* 3: 4–13.

Steynberg, R.E., Nel, P.C. & Rethman, N.F.G., 1994. Soil water use and rooting depth of Italian ryegrass (*Lolium multiflorum* Lam) in a small plot experiment. *S. A. J. Plant Soil* 11(2): 80–83.

Stobbs, T.H., 1970. The use of liveweight gain trials for pasture evaluation in the tropics: 6. A fixed stocking rate design. *J. Br. Grassld Soc.* 25: 73–77.

Stoltz, C.W. & Danckwerts, J.E., 1990. Grass species selection patterns in rotationally grazed Döhne sourveld during autumn and early winter. *J. Grassld Soc. Sth. Afr.* 7: 92–96.

Strijdom, B.W. & Wassermann, V.D., 1980. Legume inoculant and seed inoculation. *Farming S. Afr.*, Government Printer, Pretoria.

Stuart-Hill, G.C., 1999. Trees and shrubs. In: *Veld management in South Africa*, ed. N.M. Tainton, University of Natal Press, Pietermaritzburg: 109–113.

Sumner, M.E. & Farina, M.P.W., 1986. Phosphorus interactions with other nutrients and lime in field cropping systems. *Adv. Soil Sci.* 5: 201–236.

Suttle, N.F., 1987. The absorption, retention and function of minor nutrients. In: *The nutrition of herbivores*, eds. J.B. Hacker & J.H. Ternouth, Academic Press, Sydney: 333–361.

Sutton, C.D., 1969. Effect of low soil temperature on phosphate nutrition of plants: a review. *J. Sci. Fd. Agric.* 20: 1–3.

Sutton, J.D., 1980. Digestion and end product formation in the rumen from production rations. In: *Digestive physiology and metabolism in ruminants*, eds. Y. Ruckebusch & P. Thivend, AVI, Connecticut: 271–290.

Swingle, R.S., Araiza, A. & Urias, A.R., 1977. Nitrogen utilisation by lambs of wheat straw alone or with supplements containing dried poultry waste, cottonseed meal or urea. *J. Anim. Sci.* 45: 1435–1441.

Syers, J.K. & Springett, J.A., 1984. Earthworms and soil fertility. *Plant Soil* 76: 93–104.

Tainton, N.M., 1958. Studies of the growth and development of certain veld grasses. M.Sc. thesis. University of Natal, Pietermaritzburg.

Tainton, N.M., 1973. Effects of different grazing rotations on pasture production. *J. Br. Grassld Soc.* 29: 191–202.

Tainton, N.M., 1974. Milk production off pastures in New Zealand. *Proc. Grassld Soc. Sth. Afr.* 9: 185–191.

Tainton, N.M., 1976. The management of sown pastures in South Africa. Presidential address. *Proc. Grassld Soc. Sth. Afr.* 11: 15–17.

Tainton, N.M., 1979. Management of legume pastures. In: *Symposium on legumes in pastures*, ed. A. Smith, Cedara Press, Private Bag X9059, Pietermaritzburg 3200: 80–83.

Tainton, N.M., 1999. The ecology of the main grazing lands of South Africa. In: *Veld management in South Africa*, ed. N.M. Tainton, University of Natal Press, Pietermaritzburg: 23–53.

Tainton, N.M. & Booysen, P.deV., 1965. Growth and development in perennial veld grasses. 1. *Themeda triandra* tillers under various systems of defoliation. *S. Afr. J. Agric. Sci.* 8: 93–110.

Tainton, N.M., Bransby, D.I. & Booysen, P.deV., 1990. *Common veld and pasture grasses of Natal*. 2nd ed. Shuter & Shooter, Pietermaritzburg.

Tainton, N.M., Heard, C.A.H. & Nash, R.C., 1981. Response of *Eragrostis curvula* to differences in the seasonal distribution of nitrogen fertilizer. *Proc. Grassld Soc. Sth. Afr.* 16: 71–74.

Talibudeen, O., Page, M. & Mitchell, J.D.D., 1976. The interaction of nitrogen and potassium nutrition on dry matter and nitrogen yields of the graminae: perennial ryegrass. *J. Sci. Fd. Agric.* 27: 999–1004.

Taylor, C.R., 1968. The minimum water requirements of some East African bovids. *Symp. Zool. Soc.* London 21: 195–206.

Taylor, N.L. (ed), 1995. *Clover science and technology*. Agronomy series No. 25. Amer. Soc. Agron., Madison, Wisconsin, USA.

Teague, W.R., 1989. Patterns of selection of *Acacia karroo* by goats and changes in tannin levels and *in vitro* digestibility following defoliation. *J. Grassld Soc. Sth. Afr.* 6: 230–235.

Terry, R.A. & Tilley, J.M.A., 1964. The digestibility of the leaves and stems of perennial ryegrass, cocksfoot, timothy, tall fescue, lucerne and sainfoin, as measured by an *in vitro* procedure. *J. Br. Grassld Soc.* 19: 363–372.

Theiler, A., 1912. *Facts and theories about stijfziekte and lamziekte.* 2nd Report. Dir. Vet. Res. S. Afr: 7 pp.

Theiler, A., 1920. The cause and prevention of lamsiekte. *J. Dept. Agric. S. Afr.* 1: 221.

Theron, E.P. & Booysen, P.deV., 1966. Palatability in grasses. *Proc. Grassld Soc. Sth. Afr.* 1: 111–120.

Theron, E.P., Lesch, S.F. & Mappeldoram, B.D., 1974. The potential of Natal for the radical improvement of the veld and the fortification of established pastures. *Proc. Grassld Soc. Sth. Afr.* 9: 175–178.

Thomas, C. & Thomas, P.C., 1985. Factors affecting the nutritive value of grass silage. In: *Recent advances in animal nutrition,* eds. W. Haresign & D.J.A. Cole, Butterworths, London: 223–256.

Thomas, R.G., 1987. The structure of the mature plant. In: *White clover,* eds. M.J. Baker & W.M. Williams, CAB International, Wallingford, Oxon, United Kingdom: 2–4.

Thompson, D.J., 1978. Biological availability of macro elements. In: *Proceedings of the Latin American symposium on mineral nutrition research with grazing ruminants,* eds. J.H. Conrad & L.R. McDowell, University of Florida, Gainsville: 127–135.

Thompson, J.A., 1961. Studies on nodulation responses to pelleting of subterranean clover seed. *Aust. J. Agric. Res.* 12: 578–592.

Tilley, J.M.A. & Terry, R.A., 1963. A two-stage technique for the *in vitro* digestion of forage crops. *J. Br. Grassld Soc.* 18: 104–111.

Tisdale, S.L., Nelson, W.L. & Beaton, J.D., 1985. *Soil fertility and fertilizers.* Macmillan Publishing, New York.

Trlica, M.J., 1977. Distribution and utilisation of carbohydrate reserves in range plants. In: *Rangeland plant physiology,* ed. R.E. Sosebee, Soc. Range Manage., Denver, Colorado: 73–96.

Trlica, M.J. & Cook, C.W., 1971. Defoliation effects on carbohydrate reserves of desert species. *J. Range Manage.* 24: 418–425.

Trlica, M.J. & Cook, C.W., 1972. Carbohydrate reserves of crested wheatgrass and Russian wildrye as affected by development and defoliation. *J. Range Manage.* 25: 430–435.

Trollope, W.S.W., Trollope, L.A. & Bosch, O.J.H., 1990. Veld and pasture management terminology in southern Africa. *J. Grassld Soc. Sth. Afr.* 7: 52–61.

Turner, J.R. & Tainton, N.M., 1990. A comparison of four different standards of reference for the animal unit for determining stocking rate. *J. Grassld Soc. Sth. Afr.* 7: 204–207.

Turner, M.A., 1981. Dietary potassium:sodium imbalances as a factor in the aetiology of primary ruminal tympany in dairy cows. *Vet. Res. Comm.* 5: 159–164.

Turner, N.C. & Begg, J.E., 1978. Responses of pasture plants to water deficits. In: *Plant relations in pastures,* ed. J.R. Wilson, CSIRO, Melbourne.

Tyson, P.D., Kruger, F.J. & Louw, C.W., 1988. Atmospheric pollution and its implications in the Eastern Transvaal Highveld. S. Afr. Nat. Sci. Prog. Rep. No 150. CSIR, Pretoria.

Underwood, E.J., 1977. *Trace elements in human and animal nutrition.* Academic Press, New York.

Underwood, E.J., 1981. *The mineral nutrition of livestock.* CAB, London.

Van der Merwe, F.J. & Perold, I.W., 1967. Trace elements in natural pastures. *J. S. Afr. Vet. Med. Ass.* 38: 355.

Van der Merwe, F.J. & Smith, W.A., 1991. *Dierevoeding.* Anim. Sci., Pinelands.

Van Egeraat, A.W.S.M., 1975. The growth of *Rhizobium leguminosarum* on the root surface and in the rhizosphere of pea seedlings in relation to root exudates. *Plant Soil* 42: 367–379.

Van Heerden, J.M., 1986. Effect of cutting frequency on the yield and quality of legumes and grasses under irrigation. *J. Grassld Soc. Sth. Afr.* 3: 43–46.

Van Heerden, J.M., 1990. The influence of the application of grass herbicides on the production of dryland medic and lucerne pastures in the Ruens area of the southern Cape. *J. Grassld Soc. Sth. Afr.* 7(3): 152–156.

Van Heerden, J.M. & Tainton, N.M., 1987. Potential of medic and lucerne pastures in the Ruens area of the southern Cape. *J. Grassld Soc. Sth. Afr.* 4(3): 95–99.

Van Heerden, J.M. & Tainton, N.M., 1988. Influence of grazing management on the production of an irrigated grass/legume pasture in the Ruens area of the southern Cape. *J. Grassld Soc. Sth. Afr.* 5: 130–137.

Van Heerden, J.M. & Tainton, N.M., 1989. Seasonal grazing capacity of an irrigated grass/legume pasture in the Ruens area of the southern Cape. *J. Grassld Soc. Sth. Afr.* 6(4): 216–219.

Van Heerden, J.M., Tainton, N.M. & Botha, P.R., 1989. A comparison of grasses and grass/legume pastures under irrigation in the Outeniqua area of the southern Cape. *J. Grassld Soc. Sth. Afr.* 6(4): 220–224.

Van Keuren, R.W. & Hoveland, C.S., 1985. Clover management and utilisation. In: *Clover science and technology,* ed. N.L. Taylor, Amer. Soc. Agron., No. 25, Madison, USA.

Van Niekerk, A., 1990. Personal communication. Cedara Agricultural Development Institute, Private Bag X9059, Pietermaritzburg 3200.

Van Niekerk, B.D.H., 1978. Limiting nutrients: their identification and supplementation in grazing ruminants. In: *Proceedings of the Latin American symposium on mineral nutrition research with grazing ruminants,* eds. J.H. Conrad & L.R. McDowell, University of Florida, Gainsville: 194–200.

Van Niekerk, B.D.H. & Jacobs, G.A., 1985. Protein, energy and phosphorus supplementation of cattle fed low-quality forage. *S. Afr. J. Anim. Sci.* 15: 133–136.

Van Niekerk, J.P., Du Pisani, L.G. & Marais, A.deK., 1987. The potential for dryland planted pastures in the Free State Region. *J. Grassld Soc. Sth. Afr.* 4(4): 127–132.

Van Niekerk, W.A., 1987. Minerale. In: *Handelinge van 'n kortkursus aan veekundiges*, Dept. Veekunde, University van Pretoria, Pretoria.

Van Ryssen, J.B.J., 1990. Chicken litter as a feedstuff for ruminants. Paper delivered to the Natal Branch of the S. Afr. Soc. Anim. Prod., Howick, Natal.

Van Soest, P.J., 1967. Development of a comprehensive system of feed analyses and its application to forages. *J. Anim. Sci.* 26: 119–128.

Van Soest, P.J., 1982. *Nutritional ecology of the ruminant.* O & B Books, Corvallis, Oregon: 374.

Van Soest, P.J., 1984. Chemical procedures for estimating nutritive value. In: *Laboratory evaluation of fibrous feeds: relationships to nutrient requirements and utilisation by ruminants.* Banff., Canada: Section 3.

Van Soest, P.J., 1985. Composition, fibre quality, and nutritive value of forages. In: *Forages, the science of grassland agriculture*, eds. M.E. Heath, R.F. Barnes & D.S. Metcalfe, Iowa State University Press, Iowa: 412–421.

Vincent, J.F.V., 1983. The influence of water content on the stiffness and fracture properties of grass leaves. *Grass & Forage Science* 38: 107–114.

Vincent, J.M., Thompson, J.A. & Donovan, K.O., 1962. Death of root nodule bacteria on drying. *Aust. J. Agric. Res.* 13: 258–270.

Visser, J.H., 1966. Bemesting van veld. *Proc. Grassld Soc. Sth. Afr.* 1: 41–48.

Vogel, J.C., Fuls, A. & Ellis, R.P., 1978. The geographical distribution of Kranz grasses in South Africa. *S. Afr. J. Sci.* 74.

Volenec, J.J., 1985. Leaf area expansion and shoot elongation of diverse alfalfa germplasms. *Crop Sci.* 25: 822–827.

Vorster, F., 1976. 'n Vergelyking tussen springbokke en skape ten opsigte van sekere voedingsaspekte. M.Sc. thesis. University of Stellenbosch, Stellenbosch.

Vosloo, L.P. Personal communication. Dept. Animal Science, University of Stellenbosch, Stellenbosch.

Wagner, R.E., 1981. Interaction: name of game in maximum yields. *Solutions* March/April: 60–74.

Waldo, D.R., 1973. Extent and partition of cereal grain starch digestion in ruminants. *J. Anim. Sci.* 37: 1062–1074.

Waldo, D.R., & Jorgensen, N.A., 1981. Forages for high animal production: nutritional factors and effects of conservation. *J. Dairy Sci.* 64: 1207–1229.

Walker, T.W. & Adams, A.F.R., 1958. Competition for sulphur in grass-clover association. *Plant Soil* 9: 353–366.

Wallace, L.L., 1981. Growth, morphology and gas exchange on mycorrhizal and non-mycorrhizal *Panicum coloratum* L, a C_4 grass species, under different clipping and fertilization regimes. *Oecologia* 49: 272–278.

Wallace, L.L., 1990. Comparative photosynthetic responses of big bluestem to clipping versus grazing. *J. Range Manage.* 43: 58–61.

Wallace, L.L., McNaughton, S.J. & Coughenour, M.B., 1985. Effects of clipping and four levels of nitrogen on the gas exchange, growth and productivity of two east African graminoids. *Amer. J. Bot.* 72: 222–230.

Waller, S.S. & Lewis, J.K., 1979. Occurrence of C_3 and C_4 photosynthetic pathways in North American grasses. *J. Range Manage.* 32: 12–28.

Wand, S.J.E., Midgley, G.F. & Musil, C.F., 1996. Physiological and growth responses of two African species, *Acacia karoo* and *Themeda triandra*, to combined increases in CO_2 and UV-B radiation. *Physiol. Plant* 98: 882–890.

Ward, C.V. & Blazer, R.D., 1961. Carbohydrate reserves and leaf area in regrowth of orchard grass. *Crop Sci.* 1: 366–370.

Wasserman, V.D., 1982. *Wieke.* Farming in South Africa. Pamphlet E4.7. Government Printer, Pretoria.

Waterhouse, D.F., 1974. The biological control of dung. *Sci. Am.* 230(4): 100–109.

Watson, D.J., 1956. Leaf growth in relation to crop yield. In: *The growth of leaves*, ed. F.L. Milthorpe, Butterworths, London: 178–191.

Watson, V.H. & Ward, C.Y., 1970. Influence of intact tillers and height of cut on regrowth and carbohydrate reserves of Dallis grass (*Paspalum dilatatum*) Poir. *Crop Sci.* 10: 474–476.

Wedin, W.F., Vetter, R.L. & Carlson, I.T., 1970. The potential of tall grasses as autumn-saved forages under heavy nitrogen fertilization and intensive grazing management. *Proc. N. Z. Grassld Ass.* 32: 160–167.

Weinmann, H., 1940a. Storage of root reserves in Rhodes grass. *Plant Physiol.* 15: 467–484.

Weinmann, H., 1940b. Seasonal changes in the roots of some South African highveld grasses. *J. S. A. Bot.* 6: 131–145.

Weinmann, H., 1943. Effects of defoliation frequency and fertiliser treatment on Transvaal highveld. *Emp. J. Exp. Agric.* 11: 113–124.

Weinmann, H., 1944. Root reserves of South African highveld grasses in relation to fertilising and frequency of clipping. *J. S. A. Bot.* 10: 37–54.

Weinmann, H., 1955. The chemistry and physiology of grasses. In: *The grasses and pastures of South Africa*, ed. D. Meredith, Central News Agency, Johannesburg: 571–600.

Weinmann, H. & Reinhold, L., 1946. Reserve carbohydrates in South African grasses. *J. S. A. Bot.* 12: 57–73.

Weir, W.C. & Torell, D.T., 1959. Selective grazing by sheep as shown by a comparison of the chemical composition of range and pasture forage obtained by hand clipping and that by oesophageal fistulated sheep. *J. Anim. Sci.* 18: 641–653.

Welker, J.M., Briske, D.D. & Weaver, R.W., 1987. Nitrogen-15 partitioning within a three generation tiller sequence of the bunchgrass *Schizachyrium scoparium*: response to selected defoliation. *Oecologia* 24: 330–334.

Welker, J.M., Briske, D.D. & Weaver, R.W., 1991. Intraclonal nitrogen allocation in the bunchgrass, *Schizachyrium scoparium*: an assessment of the physiological individual. *Funct. Ecolo.* 5: 433–440.

Welker, J.M., Rykiel, E.J. Jr., Briske, D.D. & Goeschl, J.D., 1985. Carbon import among vegetative tillers within two bunchgrasses: assessment with carbon-11 labelling. *Oecologia* 67: 209–212.

Westoby, M., 1978. What are the biological bases of varied diets? *Amer. Nat.* 112: 627–631.

Weston, R.H., Hogan, J.P. & Hensley, J.H., 1970. Some aspects of the digestion of *Atriplex nummularia* by sheep. *Proc. Aust. Soc. Anim. Prod.* 8: 517–521.

Wheeler, J.L., 1987. Pastures and pasture research in southern Australia. In: *Temperate pastures: their production, use and management*, eds. J.L. Wheeler, C.J. Pearson & G.E. Robards, Australian Wool Corporation/CSIRO, Australia.

White, J.G.H., 1973. Pasture establishment. In: *Pastures and pasture plants*, ed. R.H.M. Langer, AH & HW Reed Ltd, Wellington: 129–157.

Whitehead, D.C., 1970. *The role of nitrogen in grassland productivity*. Bull. 48. Commonwealth Agricultural Bureau, Farnham Royal.

Whitehead, D.C., 1986. Sources and transformations of organic nitrogen in intensively managed grassland soils. In: *Nitrogen fluxes in intensive grassland systems*, eds. H.G. van der Meer, J.C. Ryden & G.C. Ennik, Martinus Nijhoff Publishers, Doordrecht, The Netherlands: 47–58.

Whitehead, D.C. & Raistrick, N. 1990. Ammonia volatilization from five nitrogen compounds used as fertilizers following surface applications to soils. *J. Soil Sci.* 41: 387–394.

Whitehead, D.C., Briston, A.W. & Lockyer, D.R., 1990. Organic matter and nitrogen in the unharvested fractions of grass swards in relation to the potential for nitrate leaching after ploughing. *Plant Soil* 123: 39–49.

Whitehead, D.C., Garwood, E.A. & Ryden, J.C., 1986a. The efficiency of nitrogen use in relation to grassland productivity. Annual Report 1985–86. The Grassland Res. Inst., Hurley.

Whitehead, D.C., Pain, B.F. & Ryden, J.C., 1986b. Nitrogen in UK grassland agriculture. *J. Royal Agric. Soc. England* 147: 190–201.

Wilkinson, J.M., 1985. *Beef production from silage and other conserved forages*. Longman, New York.

Wilkinson, S.R., 1973. Cycling of mineral nutrients in pasture ecosystems. In: *Chemistry and biochemistry of herbage*, eds. G.W. Butler & R.W. Bailey, Academic Press, New York.

Wilkinson, S.R. & Lowry, R.V., 1973. Cycling of mineral nutrients in pasture ecosystems. In: *Chemistry and biochemistry of herbage*, Vol. 2, eds. G.W. Butler & R.W. Bailey, Academic Press, London & New York: 248–315.

Wilkinson, S.R. & Stuedemen, J.A., 1979. Tetany hazard of grass as affected by fertilisation with nitrogen, potassium or poultry litter and methods of grass tetany prevention. In: *Grass tetany*, ed. M. Stelly, Amer. Soc. Agron., Madison, Wisconsin: 93–121.

Williams, S.N. & McDowell, L.R., 1985. Newly discovered and toxic trace elements. In: *Nutrition of grazing ruminants in warm climates*, ed. L.R. McDowell, Academic Press, Orlando: 317–338.

Wilman, D., 1975a. Nitrogen and Italian ryegrass 1. Growth up to 14 weeks: dry-matter yield and digestibility. *J. Br. Grassld Soc.* 30: 141–147.

Wilman, D., 1975b. Nitrogen and Italian ryegrass 2. Growth up to 14 weeks: nitrogen, phosphorus and potassium content and yield. *J. Br. Grassld Soc.* 30: 243–249.

Wilson, J.R., 1975. Comparative response to nitrogen deficiency of a tropical and temperate grass and the interrelation between photosynthesis, growth and accumulation of non-structural carbohydrates. *Neth. J. Agric. Sci.* 23: 104–112.

Wilson, J.R. & Ng, T.T., 1975. Influence of water stress on parameters associated with herbage quality in *Panicum maximum* var. *trichoglume*. *Aust. J. Agric. Res.* 26: 1227–1236.

Woledge, J., 1977. The effects of shading and cutting treatments on the photosynthetic rates of ryegrass leaves. *Ann. Bot.* 41: 1279–1286.

Woledge, J. & Parsons, A.J., 1986. Temperate grasslands. In: *Photosynthesis in contrasting environments*, eds. N.R. Baker & S.P. Long, Elsevier, Amsterdam, New York, Oxford.

Wolf, W.J., 1983. Nutritional supplements for animals: oilseed proteins. In: *Handbook of nutritional supplements, Vol. II*, ed. M. Rechigl Jr., CRC Press, Florida: 163–175.

Wolfson, M.M., 1989. The effect of inorganic nitrogen on growth, morphology and some aspects of the physiology of *Digitaria eriantha* (Steud). Ph.D. thesis. University of the Witwatersrand, Johannesburg.

Woolford, M.K., 1976. A preliminary investigation into the role of yeasts in the ensiling process. *J. Appl. Bact.* 41: 29–36.

Woolford, M.K., 1984. *The silage Fermentation*. Marcel Dekker, Inc., New York.

Zacharias, P.J.K., 1986. The use of the cellulase digestion procedure for indexing the dry matter digestibility of forage. *J. Grassld Soc. Sth. Afr.* 3: 117–121.

Zacharias, P.J.K., Clayton, J. & Tainton, N.M., 1991. *Leucaena leucocephala* as a quality supplement to *Pennisetum clandestinum* foggage: a preliminary study. *J. Grassld Soc. Sth. Afr.* 8(2): 59–62.

Zarrough, K.M., Nelson, C.J. & Sleper, D.A., 1984. Interrelationships between rates of leaf appearance and tillering in selected tall fescue populations. *Crop. Sci.* 24: 565–569.

Zemo, J. & Klemmedson, J.O., 1970. Behaviour of fistulated steers on a desert grassland. *J. Range Manage.* 23: 158–163.

Species index

Subject index